Lecture Notes in Computer Science 14800

Founding Editors

Gerhard Goos
Juris Hartmanis

The series Lecture Notes in Computer Science (LNCS), including its subseries Lecture Notes in Artificial Intelligence (LNAI) and Lecture Notes in Bioinformatics (LNBI), has established itself as a medium for the publication of new developments in computer science and information technology research, teaching, and education.

LNCS enjoys close cooperation with the computer science R & D community, the series counts many renowned academics among its volume editors and paper authors, and collaborates with prestigious societies. Its mission is to serve this international community by providing an invaluable service, mainly focused on the publication of conference and workshop proceedings and postproceedings. LNCS commenced publication in 1973.

Nikolaos Pitropakis · Sokratis Katsikas
Editors

Security and Privacy in Smart Environments

 Springer

Editors
Nikolaos Pitropakis ⓘ
Edinburgh Napier University
Edinburgh, UK

Sokratis Katsikas ⓘ
Norwegian University of Science
and Technology
Gjøvik, Norway

ISSN 0302-9743   ISSN 1611-3349 (electronic)
Lecture Notes in Computer Science
ISBN 978-3-031-66707-7   ISBN 978-3-031-66708-4 (eBook)
https://doi.org/10.1007/978-3-031-66708-4

This Springer imprint is published by the registered company Springer Nature Switzerland AG
The registered company address is: Gewerbestrasse 11, 6330 Cham, Switzerland

If disposing of this product, please recycle the paper.

Preface

Over the past decade, the term "smart" has permeated virtually every facet of our daily existence, exerting a profound influence on various domains. Its impact ranges from catalyzing automation and spearheading the advent of Industry 4.0 to reshaping interpersonal interactions and revolutionizing how individuals monitor their daily routines and health statuses. Initially, the concept of "smart" gained traction within the realm of the Internet of Things (IoT), where devices began to boast computational capabilities and achieved a newfound level of autonomy, departing from traditional centralized approaches. Subsequently, this transformative power extended to everyday carry items like mobile phones, and eventually to wearable devices such as smartwatches, ushering in a new era of connectivity and convenience.

However, with the heightened popularity of newly introduced technologies comes a corresponding rise in the attention garnered from malicious parties. As the attack surface of existing infrastructures expands, vulnerabilities proliferate, rendering them more susceptible to attacks compared to conventional systems. Furthermore, the enhanced computational prowess of smart devices enables the execution of increasingly sophisticated attacks that were previously inconceivable. It is essential to acknowledge that amidst these advancements, the enduring challenge of the malicious insider threat persists, remaining as pertinent as ever since the inception of Information and Communication Technology (ICT) systems.

As smart devices increasingly permeate critical infrastructures, the battle between attackers and defenders intensifies, escalating the arms race within the global cybersecurity landscape. To confront this challenge, the cybersecurity community employs a multifaceted approach, blending traditional methodologies with innovative solutions. This includes comprehensive assessments of the threat landscape, focusing on imminent dangers and potential impacts, while bolstering traditional intrusion detection techniques with cutting-edge advancements from the realm of machine learning. Furthermore, distributed ledgers are leveraged as a strategic tool to fortify existing security measures, introducing an additional layer of protection. Despite these technological advancements, the human element remains pivotal, as new methodologies demand heightened awareness and proactive engagement. With attacks against machine learning growing increasingly sophisticated, fostering a culture of vigilance and involvement becomes imperative to thwart emerging threats effectively.

In our endeavor to decipher the swiftly evolving threat landscape, and with the invaluable contributions of esteemed authors, we have compiled this book with the aim of furnishing comprehensive insights not only to the cybersecurity community but also to all stakeholders with an interest in the subject matter. To enhance readability, we have categorically organized the book into four distinct sections. The first section primarily delves into evaluating the current state of smart environments, shedding light on their complexities and vulnerabilities. The second section provides a fresh perspective on existing technologies and cyber challenges within the context of the modern era, offering novel insights and analyses. Moving forward, our third section is dedicated to exploring the intricacies of critical infrastructures. Lastly, the concluding section offers a succinct overview of distributed ledger and privacy enhancement approaches within the same purview, underscoring their potential role in fortifying cybersecurity measures. Through this structured approach, we endeavor to furnish readers with a comprehensive understanding of the multifaceted threats posed by the evolving cybersecurity landscape, while also offering strategic insights into mitigative measures and proactive solutions.

According to Romain Dagnas, Michel Barbeau, Maxime Boutin, Joaquin Garcia-Alfaro and Reda Yaich, the push for digitization emphasizes the necessity of securing interconnected and automated smart infrastructures, yet assessing their cyber resilience is complex due to layered structures. Traditional methods often lack a holistic view, risking conflicting priorities and compromised performance. This chapter reviews quantitative and qualitative assessment approaches, offering system-specific decision-making metrics and system-agnostic symbolic modeling. By exploring technological and socio-technical dimensions, it aims to enhance the resilience of smart infrastructures efficiently. Sabrina De Capitani di Vimercati, Sara Foresti and Pierangela Samarati state that in our data-driven smart society, IoT devices and sensors continuously generate vast amounts of data, crucial for complex decision-making. Often, data storage and processing are outsourced to third parties, raising concerns about trustworthiness. This chapter addresses the challenge of verifying the integrity of query computations involving external service providers. It explores deterministic approaches using authenticated data structures for full integrity guarantees, alongside probabilistic approaches using control information insertion for probabilistic integrity assurance. Finishing the first section, Marwan Lazrag, Christophe Kiennert and Joaquin Garcia-Alfaro consider that cyber-attacks pose significant threats to critical systems, impacting confidentiality, integrity, and availability, as well as mission quality and performance. Current risk assessment tools face limitations in describing enterprise infrastructure and assessing the propagation of external events' impact on business entities. This chapter surveys methods to address these limitations, focusing on financial and operational impact assessment. Operational impact assessment specifically evaluates how cyber-attacks propagate and affect infrastructure assets, company activities, and financial aspects. Additionally, a novel operational impact propagation assessment approach is presented, enhancing organizational activity definitions, and utilizing business logic modeling to quantify impact propagation probability and critical time.

The next section starts with Dennis Ivory, George Loukas and Diane Gan, who state that in cybersecurity, there's a growing recognition that users aren't the weakest link; they can actually contribute to better security, especially in intrusion detection. This "human-as-a-security-sensor" concept harnesses users' insights to detect threats accurately. While typically applied to traditional computer systems, we aim to extend this approach to IoT environments. By notifying users of new IoT device activities detected on the network, they can assess their legitimacy. This detection relies on forecasting network traffic changes, signaling new activities. Their evaluation suggests this method is effective for user notification. James Pope, Theodoros Spyridopoulos, Vijay Kumar, Francesco Raimondo, Sam Gunner, George Oikonomou, Thomas Pasquier, Ryan McConville, Pietro Carnelli, Adrian Sanchez-Mompo, Ioannis Mavromatis and Aftab

Khan consider that the integration of Edge processing nodes and IoT technologies in urban environments presents security challenges due to scalability issues in existing Intrusion Detection techniques and privacy concerns. This chapter proposes a semi-supervised distributed learning approach using Federated Learning to detect anomalies in resource-constrained IoT systems. By analyzing Linux system call data through this framework, the need for central data processing is reduced. The chapter provides an architectural overview, methodology for model deployment, and performance evaluation, indicating that dataset size and pre-trained models significantly impact the performance of FL models in intrusion detection for large-scale IoT environments. Last, Pankaj Pandey expresses that the evolution of conventional crimes into cybercrime within the metaverse underscores the challenges posed by disruptive technologies. This shift highlights issues such as avatar liability, privacy breaches, and identity theft, reflecting the complex interplay between legal frameworks and virtual environments. Balancing interaction freedom with security concerns in the metaverse necessitates robust regulations and ethical frameworks to prevent abuses and ensure its transformative potential is realized.

In the critical infrastructure section, Argiro Anagnostopoulou, Dimitris Gritzalis, Ioannis Mavridis and Panagiotis Kantas state that the integration of Internet of Things (IoT) and Industrial Internet of Things (IIoT) technologies into the electrical grid has led to the emergence of the smart grid, enhancing automation and intelligence while reducing the need for manual intervention. This chapter offers an overview of smart grid architecture, core components, and related technologies, highlighting the role of information flow control. Various use cases are examined to illustrate different business scenarios within smart grid environments. Then Sokratis Nifakos, Krishna Chandramouli and Natalia Stathakarou discuss social engineering as a growing cyber threat targeting human vulnerabilities, particularly within critical information infrastructures. They emphasize the importance of understanding employee awareness levels and the effectiveness of cybersecurity training programs. The emergence of threats like ransomware highlights the need for robust data governance policies and a deeper understanding of the impact of human behavior on cyber risks. The authors aim to provide insights into social engineering methods and propose strategies for prevention and mitigation.

The last section starts with Konstantinos Papageorgiou, Alexandros Fakis, Georgios Spathoulas and Athanasios Kakarountas who discuss the growing importance of blockchain technology and its potential to transform industries by providing secure, decentralized environments for transactions and distributed applications. Specifically, they focus on decentralized applications (DApps) and the security challenges they face. The analysis covers security issues at various layers, including smart contracts, DApps, and user wallets, and proposes solutions for detecting and mitigating attacks targeting these layers. Then Angeliki Katsika, Lydia Negka, Georgios Spathoulas and Vassilis Plagianakos shed light on the pressing need to improve blockchain scalability to address network overload and high user fees, hindering widespread adoption. Various strategies have been proposed, including enhancements to the foundational layer like sharding and off-chain techniques such as sidechains and rollups. Ethereum, recognizing its limitations, has focused on rollups as a solution to scalability issues, aiming to balance decentralization and security. Rollups are viewed as promising solutions, offering increased capacity and speed while maintaining security and flexibility. This chapter aims to comprehensively explore the complexities of rollups and their impact on blockchain scalability. Panagiotis Rizomiliotis and Maristela Chairetaki discuss the advancements in identity management, particularly the transition from centralized to decentralized solutions. It introduces and analyzes the EUDI wallet, proposed by the European Commission and recently included in the final eIDAS 2.0 document. Finally, Costas Lambrinoudakis and Christos Kalloniatis delve into the complexities surrounding privacy and data protection in eHealth/M-Health systems, which must align with user demands and GDPR regulations. It introduces a Privacy and Data Protection Framework that integrates privacy by design principles with GDPR stipulations, offering a comprehensive approach to addressing technical security and privacy needs. Furthermore, the framework proposes a method for verifying compliance with GDPR objectives outlined in the Data Protection Impact Assessment (DPIA).

We hope you enjoy reading this book as much as we enjoyed editing it.

May 2024

Nikolaos Pitropakis
Sokratis Katsikas

Contents

Distributed Ledger And Privacy Enhancement Approaches

Editors and Contributors

About the Editors

Dr. Nikolaos Pitropakis is an Associate Professor of cybersecurity at the School of Computing, Engineering & the Built Environment of Edinburgh Napier University, and a Fellow of the HEA. He is also the director of the CINEAS research group, and a core member of the Blockpass Identity Lab. Dr. Pitropakis has a strong research background in attacks against machine learning. His current research interests include adversarial machine learning, trust and privacy using distributed ledger technology, advanced cyber attack attribution, and data science applied to cyber security and IoT device security. Dr Pitropakis is leading the integrated apprenticeship scheme BSc Cyber Security, which is the first in the UK to receive full NCSC accreditation. He is teaching Cyber-related graduate apprenticeship degrees, both running in Scotland and England. He is also the external examiner of The American College in Greece (ACG), covering the BSc (Hons) Information Technology and BSc (Hons) Cyber Security and Networking programmes provided by The Open University, and the Lead External Examiner for MSc Cyber Security (Newcastle and London campuses) of Northumbria University.

Dr Pitropakis is currently leading the Horizon Europe project Trust and Privacy-Preserving Computing Platform for Crossborder Federation of Personal Data (TRUSTEE). Prior to joining ENU in 2016, he worked as a Postdoctoral Researcher for the Georgia Institute of Technology, where he was involved in a U.S. Department of Defense project (Methods for producing standardized and transparent attribution) worth 17.3 million dollars aiming at advanced attribution of malicious parties. His work has impacted major trademark holders and especially average users, whose awareness was increased against combosquatting abuse.

His recent work has included £310,843 for the EU TRUSTEE project and over £300,000 through Edinburgh Napier's latest spinout TRUEDEPLOY. Based on the outcomes of these projects, Dr Pitropakis has published 80 quality scholarly articles. His research has been awarded multiple times and recently one of his projects as well as his last PhD graduate won the Scottish Cyber Awards. Dr Pitropakis has also been invited to serve as an organizing and program committee member of major international conferences and as an editor for prestigious journals. In addition, Dr Pitropakis has been involved in supervising Research MSc and PhD students since he accomplished his PhD in 2015. Currently, Dr Pitropakis is supervising 2 PhD students, and their research activities are being disseminated in high-quality conferences and journals. His latest PhD graduate Dr Pavlos Papadopoulos is leading the ENU spinout TRUEDEPLOY.

Sokratis K. Katsikas was born in Athens, Greece, in 1960. He is the Director of the Norwegian Centre for Cybersecurity in Critical Sectors and Professor with the Department of Information Security and Communication Technology, Norwegian University of Science and Technology. He is also Professor Emeritus of the Department of Digital Systems, University of Piraeus, Greece. In 2019 he was awarded a Doctorate Honoris

Causa from the Department of Production and Management Engineering, Democritus University of Thrace, Greece. In May-June 2023 he served as Minister of Digital Governance in the interim (caretaking) government of the Hellenic Republic. In 2023 he was listed in the Stanford University list of the top 2% most cited scientists world-wide. He has authored or co-authored more than 300 journal papers, book chapters and conference proceedings papers. He is serving on the editorial board of several scientific journals, he has co-authored/edited 52 books and conference proceedings and has served on/chaired the technical program committee of more than 900 international scientific conferences, including GLOBECOM 2006, 2012, and 2015. He is a member of the Steering Committee of the ESORICS Conference (chair 2017–2023) and he is the Editor-in-Chief of the International Journal of Information Security (Springer).

Contributors

Argiro Anagnostopoulou Department of Informatics, Athens University of Economics and Business, Athens, Greece

Michel Barbeau School of Computer Science, Carleton University, Ottawa, Canada

Maxime Boutin Institut de Recherche Techologique SystemX, Palaiseau, France

Pietro Carnelli Toshiba Research Europe, Bristol, UK

Krishna Chandramouli Multimedia and Vision Research Group, Queen Mary University of London, London, UK

Romain Dagnas Institut de Recherche Techologique SystemX, Palaiseau, France

Sabrina De Capitani di Vimercati Computer Science Department, Università degli Studi di Milano, Milan, Italy

Alexandros Fakis Department of Information and Communication Systems Engineering, University of the Aegean, Samos, Mytilene, Greece

Sara Foresti Computer Science Department, Università degli Studi di Milano, Milan, Italy

Diane Gan Centre for Sustainable Cyber Security, University of Greenwich, London, UK

Joaquin Garcia-Alfaro SAMOVAR, Télécom SudParis, Institut Polytechnique de Paris, Palaiseau, France

Dimitris Gritzalis Department of Informatics, Athens University of Economics and Business, Athens, Greece

Sam Gunner University of Bristol, Bristol, UK

Maristel Hairetaki University of Piraeus, Piraeus, Greece

Dennis Ivory Centre for Sustainable Cyber Security, University of Greenwich, London, UK

Athanasios Kakarountas Department of Computer Science and Biomedical Informatics, University of Thessaly, Lamia, Greece

Christos Kalloniatis Department of Cultural Technology and Communication, University of the Aegean, University Hill, Lesvos Island, Greece

Panagiotis Kantas Department of Informatics, Athens University of Economics and Business, Athens, Greece

Angeliki Katsika Department of Computer Science and Biomedical Informatics, University of Thessaly, Lamia, Greece

Aftab Khan Toshiba Research Europe, Bristol, UK

Christophe Kiennert SAMOVAR, Télécom SudParis, Institut Polytechnique de Paris, Palaiseau, France

Vijay Kumar Cardiff University, Cardiff, UK

Costas Lambrinoudakis Department of Digital Systems, University of Piraeus, Piraeus, Greece

Marwan Lazrag SAMOVAR, Télécom SudParis, Institut Polytechnique de Paris, Palaiseau, France

George Loukas Centre for Sustainable Cyber Security, University of Greenwich, London, UK

Ioannis Mavridis Department of Applied Informatics, University of Macedonia, Thessaloniki, Greece

Ioannis Mavromatis Digital Catapult, London, UK

Ryan McConville University of Bristol, Bristol, UK

Lydia Negka School of Forestry and Natural Environment, Aristotle University of Thessaloniki, Thessaloniki, Greece

Sokratis Nifakos Department of Learning, Informatics, Management and Ethics, Karolinska Institutet, Solna, Sweden

George Oikonomou University of Bristol, Bristol, UK

Pankaj Pandey Center for Cyber and Information Security, Department of Information Security and Communication Technology, Norwegian University of Science and Technology, Gjøvik, Norway

Konstantinos Papageorgiou Department of Computer Science and Biomedical Informatics, University of Thessaly, Lamia, Greece

Thomas Pasquier University of British Columbia, Vancouver, Canada

Vassilis Plagianakos Department of Computer Science and Biomedical Informatics, University of Thessaly, Lamia, Greece

James Pope University of Bristol, Bristol, UK

Francesco Raimondo University of Bristol, Bristol, UK

Panagiotis Rizomiliotis Harokopio University, Athens, Greece

Pierangela Samarati Computer Science Department, Università degli Studi di Milano, Milan, Italy

Adrian Sanchez-Mompo Toshiba Research Europe, Bristol, UK

Georgios Spathoulas Department of Computer Science and Biomedical Informatics, University of Thessaly, Lamia, Greece;
Department of Information Security and Communication Technology, Norwegian University of Science and Technology (NTNU), Gjøvik, Norway

Theodoros Spyridopoulos Cardiff University, Cardiff, UK

Natalia Stathakarou Department of Learning, Informatics, Management and Ethics, Karolinska Institutet, Solna, Sweden

Reda Yaich Institut de Recherche Techologique SystemX, Palaiseau, France

Current State of Smart Environments

Methodological Resilience Assessment of Smart Cyber Infrastructures

Romain Dagnas[1], Michel Barbeau[2], Maxime Boutin[1],
Joaquin Garcia-Alfaro[3]([✉]), and Reda Yaich[1]

[1] Institut de Recherche Techologique SystemX, Palaiseau, France
{romain.dagnas,maxime.boutin,reda.yaich}@irt-systemx.fr
[2] School of Computer Science, Carleton University, Ottawa, Canada
barbeau@scs.carleton.ca
[3] SAMOVAR, Télécom SudParis, Institut Polytechnique de Paris, Palaiseau, France
joaquin.garcia_alfaro@telecom-sudparis.eu

Abstract. The race for digitization created a real need to protect smart infrastructures. Environments are becoming highly connected and automated. Their growing complexity and connectivity make it hard to assure and assess their cyber resilience, i.e., protecting them from cyberattacks, failures, and errors. Traditional strategies for ensuring the cyber resilience of smart infrastructures suffer from a lack of holism. Indeed, since smart infrastructures are often structured in layers, traditional protection methods can lead to conflicting and competing goals. For instance, they may increase the resilience of specific layers at the expense of decreasing the performance of others. This chapter reviews existing methods aiming to address this problem. We focus on two leading methodological assessment families: quantitative and qualitative. The former includes numerical metrics to quantify and assist system-dependent decision-making processes. The latter builds upon symbolic modeling to offer a system-agnostic assessment. The chapter provides an in-depth exploration of quantitative and qualitative methodologies with significant potential to enhance the resilience of layered smart infrastructures. Our exploration covers classical technological aspects (e.g., cascading effects) and socio-technical factors (e.g., human-in-the-loop interaction).

Keywords: Resilience Assessment · Resilience Enhancement · Attack Remediation · Socio-Technical Theory · Cascading Effect · Smart Infrastructure · Cyber-Physical System · Human-in-the-Loop · Security Ceremony · Formal Modeling · Theorem Proving · Hypergraph

1 Introduction

We live in a constantly changing world. How we conceptualize, design, and build architectures of Cyber-Physical Systems (CPS)s are also changing. We have

© The Author(s) 2025
N. Pitropakis and S. Katsikas (Eds.): *Security and Privacy in Smart Environments*, LNCS 14800, pp. 3–24, 2025.
https://doi.org/10.1007/978-3-031-66708-4_1

experienced four industrial revolutions in the last two centuries. The fourth industrial revolution was one of digitization induced by increased competitiveness. To remain competitive, the new generation of systems faces multiple challenges. The work by Ryalat *et al.* presents the main pillars of Industry 4.0 [35], which are: CPSs, additive manufacturing, automation and industrial robots, simulation, blockchain, augmented reality, big data analysis, cloud computing, Artificial Intelligence (AI), and Internet of Things (IoT). The use of these new technologies aims at enhancing the productivity and reliability of industrial processes. However, there may be a lack of hindsight on these technologies. We may not have a sufficient perception of their potential safety risks.

Motivation. Modern systems consist of multi-layered architectures. Traditional resilience enhancement strategies could help improve the resilience capabilities of one layer but to the detriment of the others. There is a need for holism in resilience mechanisms, especially in smart infrastructures where the high degree of interconnection between its functions increases the risk of cascading effects.

Contributions. The contributions of this chapter are twofold. We discuss existing methodologies to solve the above challenge. We focus on quantitative and qualitative methodologies that could enhance layered smart infrastructures' resilience potential. We review the relevance of the presented methodologies and research challenges related to the resilience enhancement of smart systems.

The chapter is organized as follows. Section 2 presents background on smart infrastructures and preliminaries on the emergence of new technologies and relevant models. Section 3 surveys existing quantitative methodologies, including numeric metrics, to help quantify and assist system-dependent decision-making processes. Section 4 surveys qualitative methodologies, including symbolic modeling and other representative methodologies. Section 5 discusses some directions and trends for further research on the resilience of smart infrastructures. Section 6 concludes the chapter.

2   Smart Infrastructures in a Digitized World

This section presents some preliminary background on smart infrastructures and the new generation of architectures in Industry 4.0. We also provide preliminaries on the emergence of new technologies and representative models used to analyze smart systems and architectures of the future [13].

2.1   Industrial Revolutions

Historically, there have been four industrial revolutions in the last centuries. The first revolution started at the beginning of the 19th century and has led to a major industrial transformation of societies in terms of mechanization. The second revolution started around 1870 with massive technological inventions and industrial advances, fostering the emergence of new energy sources such as electricity, gas, and oil. The third revolution started in the 1970s with the emergence of nuclear energy, the first computers, and the rise of electronics. The fourth revolution is the one of Industry 4.0. However, many people disagree with the arrival of such a revolution because we do not yet know the magnitude of its impact. The transition to digitization and smart technologies is underway. With the use of highly connected devices in cyberspace, we are becoming aware of new threats and vulnerabilities to which critical systems are subjected. This revolution includes advances in AI, IoT, blockchain, big data, and other technologies that we present in a more detailed way in the next section.

2.2 Emergence of New Technologies and Initiatives

The adjective *smart* describing infrastructures, facilities, and critical systems indicates a high degree of connectivity and digitization in the cyber world. Indeed, the digitization era has led to significant changes in the design of industrial systems' architecture, such as maritime port infrastructures, supply chain systems, and Industry 4.0. Cyber-physical systems are involved in this transformation because they connect the real and cyber worlds. Figure 1, which has been realized within the Secure Ports of the Future (PFS) project [39], presents new technologies used in smart infrastructures.

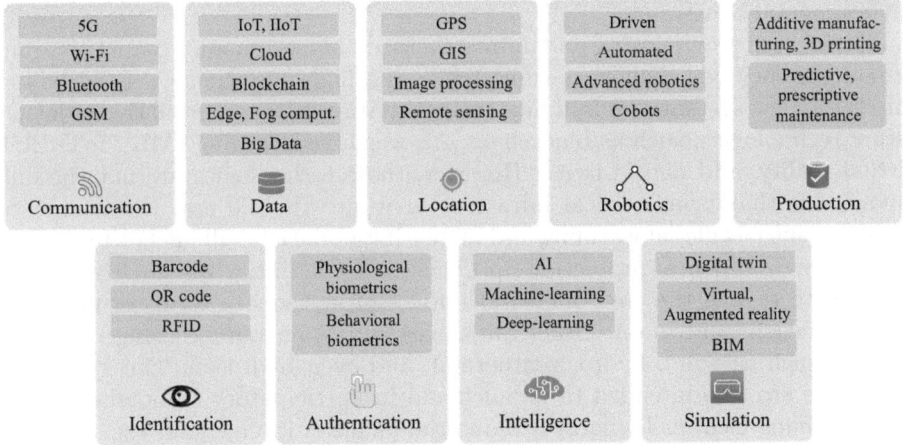

Fig. 1. Technologies used in smart infrastructures.

The new generation of complex systems comprises the integration of communication technologies for enabling communication between a larger set of connected components (including 5G networks, WiFi, Bluetooth, and GSM to enable connectivity between mobile devices); digitization, involving the use of many connected devices highlights a need to analyze, store and protect these data. These data technologies include IoT and Industrial Internet of Things (IIoT) components, Cloud, Blockchain, Big Data, and also Edge Computing (modern computing paradigm located at the edge of the network, which *allows client data to be processed closer to the data source instead of far-off centralized locations such as huge cloud data centers* [43]) and Fog Computing (located between the Cloud and the Edge, it is based on *locating certain resources and transactions closer to the edge of a network* [43]). Smart systems also integrate various location technologies to find merchandise and external or internal human employees during sensitive missions. Robotic technologies in handling or transportation systems are also a part of smart systems. In the industrial field, predictive maintenance algorithms are used to anticipate malfunctions. Recently, prescriptive algorithms have been created to provide not only the failure date of a component but also prescription recommendations to optimize a component's lifetime. Identification technologies such as barcodes, Radio Frequency Identification (RFID), and QR codes are used in smart infrastructures to obtain a detailed information sheet about products involved in the supply chain. In such smart infrastructures, there is a need to improve the authentication processes. Organizations use physiological and behavioral biometrics to guarantee an accurate and secure authentication process. Many machine-learning and deep-learning strategies are used to identify patterns in data and make appropriate decisions. Virtual, augmented reality and digital twins are also part of Industry 4.0. These advanced simulation techniques allow us to anticipate the exploitation of vulnerabilities by an adversary and analyze security scenarios in specific environmental constraints.

These new technologies and infrastructures must remain competitive and respond to society's challenges while remaining high-performance. For this reason, these new systems are building their activities on smart initiatives that use future technologies, such as blockchain, AI, Machine Learning (ML), IoT, IIoT, virtual reality, and digital twins. However, these technologies are not the only innovations that bring critical infrastructures into the 4.0 era. Environmental problems such as global warming and increasing pollution will anchor future systems in environmental protection approaches. Some initiatives are *eco-friendly* and aim to reduce the environmental impact of systems in terms of pollution, noise pollution, and so on. Other initiatives aim to encourage using renewable energies such as wind, hydro, geothermal, and even hydrogen. The systems of the future are also intended to be anchored in participatory democracy strategies, involving citizens living near industrial facilities in new land use and other projects.

In Wavestone's report dedicated to smart ports [38], Sinibaldi has presented three main families of smart solutions. They include ecological solutions, which are related to the use of renewable energies and focused on protecting the environment from the impact of facilities; economic solutions associated with the supporting functions, the core business, and the open innovation strategies of a company; and participatory which includes solutions involving citizen in new projects.

We have analyzed the evolutionary trends of the ten smartest ports in the world. Figure 2 [39] presents the results of our analysis. We can see that in mar-

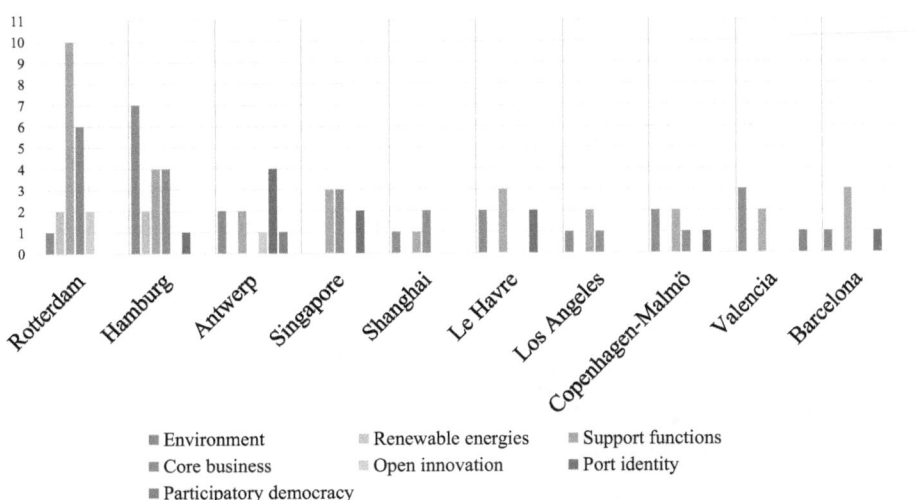

Fig. 2. Number of smart initiatives per smart port. (Color figure online)

itime port infrastructures, the emphasis is put on support functions (orange) and core business-related smart solutions (brown). Indeed, these solutions aim to enhance the port's competitiveness and facilitate operations and missions. We also see that most of the smartest ports apply environment-related solutions to reduce noise nuisance and air pollution and use automated vehicles. These diversified initiatives show that there's a shift in the way we design the architectures of the future and modernize existing ones. Industry 4.0 systems must be anchored in a world where environmental protection and the use of renewable energies are becoming a real issue.

3 Quantitative Methodologies

Quantifying the resilience of complex cyber-physical systems is a vital research axis of growing interest. Quantifying approaches aim to design complex systems considered resilient-by-design and assist designers in improving and upgrading existing infrastructures to bring resilience capacities. This section reviews relevant existing methodologies to quantify and help system-dependent decision-making processes.

Quantitative deterministic metrics are built upon intrinsic properties of a system such as performance, mission delivery, reliability, and accuracy. These metrics provide either a numerical estimation of resilience based on certain properties or a numerical score of different parameters that compose resilience. Metric-based strategies involve numerical indicators that build upon certain system properties. As resilience is highly related to performance, many metrics are used to quantify a system's ability to complete a certain mission, deliver a specific rendering, or remain secure during an adversarial event. Linkov and Kott present two different families of strategies for assessing the resilience of systems, which are metrics-based and model-based [23]. Clédel *et al.* [11] present methodologies for measuring resilience capacities. These approaches are twofold: (i) quantitative deterministic, which involves indicators to analyze performance losses of a system facing known disruptive events; (ii) quantitative probabilistic, based on uncertainties. Thus, stochastic aspects are added to resilience assessment. Another way to classify metrics is by their empirical or analytic nature. They are discussed further in the following.

Date	Duration (Minutes)
04-10-2019	124
21-10-2019	116
22-10-2019	138
23-10-2019	115
01-11-2019	191

Fig. 3. Ottawa O-Train transit system.   **Fig. 4.** Disruptions (Data source: [16]).

3.1   Empirical Metrics

Empirical metrics rely on data collected by observing cyber-physical systems during a time interval. Their use in resilience analysis of cyber-physical systems has been highlighted by Lewis [25].

We review some empirical metrics using sample data collected during the operation of the Ottawa O-Train transit system, Figs. 3 and 4. The table lists disruptions exceeding 100 min that occurred in October and November 2019. For this example, the disruption time threshold is 100 min. During that two-month observation period (T_0), five (N) disruptions exceeded 100 min. The average monthly occurrences is:

$$N_m = N/T_0, \text{ i.e., } 2.5. \tag{1}$$

A probability of exceedance model can be constructed using this parameter to predict the future. The probability of exceedence is defined as:

$$P = 1 - e^{-N_m T_p} \tag{2}$$

where T_p is the prediction period.

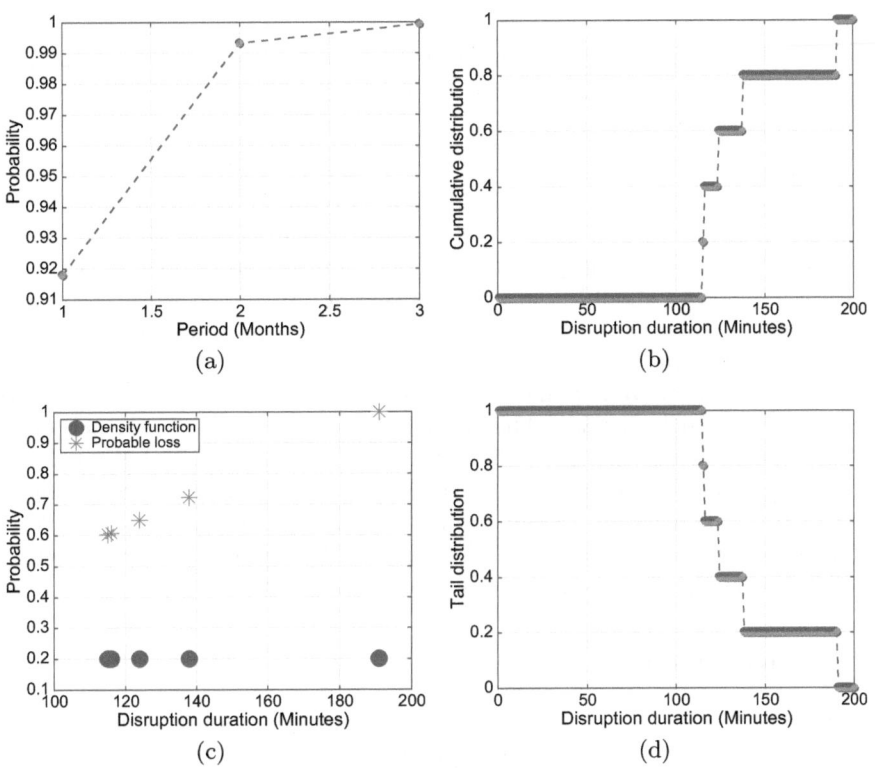

Fig. 5. Example of empirical metrics usage associated to sample data collected during the operation of the Ottawa O-Train transit system. (a) 100-min disruption exceedance. (b) Cumulative distribution. (c) Density function and probable loss. (d) Tail distribution.

Figure 5(a) plots the exceedance probability for a 100 min disruption or longer versus time (in months). The random variable is time. For this example, the probability is one for three months and longer. In such a case, the metric to minimize is the probability of a disruption. Another interesting number that can be derived is the *mean recurrence interval*, which corresponds to the ratio:

$$T_x = 1/N_m \tag{3}$$

In this example, it is 0.4 month. It roughly means that a 100-min disruption can be expected biweekly. The data can be examined from several different perspectives. Figure 5(b) plots the cumulative duration of a disruption versus the duration of a disruption (minutes). In this case, the random variable X is the disruption duration. The diagram plots

$$F(x) = Pr[X \le x], \text{ for } x = 1, 2, \ldots, 200 \text{ min} \tag{4}$$

When a disruption occurs, the plot indicates the probability that it is at least that time for every possible duration. The $F(x)$ derivative yields the probability density function of X, that is, $f(x)$. The product $xf(x)$ is interpreted as the probable loss, Fig. 5(c). This probable loss has a maximum. The ratio $xf(x)$ over the maximum indicates resilience. The complementary of $F(x)$ is:

$$\bar{F}(x) = Pr[X > x] = 1 - F(x) \qquad (5)$$

It is interpreted as the tail distribution, Fig. 5(d). When alternatives are available, resilience-wise, a thin tail distribution is preferable to a fat tail distribution. The examples presented in this section are built using a small data set. Ideally, analyses should built on large sets of data. Such data sets may not be available for new systems. In such cases, analytic metrics discussed in the upcoming section can be used.

Quantitative probabilistic metrics include stochastic strategies. Some uncertainties exist in such metrics. The probability considered in resilience evaluation includes the stochasticity of the occurrence of an adversarial event. In such quantitative probabilistic approaches, there exist event-specific metrics. For such metrics, the resilience of a system facing a specific event is considered. Certain approaches claim that resilience can be quantified only once a threat scenario is established [11,18]. This approach is called event-specific.

3.2 Analytic Metrics

Analytic approaches rely on mathematical or logical reasoning to predict cyber-physical systems' resilience. Analytic metrics may be considered in a graph modeling framework. Graph modeling is a rigorous framework that captures entities and relationships between them [21]. In contrast to classical databases, the emphasis is on the relationships and how entities interact. They are particularly well adapted for answering queries involving several relationships chained together. Graphs are well-suited for capturing the design aspect of complex cyber-physical systems. For example, functionality can be defined as the percentage of functioning graph nodes. Links from node to node can capture cascading effects due to fault propagation.

The issue of cascading effects is of paramount importance. Indeed, smart infrastructures are, due to their complexity, more prone to the damaging consequences of cascading effects, i.e., their functions, components, and sub-systems have high link densities between elements of the systems, which increase the risk. A cascading attack describes an adversarial event during which an adversary attacks a specific point of an infrastructure, gains access to another system, and compromises it due to their connectivity. According to this definition, a cascading effect corresponds to a domino effect through interconnected systems with severe and unexpected consequences.

For cyber-physical system modeling, graph nodes may correspond to system components, such as pumps and valves. Links represent connections between components, such as conduits. The interconnections can be specified with an adjacency matrix where rows and columns correspond to components. *Spectral radius* has been considered to characterize the resilience of a cyber-physical system modeled as a graph. The spectral radius is the largest eigenvalue of the adjacency matrix [40]. When there are nodes that tend to play the roles of centers, the spectral radius reflects their number of connections to other neighbor nodes. It is indicative of the potential for fault propagation through cascading effects. A graph can also be analyzed to highlight the presence of *blocking nodes* versus the number of links. Removing a blocking node partitions the graph, breaking connections from one partition to another. Resilience-wise, a low blocking nodes-number of links ratio is preferable.

Wang *et al.* [42] studied six different graph spectral metrics and highlighted that their interpretation regarding resilience may be contradictory. In a recent paper [14], we investigated the use of the spectral radius and (k, ℓ)-resilience to evaluate the resilience of a water treatment system [2,3]. We arrived at similar conclusions.

3.3 Multi-layered Frameworks

Frameworks, such as the Industrial Internet Reference Architecture (IIRA) [26] and Reference Architectural Model Industrie 4.0 (RAMI 4.0) [19], are dedicated to helping model architectures of the Industry 4.0 as multi-layered architectures. Figure 6 represents the RAMI 4.0 framework.

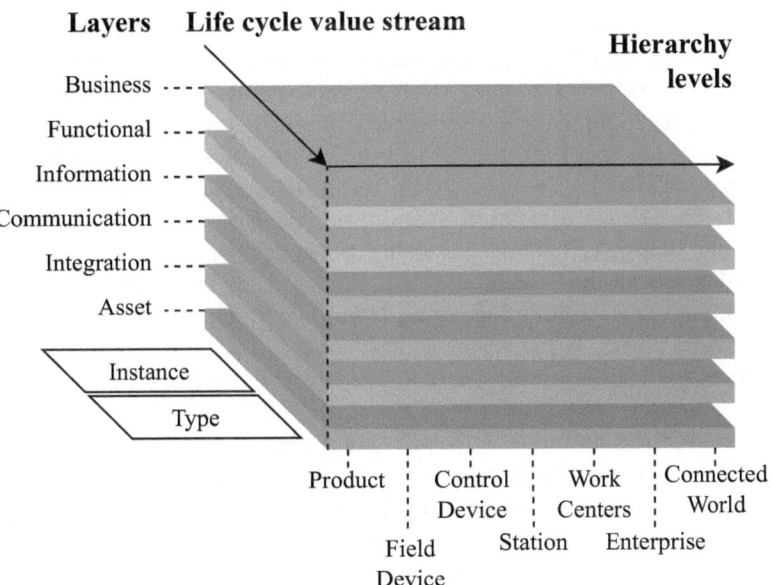

Fig. 6. RAMI 4.0 model.

The RAMI 4.0 framework works with a 3-D model by representing an architecture with the following layers: asset, integration, communication, information, functional, and business. The two other axes are the *life cycle value stream* and *hierarchy levels*. RAMI 4.0 is made to ensure that all the participants involved have a common understanding of a specific system.

We have applied RAMI 4.0 to a water treatment system architecture. Figure 8 presents the obtained model divided according to the RAMI 4.0 layers [44]. Indeed, the asset layer is dedicated to physical components such as sensors and actuators. The integration layer makes a transition between the physical and cyber-world. It deals with easy information processing. The communication layer provides communication elements between the integration and information layers. The information layer is related to all the information about materials manufactured in an industry or, in our case, information related to the water treatment process. The functional layer deals with system control and processing rules. The business layer integrates business strategies or business models used during the life cycle of a system. In the case of a water treatment system controlled by a state and healthcare agencies, business models can be summarized as initiatives to launch inter-sectoral water swaps, for example, [34].

We must highlight that the RAMI 4.0 model combines the vital elements of Industry 4.0 in a layered model. Such a structure is useful to systematically organize and flourish the technologies used in Industry 4.0.

3.4   Knowledge Graphs

The knowledge graph concept gained interest in various areas, especially cyber security. Indeed, knowledge graphs can structure and process high volumes of data generated from cyberspace. Using ontology-based knowledge representations, they capture information's complexity and heterogeneous nature [37,45].

The knowledge graph presented in Fig. 7 represents the Secure Water Treatment System (SWaT) [12]. It is a CPS test bed reproducing the behavior of an actual water treatment station in Singapore. SWaT consists of six sub-functions: (1) pumping phase; (2) chemical dosing phase; (3) Ultrafiltration (UF) phase phase, which works by alternating filtration and backwash cycles to clean the UF membrane until a manual cleaning is required; (4) ultraviolet treatment and dechlorination phase; (5) reverse osmosis phase; (6) final stage and backwash sending flow for the membrane of the UF unit in the third phase. Each of the six sub-functions is controlled by a Programmable-Logic Controller (PLC), and each of these PLCs communicate with each other during the water treatment process. The nodes and links related to the system are put into a dark square. Other entities are considered outside the scope of the system. Indeed, the system requires non-potable water from the groundwater or rivers. At the level of the last stage, the system produces potable water sent through water distribution networks to residential areas. The produced water must also comply with the regulations established by organizations responsible for implementing health policy.

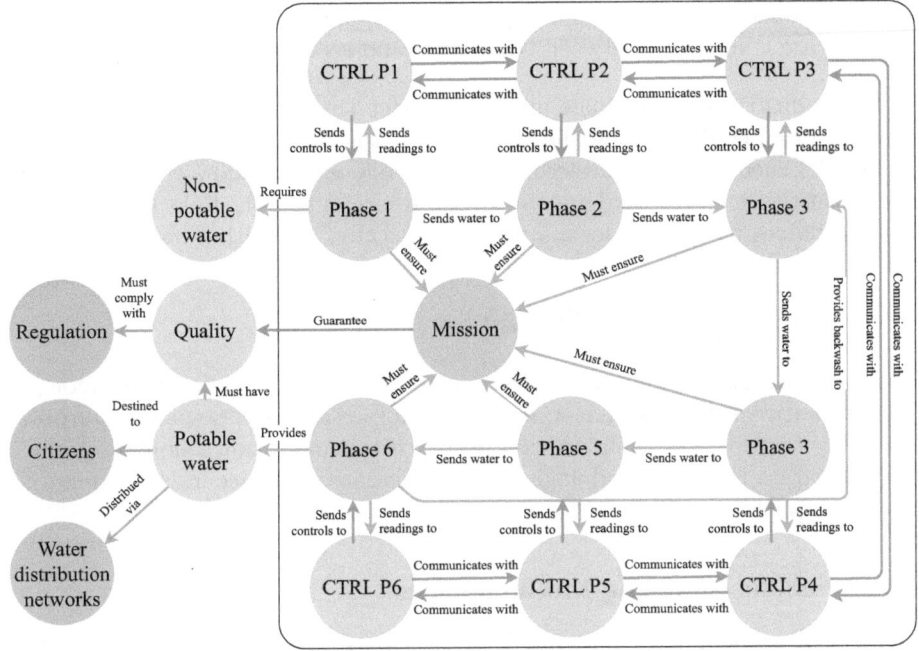

Fig. 7. Knowledge graph of a water treatment system.

3.5 Summary

We have presented multiple approaches for resilience assessment purposes. Given the wide variety of approaches in the literature to quantify the resilience of CPSs, we refer the reader to the surveys [1, 7, 11] for further details on existing resilience techniques.

We must highlight that the knowledge graph and multi-layer approaches presented in Sects. 3.3 and 3.4 are considered quantitative because the metrics applied to such models for resilience assessment purposes yield a quantitative estimation of the resilience.

Some of the techniques we presented are based on mathematical indicators, while others use probabilistic reasoning, graph representations, or disrupting events analysis. The main takeaway is that due to their new intrinsic structures, resilience applied to smart infrastructures must be considered with a new look. Indeed, the new principles of Industry 4.0 can be summarized as follows:

- More widespread Internet availability.
- Devices become smart and connected to the cyber-space.
- New services and functions are involved.
- All parties involved in business processes, in the manufacturing and process industry, are mutually connected.
- Information from suppliers, customers, and within organizations are connected and transparently available for the stakeholders.

- The production is managed autonomously by machines.
- There are transitions between companies and sectors.

The Industry 4.0 principles aim to connect the stakeholders involved in a system and allow data to be shared. Some of the metrics presented previously help model such smart architectures. Knowledge graphs allow the modeling of several entities, e.g., physical or conceptual. Representing the links between these entities enables us to consider a system holistically, i.e., within its environment.

4 Qualitative Methodologies

We review qualitative methodologies, such as symbolic modeling methodologies and alternative solutions that can be used for resilience assessment purposes. Symbolic modeling can help to examine architectures from a holistic point of view. However, these methodologies may be undated or inaccurate for quantitative assessment.

4.1 Symbolic Modeling Methodologies

We discuss how symbolic modeling can contribute to resilience evaluation. It is an interesting approach because it complements the spectral radius and (k, ℓ)-resilience metrics, focusing on system structure [2,3]. Loss scenarios, a symbolic approach, consider threats from the environment and control structures (physical controller or human employee) interacting with a system. However, due to abstraction and several assumptions about the examined system and its environment, obtaining results perfectly reflecting reality is challenging.

Modeling threat scenarios in propositional logic leads to obtaining the truth assignment of axioms that validate them. For example, consider the following scenario: *The RAW_WATER_TANK is full. CTRL-P1 does not produce the Start Pump Control Action because CTRL-P1 incorrectly believes that the tank is not full.* This flawed condition occurs when there is a delay between the data coming from the level sensors of the tank, a wrong conversion of data units from the sensors, or readings are not received from the level sensor. This may be due, for instance, to a sensor failure or a conflict between the sensor readings and the conversion tool used to adapt to the global data units. In this scenario, we distinguish between error, failure, and fault. According to ISO/IEC/IEEE 24765:2017, an error can be an erroneous system state or a difference between a computed, observed, or measured value or condition and the true, specified, or theoretically correct value or condition. A failure is the nonperformance or inability of a component (or a system) to perform its intended function. A fault is a defect in a hardware device or component that causes errors. In the scenario, it is implied that the fault *conflict between readings* causes the error *a wrong conversion of data unit from the sensors*, which in the end causes the failure *Start Pump Control Action is not produced*. This scenario can be translated into the following formula $(((FAULT\_OCCURED) \longleftrightarrow$

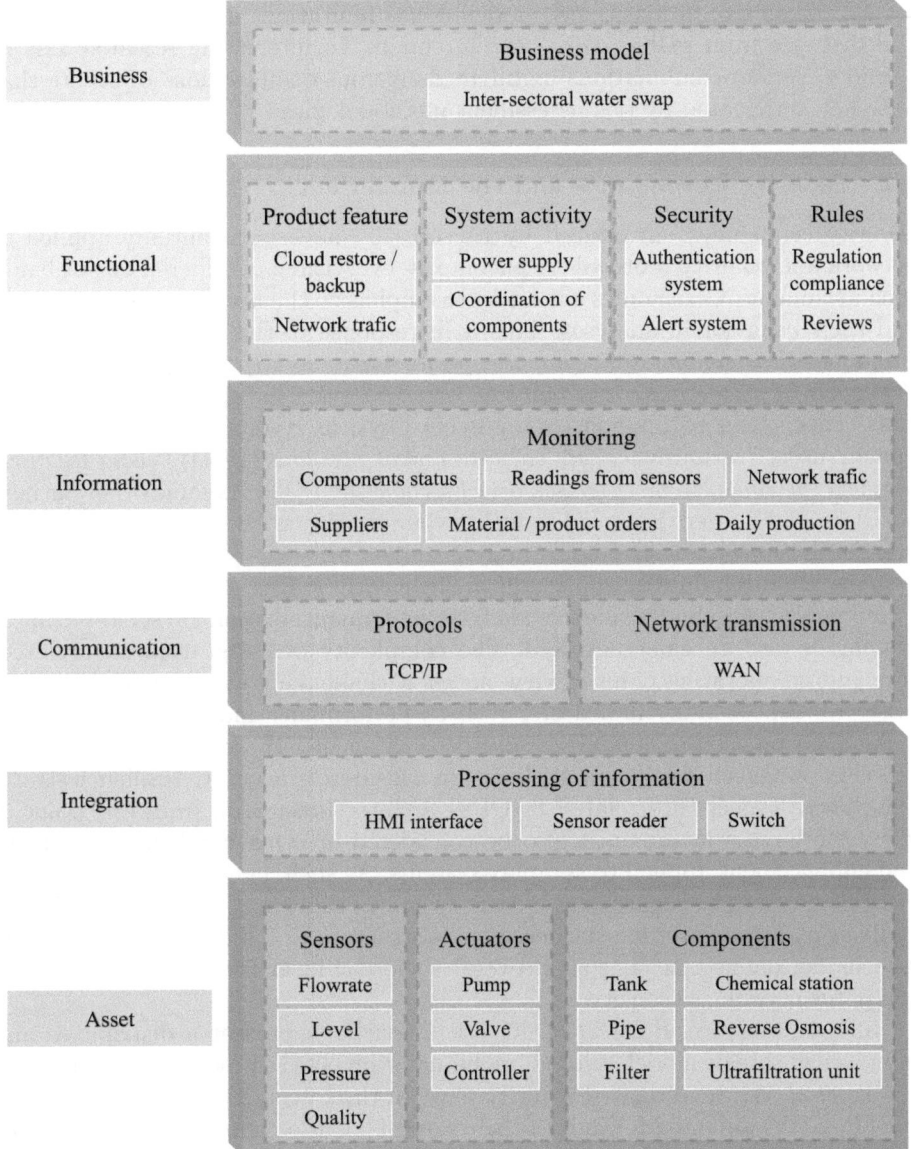

Fig. 8. RAMI 4.0 model of a water treatment system.

$(ERROR\_OCCURED)) \wedge FULL\_WATER\_TANK)$. The result of such a formula is either a loss or not. However, we could change the previous formula if we consider that an error can happen without fault. Furthermore, this level of abstraction makes it impossible to know which kind of loss occurs, such as loss of life, loss of regulatory conformity, or loss of customer satisfaction.

As we can see, propositional modeling has limitations due to its abstraction and distance from reality. However, it remains an interesting research axis to consider. Axiom combinations highlight dangerous combinations of events that were not anticipated by risk assessment methodologies.

4.2 Security Ceremonies and the Human Factor

Security ceremonies, introduced by Ellison [15], have been initially applied to network and security protocols to encompass everything considered out-of-band, such as humans. Ceremonies are a way to emphasize the need to include humans and their behavior in analyses. This is in response to the fact that, as Ellison [15] noted: *It is common for computer professionals to disparage human users as the source of all the flaws that make an excellently designed product malfunction.* This concept can be applied to any cyber-physical system that interacts with human users, variously called *socio-technical systems* [17,41], *socio-technical physical systems* [24], or *cyber-socio-technical system* [32]. As some other authors noted [5,30,33], there is a need to meld cyber security with social sciences. However, both the security ceremony analysis and security analysis of socio-technical systems are disciplines that are still in their early stages. The interdisciplinary combination of cyber resilience analysis and quantification of socio-technical systems is also an emergent field. The complexity of CPSs from an architecture and an operating point of view makes it challenging to quantify and assess cyber resilience. Many indicators, such as performance indicators and robustness, could be used to conduct resilience assessments. However, the abstraction problem remains a significant challenge in considering accurate resilience strategies. These aspects are related to CPSs architectures. Still, from our point of view, the human factor must also be considered in cyber resilience assessment and enhancement applied to complex systems. As such, cyber resilience analysis of CPSs needs to be regarded as a socio-technical challenge. However, the formal analysis requires clarifying the ceremony's structure to build a ceremony model. The model defined in [4] spans several socio-technical layers, ranging from a computer network to society.

In the literature, Bella *et al.* [6] use Tamarin to present a distributed and interacting threat model related to humans involved in security ceremonies. Radke *et al.* [33] investigated security flaws in three security ceremonies, named HTTP, EMV, and Opera Mini. In their work [9], Carlos *et al.* present a dynamic threat model for a dynamic analysis of the corresponding ceremonies. Johansen and Jøsang [22] follow a different approach and present a model based on user-interaction information obtained by sociologists to model human actions as a probabilistic process. Martina and Carlos [27] proposed an extended abstract that formalizes human-cognitive processes in security ceremonies for protocol verification. From the resilience perspective, Haring *et al.* [20] explore the resilience quantification of socio-technical CPSs.

We consider the SWaT system presented with more details in Sect. 3.4. The Fig. 9 shows two exchange diagrams of the third phase of SWaT dedicated to the UF process, namely normal operating (Fig. 9(a)) and facing a replay-adversary

(Fig. 9(b)). Seven roles have been identified, namely: *Operator* (the human supervisor having access to the monitoring system of the plant); *IHM* (the human-machine interface itself); *Controller P3* (the controller assigned to the UF phase of SWaT); *Pump P6* (the backwash pump involved after a filtration cycle); *Pump P3* (the pump sending water through the UF membrane during a filtration cycle); *Tank level* (the ultrasonic level sensor of the feed tank); *TMP* (the Transmembrane Pressure (TMP) sensor of the UF unit); plus an additional role dedicated to the replay-adversary in the diagram presented in Fig. 9(b).

The SWaT test bed has been studied and exposed to many different attacks to analyze its behavior under constraints. An example of a powerful attack perpetrated against the system is the one compromising the readings sent by the TMP sensor (located on the UF membrane, in the third sub-function) to the corresponding controller [29]. Indeed, this pressure sensor, measuring the cake layer formation pressure on the membrane, tells the controller to switch from a filtration cycle to a backwash cycle. The study has shown that an adversary attempting to compromise the readings made by the sensor and sent to the controller by increasing the measured pressure values could put the system in a perpetual backwash cycle, interrupting the production of purified water. The attack is impactful, leading to different kinds of losses such as financial loss (due to the production lack of clean water), reputation loss (due to the impact on the mission of the system), loss or damage to the material of the system (a membrane submitted of a high-pressure backwash during too much time could be degraded), and loss of mission and loss of customer satisfaction (due to the incapacity to clean the water).

According to the literature, an adversary attempting to perpetrate such an attack can do so in four different ways: (i) man-in-the-middle attack: the adversary must intercept the legitimate traffic through the network, then send corrupted data to the intended destination, i.e., the controller; (ii) replay attack (a kind of man-in-the-middle attack): the adversary must eavesdrop on the network, mimic legitimate data and replay it to the controller; (iii) jamming attack: the adversary must know the transmission power and modulation scheme of the communicating parties; and (iv) hardware Trojan attack: it requires modification of an Integrated Circuit (IC).

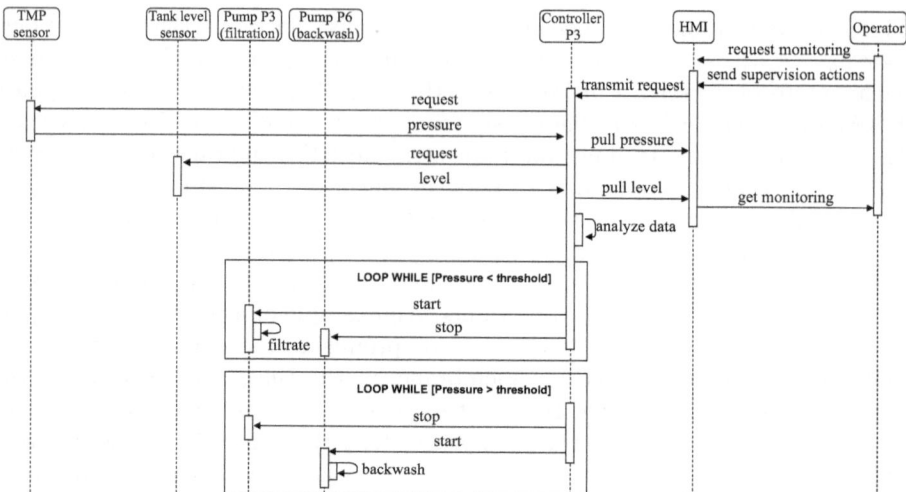

(a) Exchange diagram of the ultrafiltration phase of SWaT in normal condition.

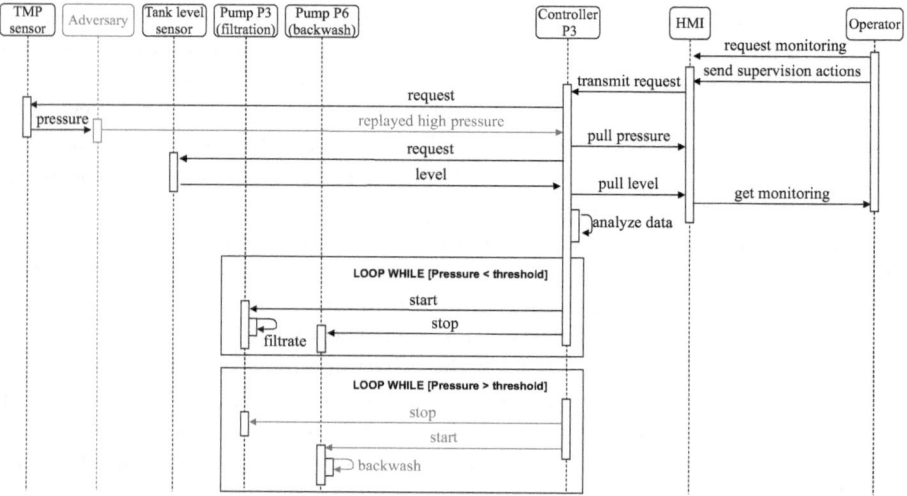

(b) Exchange diagram of the ultrafiltration phase of SWaT facing a replay attack.

Fig. 9. Exchange diagrams of the ultrafiltration phase of SWaT [12].

4.3   Summary

We have presented several approaches that consider qualitative resilience assessment. Other approaches are described in [11], including methodologies based on events handling, fuzzy rules, frameworks, and guidelines.

The approaches that we have presented are based on symbolic modeling, which offers interesting research axes, such as the theorem-proving strategies applied to the top layers of a multi-layered architecture, associated with model-checking strategies applied to the lower levels, which are dedicated to the physical, and component views of a given system. It is important to note that symbolic modeling methodologies could be complex to apply to models. However, some approaches such as the one applied by Sempreboni *et al.* [36] are interesting to consider due to their ability to capture unexpected human actions.

5   Further Research Directions

Through the previous sections, we reviewed various techniques and methodologies for resilience assessment and enhancement purposes. A new comparison between these approaches is presented in Table 1. To go beyond this comparison, we must highlight that choosing an appropriate resilience enhancement strategy depends on the system. Indeed, symbolic methodologies based on model-checking and theorem proving are suitable to conduct a resilience analysis of an IC component. On the other hand, multi-layer modeling and knowledge graphs are suitable for complex systems such as maritime ports or water treatment stations involving various families of components or many stakeholders. Graph models are adapted to network architectures or smart-grid systems.

Some of the approaches presented provide answers to specific questions. Indeed, dealing with the resilience quantification of complex systems in the context of Industry 4.0 implies using models and frameworks adapted to complex architectures. In this context, as covered in Sect. 3.3, multi-layered models are attractive for two reasons. They can represent an architecture consisting of an overlay of layers that captures how the components, functions, and sub-systems work together and communicate. Such a multi-layered representation constitutes a foundation for introducing resilience strategies at each system level. The main challenge is to make each layer resilient without detrimental effects on the others.

Particular attention must also be paid to the interpretation attributed to metrics. Indeed, it has been established that the results obtained for a given metric could have several interpretations. These interpretations may be contradictory [14,25]. An example is the spectral radius applied to graph models. This notion of graph models related to analytic metrics has been discussed in Sect. 3.2. On the one hand, a value increase from one design to another of the same system represented as graphs could reflect higher resilience. This is due to the increase in the density of links between two states, reflecting a lower risk of unreachable states in case of a node deletion, e.g., a component failure. On the other hand, a value increase can also be interpreted as lower resilience because a higher density of links between the states reflects a higher risk of producing cascading effects due to an attack. Two possible interpretations do not mean that a metric is useless for quantifying the resilience of a system. The spectral radius illustrates very well the fact that there is a delicate balance between increasing the resilience potential and mitigating the increase in the security risk [14].

Another research axis concerns the human factor in modeling and resilience assessment strategies, as covered in Sect. 4.2. It is well-known that many incidents are due to human factors. Ceremonies, which consider a system from a holistic point of view, can help evaluate the human factor. The underlying problem is the unpredictability of human actions. Risk analysis can help anticipate

Table 1. Comparison of the presented approaches.

Approach	Scope	Limitations
Empirical	Based on the study of testbed and real system data.	Does not consider the constraints of real environments.
Analytical	Mathematical modeling for predicting a system's behavior.	Difficult to apply to complex systems.
Multi-layer	Infrastructures involving stakeholders or architectural levels.	Requires a level of abstraction to build all the layers.
Knowledge graph	Complex systems with complex interactions, difficult to model.	The way the graph is built impact the assessment results.
Symbolic	Model-checking and theorem-proving analysis.	Difficult to apply to complex systems. Difficult to use.
Ceremonies	Include the human factor and how complex relationships impact a system.	Requires a well-understanding of the way a system operates.

the wrong or dangerous actions that human people can perpetrate, but an accurate knowledge of the system is necessary. This problem is related to the prediction of cascading effects on complex systems. We have shown an example of an exchange diagram in Sect. 4.1 that can be used to predict the adversary's actions. However, there is a lack of tools to accurately quantify the impact of the human factor on resilience strategies.

What we know about resilience assessment strategies is that they are based on simulation. These simulation models can be applied to data provided by test beds, as covered in Sect. 3.1. We must highlight an intrinsic limitation in the data obtained with test beds. In a resilience assessment context, test bed-generated data do not capture the stress present in live scenarios, i.e., the actual operating conditions and characteristics of a specific environment. These particular features cannot be imagined or anticipated with test beds because they do not operate in a real environment. Thus, metrics applied to test bed data may yield erroneous or inaccurate results. Using live scenarios and stochastic engineering (e.g., use of chaos monkeys [28]) can address the limitations of test beds.

This notion of environment is also related to knowledge graphs, as discussed in Sect. 3.4. Indeed, knowledge graphs allow us to consider a higher diversity of entities than traditional graphs, such as abstract entities, e.g., missions, regulations, and stakeholders involved in the life cycle of a system. In our past research [14], we used traditional graphs mapped to square adjacency matrices to model physical and logical links between the components of a use case, a water treatment system. However, this model only considers physical elements such as controllers, valves, pumps, and sensors. The knowledge graph that we have presented in Sect. 3.4 shows the potential of the model in comparison to traditional graphs. With this supplementary knowledge, the anticipation of cascading effects is more accurate.

Advances in these research axes are essential to implement the resilience principles in smart infrastructures. However, we must note that the resilience concept can conflict with other objectives. Similar to the cyber security principle that was not taken into account by all organizations during the last decades, the resilience concept is not sufficiently valued and considered. This can be explained by the fact that increasing the resilience potential of an architecture is related to augmenting the diversity of the controllability and observability aspects [14]. This augmentation has a cost. Several organizations are not prepared to spend money to increase the resilience potential of their systems. Resilience also increases the complexity of an architecture. It is known that adding more components connected to the cyber world increases the attack surface. Increasing the complexity also increases the risk of cascading effects due to the high density of links between each component or function of a system.

In our opinion, accepting resilience is highly related to regulation. Indeed, a common basis must be established to ensure that resilience principles have been officially analyzed and proven to be of paramount importance to organizations. The Cyber Resilience Act [8,10] is an initiative that works on three guarantees: (i) providing harmonized rules when products or software including a digital component are brought to market; (ii) providing a cyber security requirements framework for the governance of the planning, design, and development and the maintenance of such products. Obligations must also be met at each stage of the value chain; (iii) providing a duty of care for the whole lifecycle of these products. Other initiatives, such as the General Data Protection Regulation (GDPR) [31], which deals with data protection across the European Union, highlight the importance of regulations across several countries. These initiatives are very important to know which entity, organization, state or country is responsible in case of a problem with the conformity of products to resilience principles.

6   Conclusion

As the main conclusion, we can state that considering and applying resilience principles are based on several pillars. The first pillar is understanding the systems to be considered and acquiring sufficient knowledge about their behavior and missions to model them accurately. The second pillar is layered modeling for complex systems because the physical and cyber views depend on other levels, such as the mission, stakeholders, and business processes. The third pillar is the need to consider systems holistically, which involves threats from everywhere. The more we understand an architecture, the more we can avoid cascading effects with damaging consequences. In the context of such a holistic view, knowledge graphs have an exciting potential. They allow the modeling of various entities, their relationships, and structures. The absence of international regulations indicates that work remains to ensure that resilience principles are understood, accepted, and applied.

Acknowledgments. Authors acknowledge support from the European Commission (Horizon Europe projects DYNAMO and AI4CCAM, under grant agreements 101069601 and 101076911), the French Government (PFS project, under the "France 2030" program) and the Natural Sciences and Engineering Research Council of Canada (NSERC).

References

1. AlHidaifi, S.M., Asghar, M.R., Ansari, I.S.: A survey on cyber resilience: key strategies, research challenges, and future directions. ACM Comput. Surv. (2024). https://doi.org/10.1145/3649218
2. Barbeau, M., Cuppens, F., Cuppens, N., Dagnas, R., Garcia-Alfaro, J.: Metrics to enhance the resilience of cyber-physical systems. In: 2020 IEEE 19th International Conference on Trust, Security and Privacy in Computing and Communications (trustCom), pp. 1167–1172 (2020). https://doi.org/10.1109/TrustCom50675.2020.00156
3. Barbeau, M., Cuppens, F., Cuppens, N., Dagnas, R., Garcia-Alfaro, J.: Resilience estimation of cyber-physical systems via quantitative metrics. IEEE Access **9**, 46462–46475 (2021). https://doi.org/10.1109/ACCESS.2021.3066108
4. Bella, G., Coles-Kemp, L.: Layered analysis of security ceremonies. In: Gritzalis, D., Furnell, S., Theoharidou, M. (eds.) SEC 2012. IAICT, vol. 376, pp. 273–286. Springer, Heidelberg (2012). https://doi.org/10.1007/978-3-642-30436-1_23
5. Bella, G., Curzon, P., Lenzini, G.: Service security and privacy as a socio-technical problem. J. Comput. Secur. **23**(5), 563–585 (2015)
6. Bella, G., Giustolisi, R., Schürmann, C.: Modelling human threats in security ceremonies. J. Comput. Secur. **30**(3), 411–433 (2022)
7. Berger, C., Eichhammer, P., Reiser, H.P., Domaschka, J., Hauck, F.J., Habiger, G.: A survey on resilience in the IOT: taxonomy, classification, and discussion of resilience mechanisms. ACM Comput. Surv. **54**(7), 1–39 (2021). https://doi.org/10.1145/3462513
8. Car, P., De Luca, S.: EU Cyber resilience act. EPRS, European Parliament (2022)
9. Carlos, M.C., Martina, J.E., Price, G., Custódio, R.F.: An updated threat model for security ceremonies. In: Proceedings of the 28th Annual ACM Symposium on Applied Computing, pp. 1836–1843. SAC 2013, Association for Computing Machinery, New York, NY, USA (2013). https://doi.org/10.1145/2480362.2480705
10. Chiara, P.G.: The cyber resilience act: the EU commission's proposal for a horizontal regulation on cybersecurity for products with digital elements: an introduction. Int. Cybersecur. Law Rev. **3**(2), 255–272 (2022)
11. Clédel, T., Boulahia Cuppens, N., Cuppens, F., Dagnas, R.: Resilience properties and metrics: how far have we gone? J. Surveill. Secur. Saf. **1**(2), 119–139 (2020)
12. iTrust (Center for Research in Cyber Security): Secure Water Treatment (SWaT Testbed). Technical report, SUTD (Singapore University of Technology and Design) (2021). version 4.4
13. Dagnas, R., Arabi, W., Yaich, R.: PFS L1.2 Étude prospective, description des métiers du port et navire du futur et architectures associées. Technical report, IRT SystemX (2023)

14. Dagnas, R., Barbeau, M., Boutin, M., Garcia-Alfaro, J., Yaich, R.: Exploring the quantitative resilience analysis of cyber-physical systems. In: 2023 IFIP Networking Conference (IFIP Networking), pp.1–6 (2023). https://doi.org/10.23919/IFIPNetworking57963.2023.10186355

15. Ellison, C.: Ceremony Design and Analysis. Cryptology EPrint Archive (2007)

16. Foote, A.: Woe-train: Ottawa's LRT troubles, by the numbers. CBC (2019)

17. Giustolisi, R.: Free rides in Denmark: lessons from improperly generated mobile transport tickets. In: Lipmaa, H., Mitrokotsa, A., Matulevičius, R. (eds.) NordSec 2017. LNCS, vol. 10674, pp. 159–174. Springer, Cham (2017). https://doi.org/10.1007/978-3-319-70290-2_10

18. Haimes, Y.Y.: On the definition of resilience in systems. Risk Anal. Int. J. **29**(4), 498–501 (2009)

19. Hankel, M., Rexroth, B.: The reference architectural model industrie 4.0 (rami 4.0). Zvei **2**(2), 4–9 (2015)

20. Häring, I., Ebenhöch, S., Stolz, A.: Quantifying resilience for resilience engineering of socio technical systems. Eur. J. Secur. Res. **1**, 21–58 (2016)

21. Hutson, G., Jackson, M.: Graph Data Modeling in Python: A Practical Guide to Curating, Analyzing, and Modeling Data with Graphs. Packt Publishing, Birmingham (2023)

22. Johansen, C., Jøsang, A.: Probabilistic modelling of humans in security ceremonies. In: Garcia-Alfaro, J., et al. (eds.) DPM/QASA/SETOP -2014. LNCS, vol. 8872, pp. 277–292. Springer, Cham (2015). https://doi.org/10.1007/978-3-319-17016-9_18

23. Kott, A., Linkov, I.: Cyber Resilience of Systems and Networks, 1st edn. Springer, Cham (2018)

24. Lenzini, G., Mauw, S., Ouchani, S.: Security analysis of socio-technical physical systems. Comput. Electr. Eng. **47**, 258–274 (2015)

25. Lewis, T.G.: The many faces of resilience. Commun. ACM **66**(1), 56–61 (2022)

26. Lin, S.W., et al.: Industrial internet reference architecture. Industrial Internet Consortium (IIC), Technical report (2015)

27. Martina, J.E., Carlos, M.C.: Why should we analyse security ceremonies. Proc. CryptoForma (2010)

28. Martinez, A.G.: Chaos Monkeys: Obscene Fortune and Random Failure in Silicon Valley. Harper Business, Manhattan (2016)

29. Mathur, A.P., Tippenhauer, N.O.: SWaT: a water treatment testbed for research and training on ICS security. In: 2016 International Workshop on Cyber-Physical Systems for Smart Water Networks (CySWater), pp. 31–36 (2016). https://doi.org/10.1109/CySWater.2016.7469060

30. Nowak, V., Ullrich, J., Weippl, E.: Cybersecurity is more than a technological matter-towards considering critical infrastructures as socio-technical systems. Appl. Cybersecur. Int. Gov. **1**, 138–143 (2023)

31. Parliament, E., Council, E.: General data protection regulation. Official J. Eur. Union **59**, 294 (2016)

32. Patriarca, R., Falegnami, A., Costantino, F., Di Gravio, G., De Nicola, A., Villani, M.L.: Wax: an integrated conceptual framework for the analysis of cyber-socio-technical systems. Saf. Sci. **136**, 105142 (2021)

33. Radke, K., Boyd, C., Gonzalez Nieto, J., Brereton, M.: Ceremony analysis: strengths and weaknesses. In: Camenisch, J., Fischer-Hübner, S., Murayama, Y., Portmann, A., Rieder, C. (eds.) SEC 2011. IAICT, vol. 354, pp. 104–115. Springer, Heidelberg (2011). https://doi.org/10.1007/978-3-642-21424-0_9

34. Rao, K., Hanjra, M.A., Drechsel, P., Danso, G.: Business models and economic approaches supporting water reuse. Wastewater: Economic asset in an Urbanizing World, pp. 195–216 (2015)
35. Ryalat, M., ElMoaqet, H., AlFaouri, M.: Design of a smart factory based on cyber-physical systems and internet of things towards industry 4.0. Appl. Sci. **13**(4), 2156 (2023)
36. Sempreboni, D., Viganò, L.: X-men: a mutation-based approach for the formal analysis of security ceremonies. In: 2020 IEEE European Symposium on Security and Privacy (EuroS&P), pp. 87–104 (2020). https://doi.org/10.1109/EuroSP48549.2020.00014
37. Sikos, L.F.: Cybersecurity knowledge graphs. Knowl. Inf. Syst. **65**, 1–21 (2023)
38. Sinibaldi, T.: Les smart ports. Radar International des Solutions Smart Appliqées aux Ports de Commerce, Wavestone (2019)
39. SystemX, I.: PFS: secure ports of the future. IRT SystemX Website (2023)
40. Van Mieghem, P., Omic, J., Kooij, R.: Virus spread in networks. IEEE/ACM Trans. Netw. **17**(1), 1–14 (2008)
41. Viganò, L.: Formal methods for socio-technical security. In: ter Beek, M.H., Sirjani, M. (eds.) COORDINATION 2022. IFIP Advances in Information and Communication Technology, vol. 13271, pp. 3–14. Springer, Cham (2022). https://doi.org/10.1007/978-3-031-08143-9_1
42. Wang, X., Feng, L., Kooij, R.E., Marzo, J.L.: Inconsistencies among spectral robustness metrics. In: Duong, T.Q., Vo, N.-S., Phan, V.C. (eds.) Qshine 2018. LNICST, vol. 272, pp. 119–136. Springer, Cham (2019). https://doi.org/10.1007/978-3-030-14413-5_10
43. Yousefpour, A., et al.: All one needs to know about fog computing and related edge computing paradigms: a complete survey. J. Syst. Architect. **98**, 289–330 (2019). https://doi.org/10.1016/j.sysarc.2019.02.009
44. Zahee, M.A.: RAMI 4.0 (part 1): smart electronic industry 4.0 architecture layers. DZone (2017)
45. Zhang, K., Liu, J.: Review on the application of knowledge graph in cyber security assessment. In: IOP Conference Series: Materials Science and Engineering, vol. 768, no. 5, p. 052103 (2020). https://doi.org/10.1088/1757-899X/768/5/052103

Query Integrity in Smart Environments

Sabrina De Capitani di Vimercati$^{(\boxtimes)}$ (ID), Sara Foresti$^{(\boxtimes)}$ (ID),
and Pierangela Samarati$^{(\boxtimes)}$ (ID)

Computer Science Department, Università degli Studi di Milano, Milan, Italy
{sabrina.decapitani,sara.foresti,pierangela.samarati}@unimi.it

Abstract. Our smart society strongly relies on data, which are continuously generated, collected, stored, and processed by millions of connected IoT devices and smart sensors. Such data are at the basis of typically complex decision-making processes that require advanced analytics.

Due to the vast and increasing amount of data, their storage and processing are often outsourced to third parties (e.g., service providers and decentralized computational services) that might be not fully trustworthy in their operating. In this chapter, we focus on the problem of assessing integrity of query computations involving external service providers, and illustrate possible approaches for enabling the verification of the integrity of query results. We will cover both deterministic approaches, based on the definition of authenticated data structures over the data and giving full integrity guarantees, and probabilistic approaches, based on the insertion of control information in the data and providing probabilistic integrity guarantees.

Keywords: Query integrity · deterministic techniques · probabilistic techniques

1 Introduction

The advancements of digital and smart technologies (e.g., Internet of Things, big data analytics, and 5G/6G connectivity) are at the basis of today's smart society that supports new applications in a variety of sectors, also thank you to the availability of a powerful hyperconnected infrastructure offering unprecedented network capacity and speed. Distributed sensors, mobile and pervasive devices, cloud/edge/fog computational and storage nodes, can all be involved in providing advanced storage and computational services and applications. At the center of such novel scenarios are *data*, gathered, generated, shared, processed, and communicated among the different components of the infrastructure at an incredible pace. The possibility of efficiently performing analysis on such data for making data-informed decisions becomes then extremely important. Often data storage, as well as data analytics, are outsourced to external providers or rely on the involvement of a distributed framework for storing and processing large

© The Author(s), under exclusive license to Springer Nature Switzerland AG 2025
N. Pitropakis and S. Katsikas (Eds.): *Security and Privacy in Smart Environments*, LNCS 14800, pp. 25–48, 2025.
https://doi.org/10.1007/978-3-031-66708-4_2

datasets (e.g., peer-to-peer networks [6], Apache Hadoop [20], and Spark [40]). Outsourcing data and data analytics brings several advantages, including cost savings, increased efficiency, and flexibility. However, such advantages come at the price of the data owners losing control over their own data and processing, and introducing therefore the problem of their proper protection [12].

The problem of protecting data and computations managed by an external service provider, allowing data owners to keep the control over them, has many facets. The service provider can be *honest-but-curious* (i.e., the service provider is trusted for the management of data but, at the same time, it is not trusted with respect to the data content, which should remain confidential), or can be *lazy* (i.e., the service provider might not be considered fully trustworthy and, for example, might delete data that are accessed rarely or omit some computations to save resources) or *malicious* (i.e., the service provider may intentionally behave improperly in the storage and processing of data) and its behavior should be controlled. Depending on the trust assumption on the service provider, there are different security problems that need to be addressed, including the confidentiality and integrity of data and computations, the management and specification of policies, the exposure to different cyber-attacks, and the reliability and availability of services (e.g., [2,12,14,17,21]). For instance, natural solutions for protecting data confidentiality are based on *encryption* [30] (i.e., data are protected by encrypting them before their storage at external service providers), and *fragmentation* [1,5] (i.e., data are split in different non-linkable fragments to protect sensitive associations). Data integrity is usually ensured through the application of solutions that rely on *encryption* such as digital signatures, Proof Of Retrievability, and Provable Data Possession (e.g., [3,22]).

When the trustworthiness of the service provider cannot be taken for granted, data owners may be concerned about the correctness of the results (retrieved data or queries) received from the service provider. As a matter of fact, the lack of control of data owners and the open nature of the adopted storage and processing platforms may open the door to possible misbehavior by the providers involved in data storage and computation. It is therefore important to design efficient and practical techniques that enable data owners and, in general, clients asking the service provider to perform a query over the data, to assess the integrity of every computation performed by the service provider.

The goal of this chapter is to provide an overview of the main techniques for verifying the *integrity* of query results. The remainder of the chapter is organized as follows. Section 2 presents the reference scenario and discusses the integrity verification objectives (i.e., correctness, completeness, and freshness) together with the main characteristics of the two categories (i.e., deterministic and probabilistic) of integrity verification techniques. Section 3 describes the main deterministic techniques, that is, solutions that provide deterministic integrity guarantees, relying on authenticated data structures. Such techniques include those based on signatures, tree-based data structures, and list-based data structures. Section 4 presents the main techniques (sentinels and twins) that provide probabilistic integrity guarantees, and illustrates how to assess the completeness of join queries. Section 5 discusses some open research directions. Finally, Sect. 6 gives our conclusions.

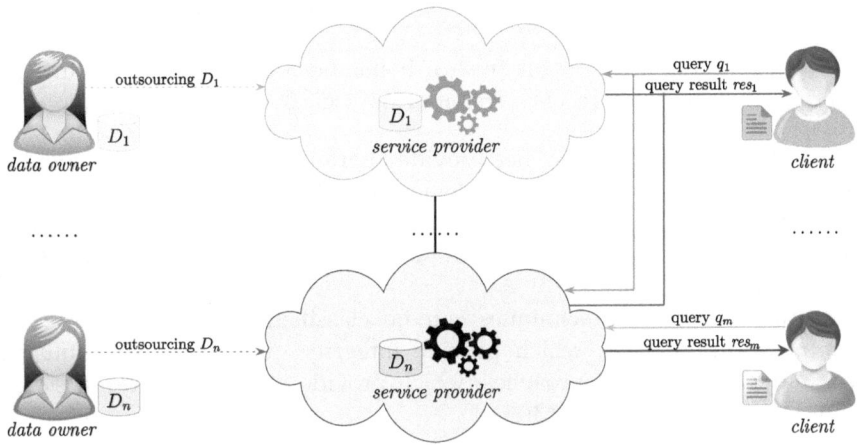

Fig. 1. Reference outsourcing scenario

2 Query Integrity Verification

We consider a data outsourcing scenario that is usually characterized by three interacting parties: data owners, service providers, and clients (see Fig. 1). The *data owner* is a company or an individual who outsources their data to an external *service provider*. The service provider offers storage and/or computation resources for the management of data. The *client* is a company or an individual who can require the execution of a computation over the outsourced data. For concreteness, we will consider the evaluation of queries over an (outsourced) dataset composed of one or more relations. Each query can involve the data managed by a service provider or can also involve data stored and managed by different service providers (e.g., a join query), which have to collaborate for query execution. If the service providers are not trustworthy, they can return incorrect query results. There can be multiple reasons for a query result to be incorrect, including a temporary misconfiguration, or a malicious action by the service provider (which may want to either sabotage the execution of the query or get the reward for computing queries while omitting to do so). Note that there is an integrity issue regardless of whether the incorrect result has been caused by failure, malfunctioning, sloppiness, or intentional opportunistic behavior since they all have the effect of the service provider not correctly computing the queries submitted to it. The problem then arises of providing clients with the ability to assess the integrity of query results, that is, to assess whether, given a query q, the result *res* obtained by the execution of q over the outsourced dataset D, denoted $q(D)$, (i.e., $res = q(D)$) is *correct*, *complete*, and *fresh*.

- *Correct.* The query result *res* is correct if the data items in *res* have been obtained from the execution of q on dataset D, and have not been tampered with.
- *Complete.* The query result *res* is complete if no valid data is missing from the result.
- *Fresh.* The query result *res* is fresh if it has been obtained by executing the query q over the most recent version of dataset D.

Most of the current approaches provide guarantees of completeness and correctness, with a few proposals complementing them with timestamps or periodical refreshing to provide freshness guarantees. In the following, we present some of the most well-known solutions proposed for ensuring completeness and correctness of query results.

Integrity verification techniques can be classified in two main categories: *deterministic* approaches, which provide integrity guarantees with full confidence, and *probabilistic* approaches, which provide integrity guarantees with a certain degree of confidence [13].

Deterministic Approaches (Sect. 3). Deterministic approaches are typically based on the generation of a proof, called *Verification Object* (VO), which provides deterministic integrity guarantees. Intuitively, given a query q, the service provider executes the query over the outsourced dataset D and constructs a VO for the query result $res = q(D)$. The service provider then returns to the client the query result *res* together with the verification object. The construction of the verification object is based on an *authenticated data structure* that is built over D by the owner of the data and is stored together with the data at the external service provider. In addition to the authenticated data structure, other pieces of information can be distributed to the client. The client then uses the VO and the information possibly received from the owner of the data to verify whether *res* satisfies certain properties (i.e., correctness, completeness, and freshness). Deterministic techniques can assess with full confidence the integrity of the result of queries defined over the attribute(s) on which the authenticated data structure has been defined; no guarantee is instead provided for queries using other attributes.

Other deterministic solutions are based on advanced cryptographic protocols (e.g., [24]) and trusted hardware (e.g., [42]). This chapter, however, focuses on those based on the definition of an authenticated data structure mainly because they are largely used and practical.

Probabilistic Approaches (Sect. 4). Probabilistic approaches provide probabilistic integrity guarantees, meaning that a query result that violates the integrity property passes the integrity check with a given probability. Such techniques enable the assessment of query integrity by injecting *control information* in the dataset. The advantage of the probabilistic approaches is a larger applicability than deterministic approaches, as they are not limited to operate considering queries defined over a specific attribute. The offered integrity guarantee is probabilistic because an integrity compromise can be detected only if it

HospitalPatient (HP)

	PId	PName	YoB	Doc
t_1	12	Amos	1961	d50
t_2	14	Bea	2000	d52
t_3	16	Cal	1985	d52
t_4	20	Dennis	1933	d54
t_5	22	Ethan	1973	d56
t_6	24	Frank	1965	d58
t_7	26	Grady	1953	d60
t_8	28	Helen	1987	d62
t_9	30	Ian	1987	d60
t_{10}	32	Loretta	1961	d60

stored at $\mathbb{H}$

FamilyDoctor (FD)

	DId	Name	Phone	Specialty
f_1	d50	Ann	123876	Cardiology
f_2	d52	Bart	784309	Allergy
f_3	d54	Carl	619345	Dermatology
f_4	d56	Dexter	914382	Nephrology
f_5	d58	Elen	903658	Cardiology
f_6	d60	Frank	814309	Urology
f_7	d62	George	357823	Psychiatry
f_8	d64	Hal	813456	Neurology

stored at $\mathbb{M}$

Fig. 2. Relations of the running example

affects the control information. Control information is of two types: non-genuine data (called *sentinels* or *markers*) injected in the original dataset, or controlled replication of data (*twinning*). Absence of a sentinel or of one replica (in the presence of the other) from a query result signals that the query result is not complete. The incorrect value for a sentinel or different results over twins signal that the query result is not correct. The guarantee is probabilistic because omissions or incorrect values of data items that are neither sentinels nor twins cannot be detected. The probability of detecting an integrity violation depends on the amount of sentinels and twins injected, as we elaborate next.

In the remainder of this chapter, we will describe the main deterministic and probabilistic integrity verification techniques. To fix ideas and make the discussion clear, the examples will refer to queries operating over the relations in Fig. 2. Relation HospitalPatient (HP) contains the identifier (attribute PId), name (attribute PName), year of birth (attribute YoB), and the identifier of the family doctor (attribute Doc) of the patients of a hospital $\mathbb{H}$, which is the owner of the relation. Relation FamilyDoctor (FD) contains the identifier (attribute DId), name (attribute Name), phone number (attribute Phone), and specialty (attribute Specialty) of family doctors who practice in Milan (Italy), and the municipality $\mathbb{M}$ of Milan is the owner of the relation.

3 Deterministic Approaches

Deterministic integrity verification techniques are based on the definition, and storage at the service provider, of an authenticated data structure built on one of the attributes (or set thereof) in the outsourced dataset. Deterministic approaches can be classified based on the kind of authenticated data structure used for integrity verification, which can be a chain of *signatures*, a *tree-based* structure, or a *list-based* structure, as illustrated in the following.

3.1 Signatures-Based Approaches

Signature-based techniques rely on the *digital signature* of tuples for generating a verification object (e.g., [4, 26, 41]). The data owner signs (e.g., using an RSA signature) each tuple of a relation, and outsources the relation where each tuple is complemented with its signature (e.g., [26]). When the service provider performs a query over the outsourced relation, the service provider returns to the client the query result together with a signature obtained by aggregating the signatures of the tuples in the query result. This aggregate signature forms the verification object of the query result. The aggregate signature can be computed according to different approaches (e.g., condensed RSA [13]). The client can then verify the correctness of the query result by recalculating the single aggregate signature combining the signatures associated with the returned tuples. The computation by the client of the aggregate signature requires a number of operations that is linear in the number of tuples in the query result. This technique permits to assess the correctness of the query result and the non-tampering of each tuple singularly taken but not the completeness of the query result. To provide completeness guarantees to query results, this technique can be extended by constructing an *authenticated chain* over the signatures associated with the tuples in the relation. Intuitively, tuples in the outsourced relation are ordered according to the values of an attribute a that is defined over a totally ordered domain. Let $\{t_0^{-\infty}, t_1, \ldots, t_n, t_{n+1}^{\infty}\}$ be the ordered list of tuples in the outsourced relation, with $t_0^{-\infty}$ and t_{n+1}^{∞} two fictitious tuples: $t_0^{-\infty}$ is a left delimiter and t_{n+1}^{∞} is a right delimiter. The owner signs each pair (t_i, t_{i+1}) of tuples, $i = 0, \ldots, n$, and stores the signature of each pair (t_i, t_{i+1}) together with t_{i+1}. The signature, denoted $s(t_{i+1})$, associated with tuple t_{i+1}, $i = 1, \ldots, n + 1$, is computed as: $s(t_{i+1}) = sign(h(t_i) \| h(t_{i+1}), sk)$, where h is a cryptographic hash function, $\|$ is the concatenation operator, $sign$ is a signature algorithm, and sk is the private key of the data owner.

Figure 3 illustrates an example of the chain of tuples built over attribute PId of relation HospitalPatient in Fig. 2. Tuples are first ordered according to the values of attribute PId, the left and right delimiters ($t_0^{-\infty}$ and t_{n+1}^{∞}) are added to the chain, and then the data owner computes the signatures as described above. Suppose now that a client submits query "SELECT * FROM HospitalPatient WHERE PId BETWEEN 15 AND 25". The result of this range query includes all the tuples in the relation with a value for attribute PId that falls in the query range [15, 25] (i.e., tuples $\{t_3, t_4, t_5, t_6\}$ with a green - dark gray on a b/w printout - background in Fig. 3) together with two left and right tuples, that is, the tuple (left) preceding the first tuple satisfying the range query and the tuple (right) following the last tuple satisfying the range query (i.e., tuples t_2 and t_7 with a light blue - light gray on a b/w printout - background in Fig. 3). The VO includes instead the signature of the tuples between t_3 and t_7 (the signature of tuple t_2 is not needed since this tuple is included in the signature of t_3), and an aggregate signature resulting from the combination of all the signatures of tuples between t_3 and t_7. The client can then verify the correctness and completeness of the query result by: *i)* checking if the set of returned tuples, together with their signatures, form a valid chain; *ii)* if the values for attribute `PId` of the returned tuples are within the query range (in our example, `PId` is between 15 and 25), and *iii)* if the boundary tuples are outside the query range (in our example, $14 < 15$ and $26 > 25$). If the signatures do not form a valid chain, the client can conclude that the query result is not correct or not complete. As an example, suppose that tuple t_4 has been omitted from the query result, meaning that the client receives the sequence t_3, t_5, t_6, t_7. By checking the signature associated with the returned tuple t_5, the client would discover that its value obtained removing the encryption layer of the signature (i.e., $h(t_4) \| h(t_5)$) is different from the one that the client can compute with the returned tuples (i.e., $h(t_3) \| h(t_5)$).

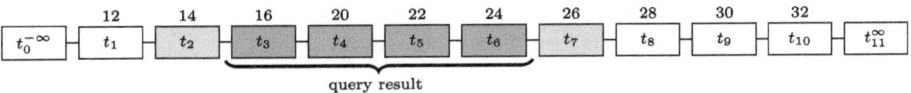

Fig. 3. An example of a chain of tuples of relation `HospitalPatient` in Fig. 2 ordered on attribute `PId` (for convenience of the reader the attribute value is reported above each tuple) (Color figure online)

3.2 Tree-Based Authenticated Data Structures

Most authenticated tree-based structures are based on variants of the Merkle Hash Tree (Merkle Tree, for short) structure. A Merkle Tree over a relation R is a binary tree that stores, in each leaf, the result of a one-way hash function h applied over a tuple of the original relation (e.g., h can be a collision-resistant hash function such as SHA-1). The tuples in the leaves of the Merkle Tree are ordered according to the values of an attribute a, defined over a totally ordered domain. The internal nodes store the result of the hash function applied over the concatenation of the values stored at their children. In other words, given an internal node n with children n_x and n_y, its hash value h_n is $h(h_{n_x} \| h_{n_y})$, where h_{n_x} and h_{n_y} are the hash values of n_x and n_y, respectively, and $\|$ is the concatenation operator. The root of the Merkle Tree is signed by the data owner. Figure 4 illustrates an example of a Merkle Tree defined over attribute `DId` of relation `FamilyDoctor` in Fig. 2. The hash of the root (i.e., $h_{12345678}$) is signed by the data owner with their private key (i.e., sk).

To verify the correctness and completeness of a point or range query over attribute a, the service provider returns to the client (together with the tuples resulting from the evaluation of the query) a VO that includes the values of the nodes in the Merkle Tree needed by the client to compute the hash value associated with the root of the tree. The client then computes the hash value of the root using the VO and the tuples in the query result, and checks whether such a value corresponds to the hash value of the root initially computed (and signed) by the data owner [15]. Note that, while being stored at the service provider, the Merkle Tree cannot be modified by the provider itself, since any

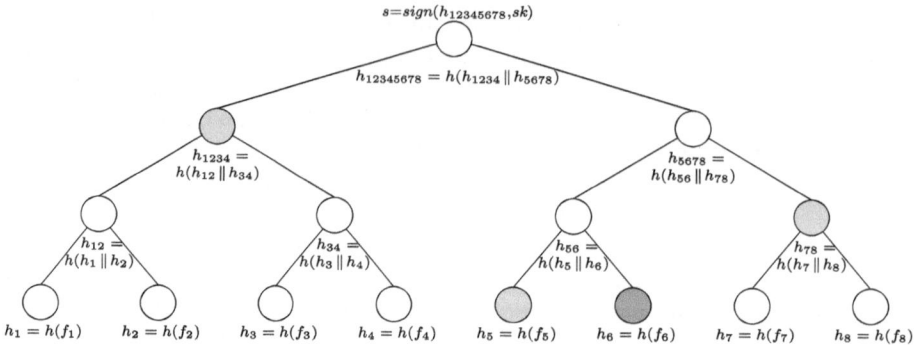

Fig. 4. An example of Merkle Tree built over the tuples of relation `FamilyDoctor` in Fig. 2 ordered according to attribute `DId` (Color figure online)

update to the structure would imply a change in the hash of the root and hence would invalidate its signature.

The computation of the VO depends on the kind of query evaluated (and hence on the set of tuples returned) and on the integrity guarantee (correctness and/or completeness) to be verified. In case of a point query (i.e., a query with a select condition of the form $a = $ 'v', with v a value in the domain of attribute a), the result includes one tuple only, and the VO contains the values of all the nodes being sibling of those in the path from the root to the leaf corresponding to the returned tuple. The use of a collision-resistant hash function for computing the hash values guarantees that it is computationally infeasible for a malicious service provider to return a fake tuple since this would imply the ability of finding a collision in the tree to pass integrity verification. In fact, the root of the Merkle Tree, being signed, should remain unchanged. The cost for verifying the correctness of a point query then corresponds to the computation of $log\ n$ hash values, where n is the number of nodes in the Merkle Tree. As an example of point query, consider relation `FamilyDoctor` in Fig. 2 and the Merkle Tree in Fig. 4 built over attribute `DId`. Suppose that a client submits query "SELECT * FROM `FamilyDoctor` WHERE `DId` = d60" that returns tuple f_6. To verify the correctness of the query result, tuple f_6 is returned together with a VO that contains the light blue nodes (light gray in b/w printout) in Fig. 4. The client can then compute the hash of tuple f_6 (green - dark gray in b/w printout - node) and combine it with the VO, as illustrated in the figure, to compute the hash of the root of the tree, which is then compared to the one computed (and signed) by the data owner. To verify the completeness of a query result, the VO must include the tuples corresponding to the two left and right nodes of the range of the returned tuple(s), together with the values of the nodes needed to compute the hash value of the root. When relying on a Merkle Tree, the size of the VO depends then on the size of the outsourced relation and is independent from both the number of tuples in the query result and from the width of the range covered by the query. Therefore, the approach based on the construction of a signature chain returns a smaller VO than a Merkle Tree for point queries, while the Merkle Tree returns a smaller VO for range queries covering more than *log n* tuples.

Variations of this basic tree-based integrity verification technique have been proposed to improve the efficiency of the verification process (e.g., [25,27]) and to support integrity verification of more complex (e.g., join) queries (e.g., [23,39]). In [23] the authors have proposed the *Merkle B-tree* (MB-tree) structure for supporting the efficient execution and verification of range queries over a single attribute. A MB-tree combines a Merkle Tree with a B+-tree, meaning that the nodes of the B+-tree are extended with a hash value associated with every pointer entry in the node. More precisely, in the leaf nodes, each tuple t is associated with a hash value $h(t)$ computed on the tuple itself. In the internal nodes, each pointer p_i to a child node is associated with hash value $h_i = h(h_{i1} \|| h_{i2} \|| \ldots \|| h_{im-1})$, with $h_{i1}, \ldots, h_{im-1}$ the hash values in the child node pointed by p_i, assuming a B+-tree of order m (i.e., a tree where the number of children of an internal node is at most m). Figure 5 illustrates an example of internal and leaf nodes of a MB-tree with order m. Similarly to Merkle Tree, the data owner signs the concatenation of the hash values stored in the root.

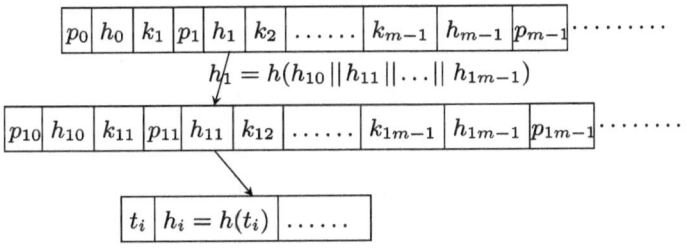

Fig. 5. An example of nodes in a MB-tree with order m

When a client submits a range query, the service provider executes the query over the dataset and returns the set of tuples that satisfy the query together with a VO. The VO is computed by visiting the MB-tree twice, to find the tuples at the left and at the right of the range, respectively. The VO then includes all the information needed to the client to reconstruct all the hash values that appear in the sub-tree whose leaf nodes contain the tuples that satisfy the range query.

Figure 6 illustrates an example of MB-tree that has been built over attribute PId of the tuples of relation HospitalPatient in Fig. 2. Suppose that the client submits query "SELECT * FROM HospitalPatient WHERE PId BETWEEN 15 AND 25". The service provider returns the query result $res = \{t_3, t_4, t_5, t_6\}$ together with the VO that the client uses to verify the completeness and correctness of the query result. In particular, the service provider includes in the VO the tuple on the left (t_2 in our example) and the tuple on the right (t_7 in our example) of those belonging to the query result, respectively. These left and right tuples allow the client to verify the completeness of the query result. The VO must also

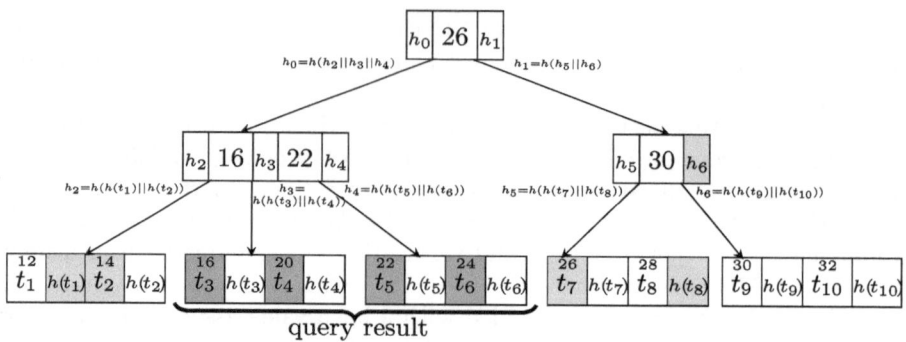

Fig. 6. An example of MB-tree with order $m = 4$ built over attribute PId of the tuples in relation HospitalPatient of Fig. 2 (Color figure online)

include all the hash values needed to the client for recomputing the hash values in the root. For the leaf nodes, the service provider includes in the VO the hash values of all the residual tuples (i.e., tuples that are not returned to the client) in the left and right leaves (i.e., the leaves that contain the left and right tuples). In the example, the VO includes $h(t_1)$ and $h(t_8)$. For the internal nodes, the service provider includes in the VO the hash values associated with all the pointers that appear in the nodes visited when searching for the left and right tuples except the hash value of the pointers that are on the right (left) of the pointer traversed when searching for the left (right) tuple. In the example, the service provider includes in the VO hash value h_6. Figure 6 illustrates the elements forming the VO with a light blue (light gray in b/w printout) background.

In [23] the authors compared a MB-tree structure defined over attribute a, with the adoption of a B+-tree structure over relation R for attribute a combined with a signature chain over a for R (Sect. 3.1). Indeed, both solutions efficiently support the evaluation of range queries and their integrity verification. While the adoption of the MB-tree structure implies that the size of the VO depends on the order of the MB-tree (and can therefore be larger than the VO defined when using a signature chain), the construction of the authenticated data structure is expected to be lower. To combine the advantages of the MB-tree structure with the ones of using a B+-tree with a signature chain over the tuples in the relation, in [23] the authors propose an alternative authenticated data structure (the embedded MB-tree).

3.3   List-Based Authenticated Data Structures

Besides tree-based structures, integrity verification approaches can rely on list-based authenticated data structures such as *skip lists* (e.g., [29]). A skip list defined for a set $\mathcal{K}$ of distinct key values includes a set of lists $L_0, L_1, \ldots, L_n$ such that: *i)* L_0 contains all the keys in $\mathcal{K}$ in non-decreasing order, together with special values $-\infty$ and $+\infty$ as first and last element in the list, respectively; *ii)* each list $L_i, i = 1, \ldots, n$, contains an arbitrary subset of the keys in L_{i-1};

iii) all lists $L_0, L_1, \ldots, L_n$ include values $-\infty$ and $+\infty$. Figure 7(a) illustrates an example of skip list defined over the set $\mathcal{K} = \{\infty\in, \infty\triangle, \infty\!\!\not\in\in, \in\!\!\not\}$ of values for attribute `PId` (i.e., patient identifier) of relation `HospitalPatient` in Fig. 2. This skip list includes three lists L_0, L_1, and L_2.

Skip lists have been designed in such a way to efficiently support search operations. The search for a key value v starts from $-\infty$ in the top list (i.e., L_n), and proceeds through the application of two operations: *hop forward*, and *drop down*. Hop forward means that the search proceeds right along the current list until the visited key value v_i is the largest value lower than or equal to v. Then, the search moves down of one step (i.e., from the current list L_j to L_{j-1}). The search iteratively applies the hops forward and drops down operations until it reaches the bottom list L_0. Figure 7(b) illustrates an example of the search process for value 22 in the skip list in Fig. 7(a). In the figure, visited nodes are denoted with bold lines, while bold arrows denote hop forward (horizontal) and drop down (vertical) operations.

Skip lists can be efficiently used to verify the integrity of point queries defined over an attribute with actual domain corresponding to $\mathcal{K}$. In this case, each node of the skip list defined over $\mathcal{K}$ is enriched with a *label*, computed through a *commutative* and *collision-resistant* hash function (i.e., a hash function h such that $h(x,y) = h(y,x)$ and such that its application to different inputs always returns different values). Given a node v of the skip list and the node w at its right, the label for v is computed as follows for L_0:

$$\ell(v, L_0) = \begin{cases} h(v,w) & \text{if } w \in L_1 \\ h(v, \ell(w, L_0)) & \text{otherwise} \end{cases} \tag{1}$$

and as follows for L_i with $i = 1 \ldots n$:

$$\ell(v, L_i) = \begin{cases} \ell(v, L_{i-1}) & \text{if } w \in L_{i+1} \\ h(\ell(v, L_{i-1}), \ell(w, L_i)) & \text{otherwise} \end{cases} \tag{2}$$

For instance, with respect to the skip list in Fig. 7(a), the label of node 22 on L_0 is computed as $\ell(22, L_0) = h(22, 26)$ and the label of node 14 on L_0 is computed as $\ell(14, L_0) = h(14, \ell(16, L_0))$. The label of node 12 on L_1 is instead computed as $\ell(12, L_1) = \ell(12, L_0)$ and the label of node 14 on L_1 is computed as $\ell(14, L_1) = h(\ell(14, L_0), \ell(22, L_1))$.

The label of the first node in the top list (i.e., L_n) is signed by the data owner and used for integrity check.

The verification of the integrity of a point query targeting value v consists in checking whether v is included in the skip list. In particular, if v belongs to the skip list, the integrity verification process verifies its presence. Otherwise, it verifies the existence of two values v' and v'', consecutive in list L_0, such that $v' < v < v''$. The verification process uses the information in the VO, which includes: the signed label of the starting node of the skip list and the labels of the nodes on the right and below the nodes in the path visited to reach v, which are necessary to recompute the label of the starting node of the skip list. If such recomputed value corresponds to the value signed by the data owner, the verification process succeeds. For instance, consider the skip list in Fig. 7(a) and a query targeting value 22. Figure 7(c) highlights the visited nodes with bold lines and the nodes included in the verification object with a light blue (light gray in b/w printout) background and with dashed lines. The verification object then corresponds to the list $\langle 22, 26, \ell(14, L_0), \ell(-\infty, L_1), \ell(26, L_2)\rangle$. The client formulating the query verifies the query result by hashing the values in the verification object and comparing the result with the label of the starting node of the skip list.

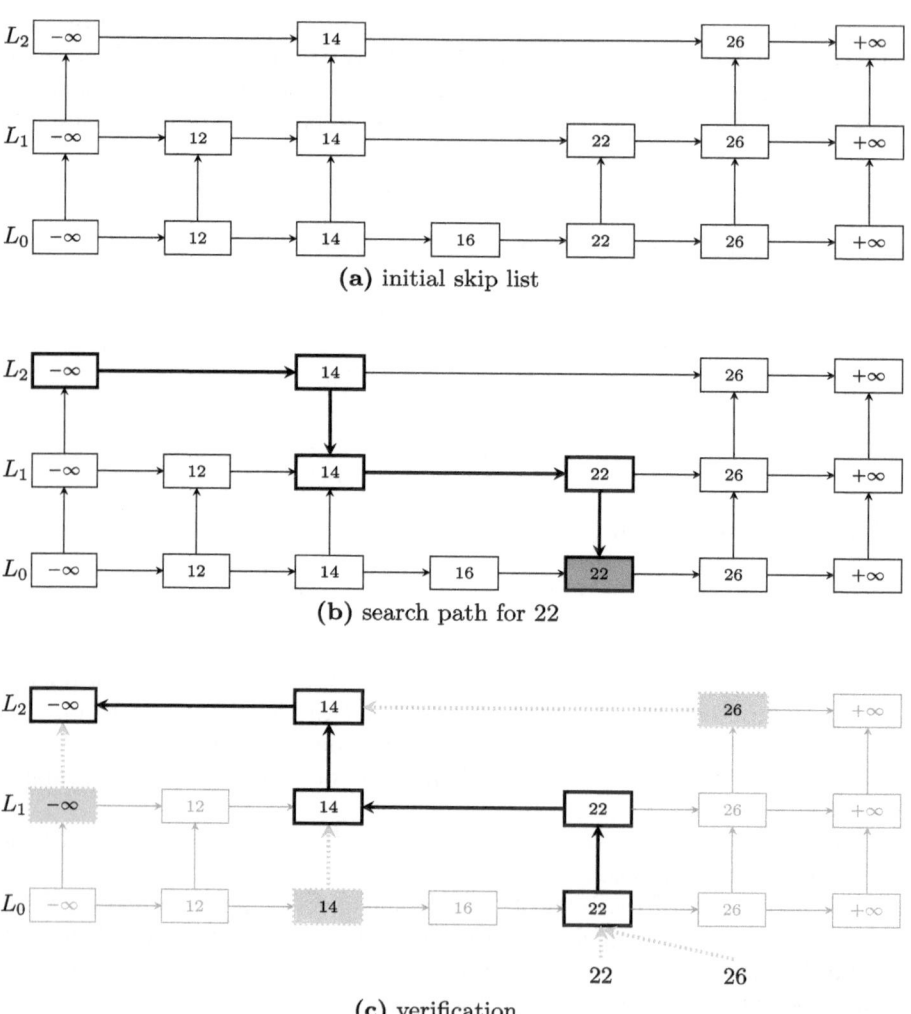

(a) initial skip list

(b) search path for 22

(c) verification

Fig. 7. A skip list for set $\mathcal{K} = \{\infty\in, \infty\triangle, \infty\notin\in, \in\}$ of keys with three lists (a), search process for value 22 (b), and verification object for a point query searching value 22 (c) (Color figure online)

The main advantage of skip lists over tree-based structures is that they can be efficiently managed by a relational database with a limited overhead at the client-side for integrity verification [16]. Such technique, however, permits the verification of the correctness and completeness of point queries only.

4   Probabilistic Approaches

Probabilistic techniques offer a probabilistic guarantee on the integrity of query results, that is, there is a (typically low) probability that an integrity viola-tion (e.g., omission of tuples in the query result) goes undetected. Probabilistic integrity verification approaches mainly focus on providing correctness and com-pleteness guarantees and most of them are based on the injection in the original dataset of non genuine data (called *sentinels* or *markers*) [19, 34–36], or the con-trolled replication of data (*twins*) [10, 32, 33]. The offered integrity guarantees are probabilistic because, while the omission of an expected sentinel or replica signals an integrity violation, its presence in the computation of the query result does not imply the integrity of the result. As a matter of fact, the service provider might have just been lucky in not missing any of the sentinels and/or twins inserted by the data owner. We now provide a more detailed description of the use of sentinels and twins for integrity verification (Sect. 4.1), and then describe their use for the verification of join queries (Sect. 4.2).

4.1   Probabilistic Integrity Controls

We describe two probabilistic integrity verification approaches (i.e., sentinels and twins) and their adoption for controlling the correctness and completeness of query results.

Sentinels. Sentinels are fake tuples inserted in a dataset before using it in a computation, and are built in such way to be indistinguishable from original data to the providers involved in the computation (e.g., [8, 36]). The insertion of a set S of sentinels is driven by the client requesting the execution of a query. Given a dataset D and a query q, the service provider executes the query over $D \cup S$ and returns to the client the query result *res*. The correctness of the query result is verified by the client by checking whether, for each sentinel in the query result *res*, the result of the evaluation of query q over the sentinel is the expected one. Absence in the query result *res* of one or more of the expected tuples (i.e., the tuples obtained by the execution of q over S, denoted $q(S)$) signals instead the fact that the query result is not complete. Otherwise, the query result is considered complete with a certain probability. As proved in [36], even a limited number of sentinels ensures high probabilistic guarantees of com-pleteness of the query result. The main drawback of the use of sentinels is that the client should store the set S of sentinels injected in the dataset to compare the query result received from the service provider with the evaluation of the query q over S. To avoid this drawback, some approaches (e.g., [36]) are based on the use of deterministic functions for sentinels generation. A client can then have knowledge and check sentinels without storing them. Another problem is related to the generation of sentinels, which should be: *i)* indistinguishable from the original tuples; and *ii)* generated to cover in a uniform way the domains of the attributes in the dataset, to maximize the probability that any query hits at least a sentinel. The approaches in [8,36] use encryption to prevent the service provider from distinguishing between sentinels and real tuples. Other approaches use specific functions that generate uniformly distributed sentinels (e.g., [19]). As an alternative, if the provider storing the data is assumed to be trusted, sentinels can be generated and injected on the fly, before sending the dataset to the provider in charge of query evaluation. In this case, sentinels can be generated in such a way that they belong to the query result. As an example of the use of sentinels, consider relation HospitalPatient in Fig. 2 and suppose to inject two sentinels (s_1 and s_2) in the relation as illustrated in Fig. 8(a), where sentinels have a yellow (light gray in b/w printout) background. Assume that a client submits query "SELECT * FROM HospitalPatient WHERE YoB ≤ 1970", asking for all patients born in 1970 or before. The client would expect sentinel s_2 to belong to the query result, that is, $res = \{t_1, t_4, t_6, t_7, t_{10}, s_2\}$. Absence of s_2 from the query result signals the incompleteness of the result. We note, however, that a result including tuples t_4, t_6, t_7, t_{10}, and s_2, while not complete (it misses t_1) would not violate the integrity check over sentinels.

HospitalPatient (HP)

	PId	Name	YoB	Doc
t_1	12	Amos	1961	d50
t_2	14	Bea	2000	d52
t_3	16	Cal	1985	d52
t_4	20	Dennis	1933	d54
t_5	22	Ethan	1973	d56
t_6	24	Frank	1965	d58
t_7	26	Grady	1953	d60
t_8	28	Helen	1987	d62
t_9	30	Ian	1987	d64
t_{10}	32	Loretta	1961	d64
s_1	18	Ben	1980	d60
s_2	25	Gloria	1970	d58

(a)

HospitalPatient (HP)

	PId	Name	YoB	Doc
t_1	12	Amos	1961	d50
t_2	14	Bea	2000	d52
t_2'	14	Bea	2000	d52
t_3	16	Cal	1985	d52
t_4	20	Dennis	1933	d54
t_5	22	Ethan	1973	d56
t_6	24	Frank	1965	d58
t_7	26	Grady	1953	d60
t_7'	26	Grady	1953	d60
t_8	28	Helen	1987	d62
t_9	30	Ian	1987	d64
t_{10}	32	Loretta	1961	d64

(b)

Fig. 8. Relation HospitalPatient in Fig. 2 enriched with sentinels (a) and twins (b) (Color figure online)

Controlled Replication (Twins). Controlled replication consists in replicating the tuples in the relation(s) involved in a query that satisfy a *replication condition* C_r. The service provider involved in the computation should not be able to identify pairs of replicated tuples. To verify the completeness and correctness of the query result, the client checks the presence of two identical copies for each tuple in the query result that satisfies the replication condition C_r. The presence of one copy only signals the incompleteness of the query result. Also receiving, for twin tuples, inconsistent results signals an incorrect result (i.e.,

at least one of the two results is incorrect). As an example, consider relation
HospitalPatient in Fig. 2 and assume to replicate tuples with PId equal to
14 or 26 (i.e., tuples t_2 and t_7), as illustrated in Fig. 8(b) where twins have a
orange (dark gray in b/w printout) background. Note that, for simplicity, in the
example we did not change twin tuples t'_2 and t'_7 to make them indistinguishable
from the corresponding twins (t_2 and t_7, respectively). Suppose that a client submits query "SELECT * FROM HospitalPatient WHERE Name = 'Bea'". The query
result should include both t_2 and t'_2. Absence of one of these tuples from the
query result signals its incompleteness. We note, however, that an empty result
would pass integrity verification.

Complementary Adoption of Sentinels and Twins. Sentinels and twins
have been proposed independently, and can be applied in isolation (i.e., either
sentinel or twins can be adopted to verify the integrity of a query result) or
can be applied in combination (e.g., [7–9]). Indeed, the complementary nature
of sentinels and twins ensures that their combined adoption provides stronger
integrity guarantees. As observed in [8,11] twins are twice as effective as sentinels in detecting omissions since the absence of any of the tuples in a twin pair
signals an integrity violation. However, twins lose effectiveness when the service
provider omits a large fraction of the tuples in the query result: the greater the
number of omitted tuples in the query result the more likely it is for the service
provider to omit twins in pairs and therefore to have the omission undetected
(e.g., the omission of all the tuples in the join result would pass the integrity
check based on twins only). This is then where sentinels come into help. In fact,
when the number of tuples omitted from the query result increases, the probability for the service provider to be undetected with respect to the sentinel
control decreases. This is due to the fact that the greater the number of omissions, the greater the probability of omitting a sentinel. The combined use of
a limited number of sentinels together with a limited number of twins ensures
then complementarity of controls and stronger integrity guarantees, as formally
illustrated by the analysis in [11].

4.2   Probabilistic Guarantees of Join Queries

In the discussion so far we have described how existing techniques can be adopted
for assessing the integrity of point and range queries. Often, however, data coming from multiple data owners need to be combined (joined) to extract useful
information. The verification of join queries is more complex than the verification
of point/range queries. In the following, we illustrate how sentinels and twins can
be also adopted for assessing the integrity of the result of join queries. In the discussion, we consider a client that wants to execute a join query over two relations,
denoted R_l and R_r, stored at two independent trusted service providers, S_l and
S_r, respectively. We distinguish service providers that offer storage resources and
that manage the relations on which the queries are performed, from *computational providers* that offer computational resources. We assume that the service
providers storing the relations are trustworthy while computational providers
are not. The execution of joins can be delegated to computational providers

because the adoption of the service providers for performing the joins (which are expensive operations) might not be the most economically viable solution or because they might not want to use their network and computational resources for performing queries on behalf of the client. In the following, we describe how to perform one-to-one or one-to-many join operations with sentinels and twins and with the help of a computational provider.

One-to-One Join. Correctness and completeness of the result of join queries performed over R_l and R_r is provided through the coordinated insertion of sentinels and twins in the relations before transmitting them to the computational provider. To have the guarantee that sentinels and twins also belong to the join result, the insertion of such checks is driven and coordinated by the client submitting the query. More precisely, the client determines the number of sentinels to be inserted in the two relations and their values for the join attribute, as well as the replication condition regulating the percentage of twins to be inserted in R_l and R_r. To avoid spurious tuples in the join result, the values of the join attribute for sentinels are chosen outside the domain of the original join attribute values. The twinning condition is defined over the join attribute, because it is the only attribute common between the two relations. The values of the join attribute for twinned tuples are combined with a random nonce. In this way, twinned tuples do not join with the original ones. The relations to be joined are encrypted before sending them to the untrusted computational provider. Note that to allow the computational provider to perform the join operation, the join attribute in the two relations R_l and R_r is encrypted separately using the same deterministic encryption algorithm and the same key. The encryption of the relations guarantees the confidentiality of the data and ensures the indistinguishability of sentinels and twins from the original tuples. The computational provider performs the join operation over the encrypted relations and returns the query result to the client. The client checks the integrity of the query result by analyzing sentinels and twins: an integrity violation is detected if a sentinel is missing or a tuple satisfying the twinning condition appears solo.

Figure 9 illustrates an example of one-to-one join query "SELECT DId, Name FROM FamilyDoctor JOIN HospitalPatient ON DId = Doc" retrieving the identifier and name of the family doctors with a patient in the hospital. For simplicity, the figure shows a simplified version of the schema of the relations in Fig. 2, and a simplified version of the instances that produces a one-to-one join. Also, we use the abbreviations FD and HP as name of the relations. Service providers $\mathbb{M}$ and $\mathbb{H}$, storing the two relations, first inject both sentinels and twins (which are the tuples that satisfy replication condition DId = d52 or Doc = d52, respectively). The resulting extended relations (FD* and HP*) are then encrypted on-the-fly and sent to the computational provider (in the figure, encrypted values are represented as Greek letters). The encrypted relations FD_k^* and HP_k^* have two attributes: I_k, the encrypted join attribute; and T_k, the encryption of all the attributes in the original relation (including the join attribute). The computational provider computes the natural join between the received encrypted relations and sends the result (J_k^*) to the client. The client decrypts J_k^* (J^*), verifies its completeness (i.e., if all the expected sentinels and twins are in J^*)

and correctness, and, if no omission is detected, projects over attributes `DId` and `Name`, and removes twins and sentinels to obtain the final join result `J` to be returned to the client.

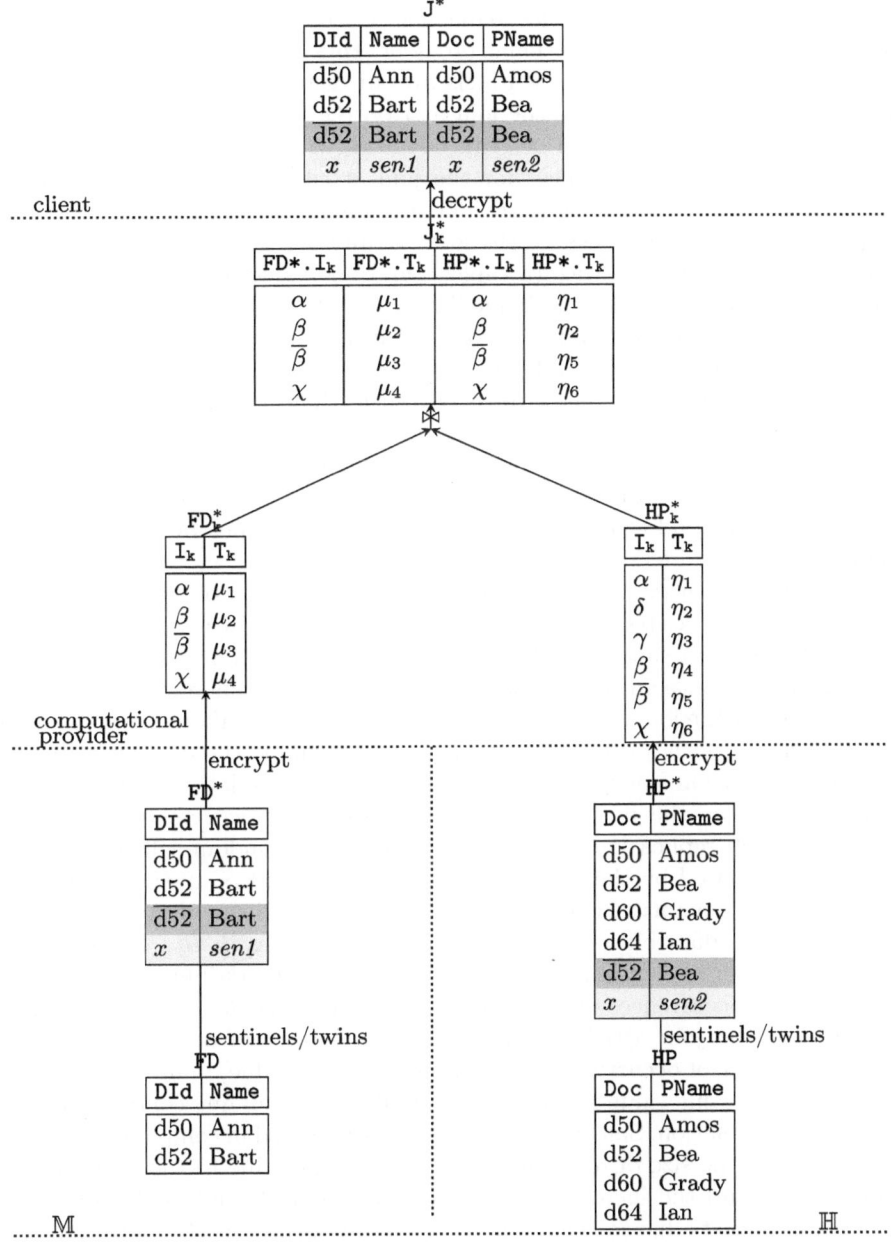

Fig. 9. An example of evaluation of a one-to-one join query with twins (orange - dark gray on a b/w printout - tuples) on d52' and one sentinel (yellow - light gray on a b/w printout - tuple) (Color figure online)

One-to-Many Join. The correctness and completeness of one-to-many joins can be verified combining sentinels and twins as illustrated for one-to-one join operations. However, when the join between relation R_l and relation R_r is a one-to-many join (we assume R_l to be the relation on side "one" and R_r to be the relation on side "many"), the frequencies of the values of the join attribute remain visible to the computational provider, thus possibly making twins and sentinels recognizable. Indeed, sentinels are all distinct and therefore multiple tuples with the same value for the join attribute cannot be sentinels. Also, the uncertainty on twin tuples can be reduced since twin pairs are characterized by the same number of occurrences for the join attribute. As a simple example, suppose that the computational provider receives a relation R_r where there are 8 tuples with 4 distinct values for the join attribute: one value occurs 3 times (3 tuples), two values occur twice (4 tuples), and one value occurs once (1 tuple). The computational provider can immediately infer that the three tuples with the same value for the join attribute cannot be sentinels, since sentinels have distinct values for the join attribute, and cannot be twins since twins are always in pairs and no other value in R_r has three occurrences. The two pairs of tuples with two occurrences each can be twins or can be genuine tuples. The only tuple with one occurrence can be a sentinel or a genuine tuple. The frequency distribution of the values of the join attribute can then compromise the indistinguishability property of twins and sentinels, and therefore it should not be revealed to the computational provider.

The approach in [8] flattens the frequency distribution of values of the join attribute of the tuples participating in a one-to-many join by using *salts*, *buckets*, or a combination of salts and buckets. Intuitively, salts aim at transforming a one-to-many join into an equivalent one-to-one join. The different occurrences of a same value for the join attribute in relation R_r are made different by combining each occurrence with a different random salt. To enable the correct evaluation of the join operation, each value of the join attribute in relation R_l is replicated as many times as the maximum expected number of occurrences of a value in R_r, and each replica is combined with a different random salt. Clearly, the values for salts used by the two service providers S_l and S_r must be the same, therefore their generation is coordinated. Buckets aim instead at guaranteeing a flat frequency distribution of the join attribute values in R_r. The idea is to make the number of occurrences of all values of the join attribute in R_r equal to the number of occurrences of the join value that appears more frequently in the relation. For the join attribute values with a number of occurrences smaller than the number of occurrences of the most frequent value, dummy tuples (i.e., tuples with the same value for the join attribute and a dummy content) are then inserted in the relation. Salts and buckets can also be used in combination to limit the increase of the size of relation R_l when using salts due to the replication of salted tuples, and the increase of the size of relation R_r and of the join result when using buckets due to the addition of dummy tuples in R_r. Note that salts and/or buckets operate on original as well as on control (i.e., sentinels and twins) tuples, to guarantee their indistinguishability. The number of salts and the size of buckets need to be coordinated among the client and the service providers, and the number of salts in particular need to be known to both S_l and S_r. The client is therefore in charge of choosing and communicating the number of salts to be used. The size of buckets can be autonomously computed by S_r based on the maximum frequency in relation R_r (i.e., by dividing such a frequency by the number of available salts).

Consider, as an example, the evaluation of the one-to-many join query "SELECT DId, Name FROM FamilyDoctor JOIN HospitalPatient ON DId = Doc" in Fig. 10, where each provider $\mathbb{H}$ and $\mathbb{M}$ injects in its relation one sentinel and twins tuples that satisfy the replication condition DId = d52 or Doc = d52. Like for the one-to-one join, the figure reports a simplified version of the schema of the relations in Fig. 2. We assume that the client sets the number of salts to 2. Buckets will have size 2, since the most frequent value in HospitalPatient (i.e., value d60) has 3 occurrences ($\lceil \frac{3}{2} \rceil = 2$). The 3 tuples with value d60 are then split in two buckets, one of which includes a dummy tuple. Similarly, sentinel x is included in a bucket with a dummy tuple, since by definition sentinels have one occurrence only. Figure 10 illustrates the evaluation of the join, which proceeds as already discussed for the one-to-one join in Fig. 9, with the addition of salts and buckets.

5   Open Issues

Although the problem of query integrity verification has been widely studied, there are still several interesting research directions that need to be further investigated, as summarized in the following.

- *Type of outsourced computation and data.* Existing integrity verification techniques mainly consider SQL queries (e.g., point, range, aggregate, and join queries), location-based range queries, and top-k queries as outsourced computations, and work on relational databases. An interesting research direction would then be the investigation of integrity verification techniques for other kinds of computations, (e.g., data mining, machine learning, data classification, and clustering) as well as for other kinds of data. Very few approaches have addressed the problem of verifying the integrity of queries formulated, for example, over spatial data or graph data (e.g., [18,31]). These solutions are based on a variation of the Merkle hash tree.
- *Distributed platforms.* The growing interest toward the use of distributed platforms (e.g., distributed cloud storage platforms) for processing large volumes of data goes along with the growing interest for designing integrity verification techniques that are able to verify the behavior of all the parties involved in the computation (e.g., [7]). These platforms are typically characterized by the presence of independent *workers* that collaboratively perform a computation. There are several aspects that need to be studied to efficiently and effectively apply the integrity verification techniques in such distributed context, including: *i)* the need to verify the behavior of all the workers; *ii)* the consideration of different trust assumptions on the workers involved in the

computation; and *iii)* the need to limit the overhead due to coordination of integrity verification. The presence of different workers also introduces the problem of collusion. As a matter of fact, the working of distributed platforms is typically based on the assumption that workers are independent and do not communicate. It would then be interesting to investigate what can happen when such independency cannot be assumed and some workers, under the control of a same subject, can communicate and collude to go undetected in their omissions. The design of probabilistic integrity verification techniques for distributed scenarios also requires the definition of a model for capturing the distribution among workers of sentinels and twins, their combined use, and their generation so to provide best effectiveness for integrity guarantees [11].

− *Freshness.* Most of the existing integrity verification techniques mainly focus on the correctness and completeness of query results. Only few proposals also address the problem of verifying the freshness of query results (e.g., [28,37]). An interesting research direction would then be the design of efficient integrity verification techniques (especially probabilistic) able to assess at the same time the correctness, completeness, and freshness of query results.

− *Combined application of integrity verification and other security approaches.* While integrity verification techniques might work well in isolation, their combined application with other approaches for protecting data/computations may open the door to new vulnerabilities that need to be addressed. As an example, the problem of combining integrity and query privacy in its different aspects (e.g., protection of the data, protection of single queries, and protection of query patterns) requires a careful analysis, investigating approaches that can balance the trade-off between protection enjoyed and performance overhead paid, allowing clients to tune the protection guarantees and overhead in different contexts and scenarios. A further example is represented by the combination of integrity and access control. In fact, existing query integrity verification techniques are typically based on the assumption that the client is authorized to access the whole outsourced datasets. Such an assumption, however, does not fit real world applications, which demand for selective access by different users (e.g., [38]).

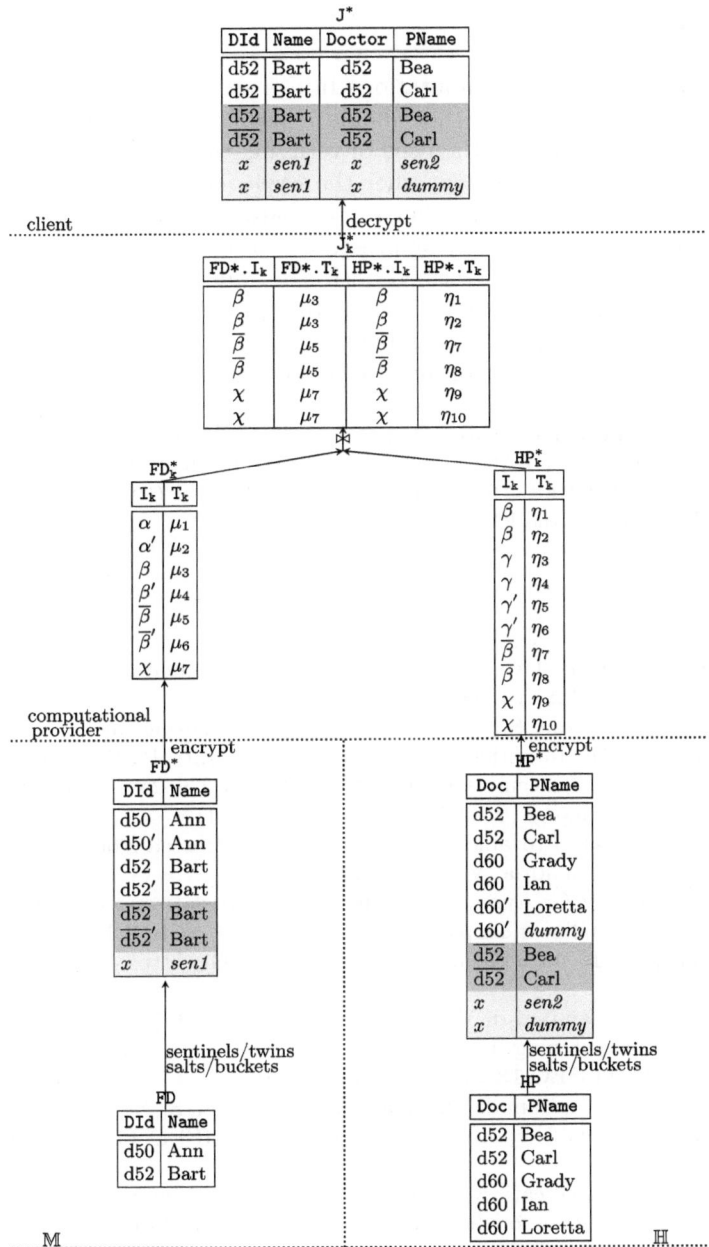

Fig. 10. An example of evaluation of a one-to-many join query with twins (orange - dark gray on a b/w printout - tuples) on 'd52', one sentinel (yellow - light gray on a b/w printout - tuple), two salts, and bucket of size two (Color figure online)

6   Conclusions

The outsourcing of query computations brings several advantages but, at the same time, requires solutions that should consider not only the protection of data confidentiality, but also the need of verifying the integrity of query results. Integrity is particularly important when the service providers involved in a query evaluation are not trustworthy, that is, they are not reliable for properly performing queries. In this chapter, we presented an overview of the main integrity verification techniques. We described both deterministic and probabilistic techniques, which differ on the kind of guarantees provided. For each category, we illustrated the main techniques, highlighting the verification cost. We concluded the chapter with a discussion on possible open research directions.

Acknowledgements. This work was supported in part by the EC under projects EdgeAI (101097300) and GLACIATION (101070141), by the Italian MUR under PRIN project POLAR (2022LA8XBH), and by project SERICS (PE00000014) under the MUR NRRP funded by the EU - NGEU.

References

1. Aggarwal, G., et al.: Two can keep a secret: a distributed architecture for secure database services. In: Proceedings of CIDR, Asilomar, CA, USA (2005)
2. Albanese, M., Jajodia, S., Jhawar, R., Piuri, V.: Securing mission-centric operations in the cloud. In: Jajodia, S., Kant, K., Samarati, P., Singhal, A., Swarup, V., Wang, C. (eds.) Secure Cloud Computing, pp. 239–259. Springer, New York (2014). https://doi.org/10.1007/978-1-4614-9278-8_11
3. Ateniese, G., et al.: Provable data possession at untrusted stores. In: Proceedings of ACM CCS, Alexandria, VA, USA (2007)
4. Boneh, D., Gentry, C., Lynn, B., Shacham, H.: Aggregate and verifiably encrypted signatures from bilinear maps. In: Biham, E. (ed.) EUROCRYPT 2003. LNCS, vol. 2656, pp. 416–432. Springer, Heidelberg (2003). https://doi.org/10.1007/3-540-39200-9_26
5. Ciriani, V., De Capitani di Vimercati, S., Foresti, S., Jajodia, S., Paraboschi, S., Samarati, P.: Combining fragmentation and encryption to protect privacy in data storage. ACM TISSEC **13**(3), 22:1–22:33 (2010)
6. Damiani, E., De Capitani di Vimercati, S., Paraboschi, S., Samarati, P.: Managing and sharing servents' reputations in P2P systems. IEEE TKDE **15**(4), 840–854 (2003)
7. De Capitani di Vimercati, S., Foresti, S., Jajodia, S., Livraga, G., Paraboschi, S., Samarati, P.: Integrity for distributed queries. In: Proceedings of IEEE CNS, San Francisco, CA, USA (2014)
8. De Capitani di Vimercati, S., Foresti, S., Jajodia, S., Paraboschi, S., Samarati, P.: Integrity for join queries in the cloud. IEEE TCC **1**(2), 187–200 (2013)
9. De Capitani di Vimercati, S., Foresti, S., Jajodia, S., Paraboschi, S., Samarati, P.: Optimizing integrity checks for join queries in the cloud. In: Atluri, V., Pernul, G. (eds.) DBSec 2014. LNCS, vol. 8566, pp. 33–48. Springer, Heidelberg (2014). https://doi.org/10.1007/978-3-662-43936-4_3

10. De Capitani di Vimercati, S., Foresti, S., Jajodia, S., Paraboschi, S., Samarati, P.: Efficient integrity checks for join queries in the cloud. JCS **24**(3), 347–378 (2016)
11. De Capitani di Vimercati, S., Foresti, S., Jajodia, S., Paraboschi, S., Sassi, R., Samarati, P.: Sentinels and twins: effective integrity assessment for distributed computation. IEEE TPDS **34**(1), 108–122 (2023)
12. De Capitani di Vimercati, S., Foresti, S., Livraga, G., Paraboschi, S., Samarati, P.: Privacy in pervasive systems: social and legal aspects and technical solutions. In: Colace, F., De Santo, M., Moscato, V., Picariello, A., Schreiber, F.A., Tanca, L. (eds.) Data Management in Pervasive Systems. DSA, pp. 43–65. Springer, Cham (2015). https://doi.org/10.1007/978-3-319-20062-0_3
13. De Capitani di Vimercati, S., Foresti, S., Livraga, G., Samarati, P.: Practical techniques building on encryption for protecting and managing data in the cloud. In: Ryan, P.Y.A., Naccache, D., Quisquater, J.-J. (eds.) The New Codebreakers. LNCS, vol. 9100, pp. 205–239. Springer, Heidelberg (2016). https://doi.org/10.1007/978-3-662-49301-4_15
14. De Capitani di Vimercati, S., Foresti, S., Samarati, P.: Protecting data and queries in cloud-based scenarios. SN Comput. Sci. **4**(5), 440 (2023)
15. Devanbu, P., Gertz, M., Martel, C., Stubblebine, S.G.: Authentic third-party data publication. In: Thuraisingham, B., van de Riet, R., Dittrich, K.R., Tari, Z. (eds.) DBSec 2000. IIFIP, vol. 73, pp. 101–112. Springer, Boston, MA (2002). https://doi.org/10.1007/0-306-47008-X_9
16. Di Battista, G., Palazzi, B.: Authenticated relational tables and authenticated skip lists. In: Barker, S., Ahn, G.-J. (eds.) DBSec 2007. LNCS, vol. 4602, pp. 31–46. Springer, Heidelberg (2007). https://doi.org/10.1007/978-3-540-73538-0_3
17. Donida Labati, R., Genovese, A., Piuri, V., Scotti, F., Vishwakarma, S.: Computational intelligence in cloud computing. In: Kovács, L., Haidegger, T., Szakál, A. (eds.) Recent Advances in Intelligent Engineering. TIEI, vol. 14, pp. 111–127. Springer, Cham (2020). https://doi.org/10.1007/978-3-030-14350-3_6
18. Fan, Z., Peng, Y., Choi, B., Xu, J., Bhowmick, S.: Towards efficient authenticated subgraph query service in outsourced graph database. IEEE TSC **7**(4), 696–713 (2014)
19. Ghazizadeh, P., Mukkamala, R., Olariu, S.: Data integrity evaluation in cloud database-as-a-service. In: Proceedings of IEEE SERVICES, Santa Clara, CA, USA (2013)
20. Apache hadoop. http://hadoop.apache.org/
21. Jhawar, R., Piuri, V.: Fault tolerance and resilience in cloud computing environments. In: Vacca, J. (ed.) Computer and Information Security Handbook, 2nd edn., pp. 125–141. Morgan Kaufmann (2013). ISBN: 978-0-1239-4397-2
22. Juels, A., Kaliski, B.: PORs: proofs of retrievability for large files. In: Proceedings of ACM CCS, Alexandria, VA, USA (2007)
23. Li, F., Hadjieleftheriou, M., Kollios, G., Reyzin, L.: Dynamic authenticated index structures for outsourced databases. In: Proceedings of SIGMOD, Chicago, IL, USA (2006)
24. Li, X., Weng, C., Xu, Y., Wang, X., Rogers, J.: ZKSQL: verifiable and efficient query evaluation with zero-knowledge proofs. PVLDB **16**(8), 1804–1816 (2023)
25. Mouratidis, K., Sacharidis, D., Pang, H.: Partially materialized digest scheme: an efficient verification method for outsourced databases. VLDB J. **18**, 363–381 (2009)
26. Pang, H., Jain, A., Ramamritham, K., Tan, K.: Verifying completeness of relational query results in data publishing. In: Proceedings of ACM SIGMOD 2005, Baltimore, MD, USA (2005)

27. Pang, H., Tan, K.: Authenticating query results in edge computing. In: Proceedings of ICDE, Boston, MA, USA (2004)
28. Pang, H., Zhang, J., Mouratidis, K.: Scalable verification for outsourced dynamic databases. PVLDB **2**(1), 802–813 (2009)
29. Pugh, W.: Skip lists: a probabilistic alternative to balanced trees. Commun. ACM **33**(6), 668–676 (1990)
30. Samarati, P., De Capitani di Vimercati, S.: Cloud security: issues and concerns. In: Murugesan, S., Bojanova, I. (eds.) Encyclopedia on Cloud Computing. Wiley (2016)
31. Tian, F., Wu, Z., Gui, X., Ni, J., Shen, X.: Fine-grained query authorization with integrity verification over encrypted spatial data in cloud storage. IEEE TCC **10**(3), 1831–1847 (2022)
32. Ulusoy, H., Kantarcioglu, M., Pattuk, E.: TrustMR: computation integrity assurance system for MapReduce. In: Proceedings of BigData, Santa Clara, CA, USA (2015)
33. Wand, H., Yin, J., Perng, C.S., Yu, P.: Dual encryption for query integrity assurance. In: Proceedings of ACM CIKM, Napa Valley, CA, USA (2008)
34. Wang, C., Chow, S., Wang, Q., Ren, K., Lou, W.: Privacy-preserving public auditing for secure cloud storage. IEEE TC **62**(2), 362–375 (2013)
35. Wong, W., Cheung, D., Kao, B., Hung, E., Mamoulis, N.: An audit environment for outsourcing of frequent itemset mining. PVLDB **2**(1), 1162–1172 (2009)
36. Xie, M., Wang, H., Yin, J., Meng, X.: Integrity auditing of outsourced data. In: Proceedings of VLDB, Vienna, Austria (2007)
37. Xie, M., Wang, H., Yin, J., Meng, X.: Providing freshness guarantees for outsourced databases. In: Proceedings of EDBT, Nantes, France (2008)
38. Xu, C., Xu, J., Hu, H., Au, M.: When query authentication meets fine-grained access control: a zero-knowledge approach. In: Proceedings of SIGMOD, Houston, TX, USA (2008)
39. Yang, Y., Papadias, D., Papadopoulos, S., Kalnis, P.: Authenticated join processing in outsourced databases. In: Proceedings of SIGMOD, Providence, RI, USA (2009)
40. Zaharia, M., Chowdhury, M., Franklin, M., Shenker, S., Stoica, I.: Spark: Cluster computing with working sets. In: Proceedings of HotCloud, Boston, MA, USA (2010)
41. Zheng, Q., Xu, S., Ateniese, G.: Efficient query integrity for outsourced dynamic databases. In: Proceedings of ACM CCSW, Raleigh, NC, USA (2012)
42. Zhou, W., Cai, Y., Peng, Y., Wang, S., Ma, K., Li, F.: VeriDB: an SGX-based verifiable database. In: Proceedings of SIGMOD (2021)

Quantifying the Impact Propagation of Cyber Attacks Using Business Logic Modeling

Marwan Lazrag⬤, Christophe Kiennert⬤, and Joaquin Garcia-Alfaro^(✉)⬤

SAMOVAR, Télécom SudParis, Institut Polytechnique de Paris, Palaiseau, France
{marwan.lazrag,christophe.kiennert,
joaquin.garcia-alfaro}@telecom-sudparis.eu

Abstract. Cyber-attacks affect the security properties of critical systems, such as confidentiality, integrity, and availability of crucial business activities. They also affect mission quality and performance. Existing risk assessment tools handling the problem still present some limitations, owing to the difficulty of describing the enterprise infrastructure, such as identifying assets, missions, and their dependencies. Furthermore, little research has been conducted to assess the impact propagation of external events on business entities.

In this chapter, we survey existing methods aiming to solve the aforementioned limitations. We focus on two main families: financial and operational impact assessment. The latter aims to specifically assess the impact propagation of cyber-attacks. For instance, cyber-attacks targeting the infrastructure assets and perturbing the execution and performance of the company's activities. It can also include the evaluation of the financial impact based on former financial assessment methodologies.

We also present a concrete operational impact propagation assessment contribution. This contribution extends previous work by enhancing the definition associated to organizational activities that might be impacted by cyber-attacks. It relies on business impact analysis via business logic modeling. It also includes metrics to quantify (i) the impact propagation probability on the business entities, and (ii) critical time (i.e., the time during which the business entity is not be impacted).

Keywords: Cybersecurity · Risk Analysis · Impact Assessment · Cyber-Attack · Impact Propagation · Resource Dependency Graph · Mission Dependency Graph

1 Introduction

Modeling the technical assets and missions of a company, as well as identifying the dependencies between them, can assist the security analyst in responding to the incident more effectively. With the discovery of new vulnerabilities and an increase in attacks aiming at compromising the confidentiality, integrity, and

N. Pitropakis and S. Katsikas (Eds.): *Security and Privacy in Smart Environments*, LNCS 14800, pp. 49–71, 2025.
https://doi.org/10.1007/978-3-031-66708-4_3

availability of business activities, as well as deteriorating mission quality and performance, assessing the impact of these external events may help the operator in determining the level of emergency and making decisions.

In addition to cyber-attacks, the limitations of existing cyber defense tools to protect missions and enterprise networks, some recent research has been conducted to quantify the impact of attacks based on attack graph tools and Common Vulnerability Scoring System (CVSS) scores. However, due to the difficulty of modeling the company's business functions and processes, little of this research has focused on assessing the operational impact of external events on business missions to protect critical infrastructure and complex enterprise architecture.

Previous work focuses on methodologies based on extending metrics such as CVSS scores, the Impact Factor (IF) to assess the attack impact [10]. For instance, Cao *et al.* [2] use CVSS scores over attack graphs, in order to compute an eventual business impact score. Operational impact assessment consists of estimating the impact of interrupting services and functionalities of missions, such as business functions and processes, due to an attack. For acquiring knowledge about the business activity of a company, we assume Business Process Model Notation (BPMN), seen as a common standard to derive the list of business functions and organizational processes.

Regarding the assessment of the propagation of impacts on business functions and processes, we evaluate the operational impact of an external event that has already occurred within the infrastructure, composed of technical assets and business entities. Our approach is not just aimed at studying the propagation of the attack within assets, but also assesses the operational impact at the level of the business entities.

We extend previous work [9,13–15] by including business logic into previous models. The added business logic model is a layered structure composed of assets and business entities, such as business functions and business processes. It consists of a graph-like structure, which links technical assets into business entities. The resulting model is used to assess the operational impact of shock events on missions, as well as to calculate the criticality of technical assets and estimate the downtime tolerance. Based on the business logic model, we implement a method for assessing the impact propagation of attacks on business entities, as well as to conduct a realistic case study on various business models to evaluate our method. To accomplish this, we define two metrics: (i) the probability of the impact on the business entities, and (ii) the critical time, which represents the time during which the business entity will not be impacted.

To identify the most critical assets and the most impacted missions, we assume graphical language for reasoning on functionalities such as: 1) calculation of the impact of events on the missions: impact probability and critical time, and 2) computation of the criticality of assets. The methodology is evaluated on realistic use cases and provides relevant results.

The focus of the contribution is on generating a business logic model, as well as demonstrating our method to assess the impact propagation of an external event into the nodes of this model. Then, using Monte-Carlo approximation, we can assess the scalability of the required computations needed by the system to evaluate the criticality of assets in some larger scenarios. We also focus on identifying the most critical assets in the infrastructure to determine which assets contribute the most to the propagation of the impact by assigning a criticality value to each asset. Our approach is based on the following assumptions:

- Generate the business logic model by creating the resource dependency model and merging it with the mission dependency model, which represents the business functions and processes of the enterprise and their interdependencies.
- Assess the criticality of the assets and determine the most critical assets in the infrastructure.
- Assess the impact probability of an external event on the different business entities of the company: *Business Functions, Business Processes* and the *Business Company*, and evaluate the impact when duplicating and backing up assets in the infrastructure.
- Calculate the critical time by deducing the shortest path from the shock event to the *Business Company*.

For the literature survey, we examined contributions from various sources of information, such as academic articles, books and case studies. We identified methodologies and works assessing the impact of cyber-attacks on companies. We conducted our search using keywords and technical terms such as *Impact Assessment, Risk Quantification, Business Logic Modeling, Financial Assessment Methodologies* and *Operational Impact Propagation Assessment*. We searched for publications published between 2004 and 2023.

The chapter is organized as follows. Section 2 surveys relevant related work. Section 3 describes our contribution. Section 4 presents a proof-of-concept tool, implementing all the concepts and models of our contribution. Section 5 discusses future directions for research. Section 6 concludes the chapter.

2   Related Work

In this section, we discuss two complementary family methodologies for assessing the impact of an cyber-attacks affecting the business functions and processes of a company given company. The first family focus on the quantification of financial aspects associated to the perpetration of attacks, while the second family focus on operational impact assessment. Tables 1 and 2 summarize the references and findings covered for each of these two families, as covered in this section.

2.1   Financial Impact Assessment

Earlier research on quantifying the impact of cyber-attacks starts with a focus on financial aspects. Some representative contributions in the literature are presented next (and summarized in Table 1).

Brian *et al.* [3] and Rainer *et al.* [1] focus on their works on economic security metrics. They conduct a survey of the state of knowledge on the cost of cyber-attacks. Two models will be discussed in our work: *Annual Loss Expectancy*

Table 1. Related work with a focus on financial impact assessment.

References	Contribution	Methodology	Metrics
– Brian et al. [3], (2004) – Rainer et al. [1], (2008)	– Survey the economic security metrics and the state of knowledge on the cost of cyber-attacks	– Use of security metrics and models for security investments and cyber-risk measurement	– Annual Loss Expectancy (ALE) – Return On Security Investment (ROSI)
– Freund and Jones [8], (2014)	– Propose the Factor Analysis of Information Risk (FAIR) metric and how to use it for impact assessment	– Discuss risk management using FAIR – Describe ontologies and terminology of the FAIR framework – Leveraging FAIR in risk decision-making and risk management	– Factor Analysis of Information Risk (FAIR)
– Dongre et al. [7], (2019)	– Develop a cost function to quantify the cost of the impact of data breaches	– Cost function as the sum of costs incurred by providers and consumers – Identify the cost components of the cost function for provider and consumers – Present two case studies: Equifax data breach (2017) and the Target data breach (2013)	– Cost function
– Orlando [18], (2021)	– Analysis of the role of Cyber Value at Risk (Cy-VaR) model in quantifying the cyber risk.	– Definition of the role of the Cy-VaR model – Highlight issues and difficulties in estimating Cy-VaR – Description of the role of Cy-VaR in supporting security investment decisions	– Cyber Value at Risk (Cy-VaR)

(ALE) and *Return On Security Investment* (ROSI). The ALE is represented as a quantitative metric for IT security, that can calculate the expected loss due to a risk in one year. Rainer [1] considers ROSI as a methodology for determining whether a firm should invest in implementing a security measure or not, i.e. it can be used to support a decision for or against implementing a security measure. Freund and Jones [8] the Factor Analysis of Information Risk (FAIR) framework. The FAIR framework aims to analyze risk and provide quantitative risk analysis. It provides a foundational understanding of risk, as well as risk assessment and analysis. Dongre *et al.* [7] quantify the cost of the impact of data breaches. In this chapter, the authors focus on data breaches that have exposed personal information. They present a mathematical function that expresses the cost impacts of data breaches. The developed cost function quantifies the cost of data breaches for providers and consumers. Orlando [18] presents the Cyber Value at Risk (Cy-VaR) model and its role in quantifying and measuring cyber-risk in the cyber security domain. The Cy-VaR model could provide an estimation and quantification of losses caused by cyber incidents, as well as support security investment decisions.

2.2   Operational Impact Assessment

Operational impact assessment aims at assessing the impact propagation of cyber-attacks and evaluating the perturbation of such attacks against the activities of a given organization. Next, we survey some representative works in the related literature. Table 2 summarizes our survey.

Liu *et al.* [11] illustrate the utility of a layered graphical model which has three layers: the upper layer, the middle layer and the lower layer that, on the one hand, model the tasks and missions and their inter-dependencies, and on the other hand, construct the attack scenarios and their inter-relationships in order to compute the impact of attacks on missions. Cao *et al.* propose in [2] a method to assess the impact of attacks on business process by generating an interconnecting graph from the attack graph showing the possible attack paths from the vulnerabilities to the target, and the entity dependency graph that contains three layers: asset layer, service layer, and business process task layer, as well as the dependencies between these three layers and on each individual layer, and calculating the impact score of the attack on the tasks that compose a business process. Musman *et al.* [17] compare the measures of effectiveness for the simulation of a mission under baseline conditions and the measures of effectiveness for the simulation of a mission under attack to evaluate the impact of an attack on a mission. The impact evaluation in this paper was based on mission models created using BPMN.

Jakobson [10] presents the impact assessment of a cyber attacks on missions by using the impact dependency graphs. In this paper, he presents a framework that quantifies the impact of attack on directly attacked assets and also calculates the cyber attack impact propagation through the nodes of the Impact Dependency Graph using operational capacity. Mukherjee and Mazumdar [16] provide a hierarchical model of a business process, and the metric *security concern* which is introduced as a new metric for measuring the security of a business process. This metric represents the impact on the business process of vulnerability exploitations in the context of a threat scenario.

Barreto *et al.* [6] propose the cyber argus framework, which helps to understand how to assess the impact of a cyber attack on missions and which critical assets contribute the most to accomplish the tasks performed in a mission. In

Table 2. Related work with a focus on operational impact assessment.

References	Contribution	Methodology	Metrics
– Liu *et al.* [11], (2017)	– Layered-graphical modeling – Impact quantifiers	– Calculate the impact score (from NIST NVD) – Assign weight to missions – Map LEGs (Logical Evidence Graph) to BPDs (Business Process Diagram) – Compute cumulative mission impact	– CMI (short for Cumulative Mission Impact)
– Cao *et al.* [2], (2018)	– Extend Ref. [19] via CVSS – Implement a tool that automatically generates an interconnected graph (interconnects the attack graph and the entity dependency graph) and compute the impact scores of an attack on tasks – Assess attack impact via business processes	– Generate the interconnected graph – Prune the interconnected graph – Compute the impact score based on the CVSS score	– Impact Scores
– Musman *et al.* [17], (2011)	– Assess the impact of a cyber attack on a mission – Compute the impact (by measuring the measures of effectiveness)	– Model creation –Compute metrics by simulating system's mission under different initial conditions – Categorize attack effects into categories and modify the mission simulation depending on the category of the attack – Compute attack impact as the difference between nominal vs. system under attack	– MoE (Measures of Effectiveness)
– Jakobson [10], (2011)	– Assess the impact of cyber attacks – Compute impact propagation	– Compute direct attack impact – Compute nodes' impact propagation sparabreak – Assess the impact of a cyber attack on a mission	– POC (Permanent Operational Capacity) – OC (Operational Capacity)
– Barreto *et al.* [6], (2013)	– Evaluate the mission impact	– Mission modeling – Collection of cyber and mission situation awareness & cyber impact assessment	– Cyber-ARGUS Framework Metrics
– Mukherjee and Mazumdar [16], (2019)	– Modeling the business process – Compute the *Security Concern* metric	– Identify the vulnerabilities – Analyze the possibility of exploiting vulnerabilities (by comparing the max_Effort with the minimum effort for exploiting a vulnerability) – Compute the impact on data items and software instances – Compute the impact on activities and information items (estimated from the impact on data items and software instances) – Calculate the Security Concern	– Security Concern
– Motzek *et al.* [9,13–15], (2015–2018)	– Define a mathematical model for mission impact modeling – Assess the mission impact	– Use the Monte-Carlo approximation to compute the conditional probability: • Find paths leading to external shock events • Monte-Carlo simulation	– Conditional Probability Metrics

their work, the mission model is designed using BPMN. To avoid developing the mission ontology from scratch, Cyber-argus integrates the previous work of D'Amico *et al.* [5] and Matheus *et al.* [12] into its own architecture. Motzek *et al.* present in [9,13–15] a mathematical mission impact assessment model. Their model takes into account external shock events. Their contribution includes a probabilistic graphical model, which is generated from mission and resource dependencies. In the sequel, we present a novel contribution expanding the work of Motzek *et al.*, specifically, expanding the theoretical background for the Business Logic Modeling in [13,14]. We also expand the approach, by generating a novel resource dependency model and automating the update of dependencies between technical assets and impact values.

3   Operational Assessment Using Business Logic Modeling

To further improve the assessment of the operational impact of external events associated to organizational missions, we assume the need of adding business logic to describe the interdependencies between technical assets to those other business functions and processes. Hence, we extend previous work in Refs. [9,13–15] with a novel Business Logic Model (BLM). More precisely, a new mission dependency model is derived, including the use of a novel resource dependency model. A more detailed description of the two models, as well as an example of the BLM of an *Online Shopping* company, are provided next.

3.1   Resource Dependency Model

The resource dependency model represents dependencies between assets. These dependencies are defined using a traffic matrix that quantifies the amount of data exchanged between each pair of assets. The traffic matrix can be generated from, e.g., NetFlow data [4], representing network traffic flows (i.e., datagrams in packet-switched network) collected from routers. The matrix is eventually processed to build a probability matrix, which contains conditional impact probabilities. The list of assets in the resource dependency model and the amount of data exchanged between assets can be retrieved and updated periodically (for instance, every 24 h).

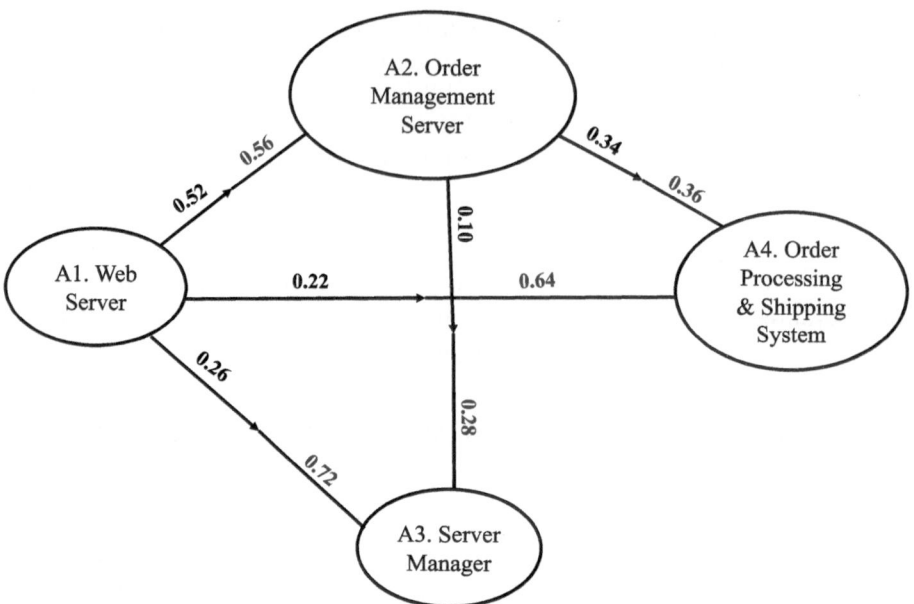

Fig. 1. Resource dependency model example. It provides a high-level representation of the interactions and dependencies between four different technical assets of the organization, depicted as vertices *A1* (*Web Server*), *A2* (*Order Management Server*), *A3* (*Server Manager*), and *A4* (*Order Processing & Shipping System*). The edges represent the interdependencies between assets. For a bidirectional impact, the value of the forward impact probability is colored in dark blue, and the value of the backward probability is colored in light blue. For instance, the impact probability from *A1* to *A3* is 0.72. The impact probability from *A3* to *A1* is 0.26.

Table 3. Matrix associated to the resource dependency model depicted in Fig. 1.

	A1	*A2*	*A3*	*A4*
A1	0.0	0.56	0.72	0.64
A2	0.52	0.0	0.28	0.36
A3	0.26	0.10	0.0	0.0
A4	0.22	0.34	0.0	0.0

Figure 1 depicts an example of our resource dependency model. We assume it is obtained from the probability matrix shown in Table 3. The example contains four representative assets (identified as *A1. Web Server*, *A2. Order Management Server*, *A3. Server Manager* and *A4. Order Processing & Shipping System*) and their interdependencies. Interdependencies between assets are represented by the edges. For a bidirectional impact, the value of the forward impact probability is colored in dark blue, and the value of the backward probability is colored in light blue. For example, the impact probability from *A1* to *A2* is 0.56, and the impact probability from *A2* to *A1* is 0.52.

3.2   Mission Dependency Model

The mission dependency model describes the interdependencies between business functions and processes, as well as the assets that directly support business functions. A Business Process Model Notation (BPMN) model and expertise in the business activity of a company are needed to derive the list of business functions and processes and build the mission dependency model. The mission dependency model can be built manually, using such an expert knowledge from the company's business activities.

Table 4. Mission Dependency Model.

	BF1	BF2	BF3	BF4	BF5	BF6	BP1	BP2	BC
A1	0.24	0.0	0.0	0.0	0.0	0.0	0.0	0.0	0.0
A2	0.0	0.76	0.61	0.0	0.0	0.0	0.0	0.0	0.0
A4	0.0	0.0	0.0	0.67	0.48	0.41	0.0	0.0	0.0
BF1	0.0	0.0	0.0	0.0	0.0	0.0	0.33	0.0	0.0
BF2	0.0	0.0	0.0	0.0	0.0	0.0	0.9	0.0	0.0
BF3	0.0	0.0	0.0	0.0	0.0	0.0	0.72	0.52	0.0
BF4	0.0	0.0	0.0	0.0	0.0	0.0	0.0	0.47	0.0
BF5	0.0	0.0	0.0	0.0	0.0	0.0	0.0	0.42	0.0
BF6	0.0	0.0	0.0	0.0	0.0	0.0	0.0	0.56	0.0
BP1	0.0	0.0	0.0	0.0	0.0	0.0	0.0	0.0	0.68
BP2	0.0	0.0	0.0	0.0	0.0	0.0	0.0	0.0	0.72
BC	0.0	0.0	0.0	0.0	0.0	0.0	0.0	0.0	0.0

Figure 2 shows an example of a mission dependency model. The model is generated from the matrix in Table 4, containing the impact probabilities. The example contains the interdependencies between business functions and processes, as well as the assets that directly support business functions. In this example, we can see that three of the initial assets already identified in the resource dependency model (cf. Fig. 1, assets *A1. Web Server, A2. Order Management Server*, and *A4. Order Processing & Shipping System*) have a direct link to six business functions (*BF1. Search product, BF2. Payment, BF3. Register Order, BF4. Confirm Order, BF5. Process Order* and *BF6. Order Shipping*). We can also see their link to two representative business processes (*BP1. Place Order* and *BP2. Order Picking & Shipping*) and one business company (identified as *BC. Online Shopping*), that represents the most important business function in the company.

The mission dependency model describes the dependencies between business processes and business functions, as well as between the business functions and the technical assets that directly support them. In Fig. 2, the first asset (*A1. Web Server*) may have an impact on one business function (*BF1. Search Product*).

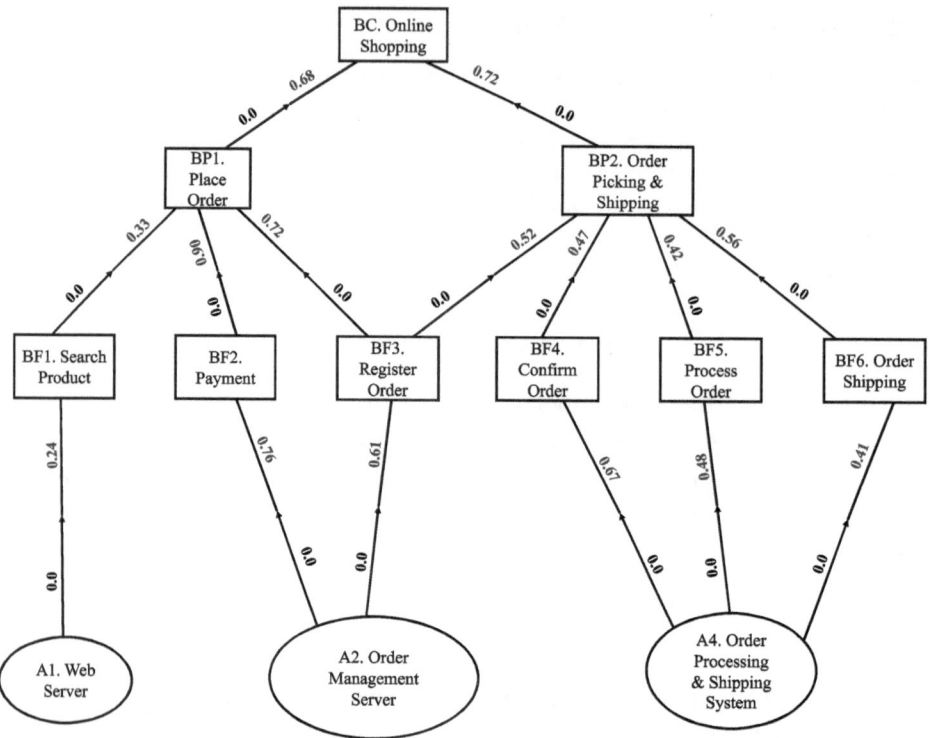

Fig. 2. Mission dependency model. It contains business functions, business processes, and assets directly supporting the business functions. The model also describes the interdependencies between those elements. In the depicted example, the model contains three assets (*A1. Web Server*, *A2. Order Management Server* and *A4. Order Processing & Shipping System*) that have a direct link with six business functions (from *BF1. Search Product* to *BF6. Order Shipping*), two business processes (*BP1. Place Order* and *BP2. Order Picking & Shipping*) and the global business of the company (*BC. Online Shopping*).

Similarly, the second asset (*A2. Order Management Server*) may have an impact on two business functions (*BF2. Payment* and *BF3. Register Order*). Finally, the last asset (*A4. Order Processing & Shipping System*) can have an impact on three business functions (*BF4. Confirm Order*, *BF5. Process Order* and *BF6. Order Shipping*). Figure 2 also shows how business functions may impact business processes (e.g., impact of *BF1* over *BP1*, and *BF3,BF4* over *BP2*). It also shows that business processes *BP1* and *BP2* may have an impact on the company's mission (identified in our example as *BC*).

3.3   BLM Generation

Once the resource dependency model and the mission dependency model have been generated, they get fused into a single adjacency matrix representing the

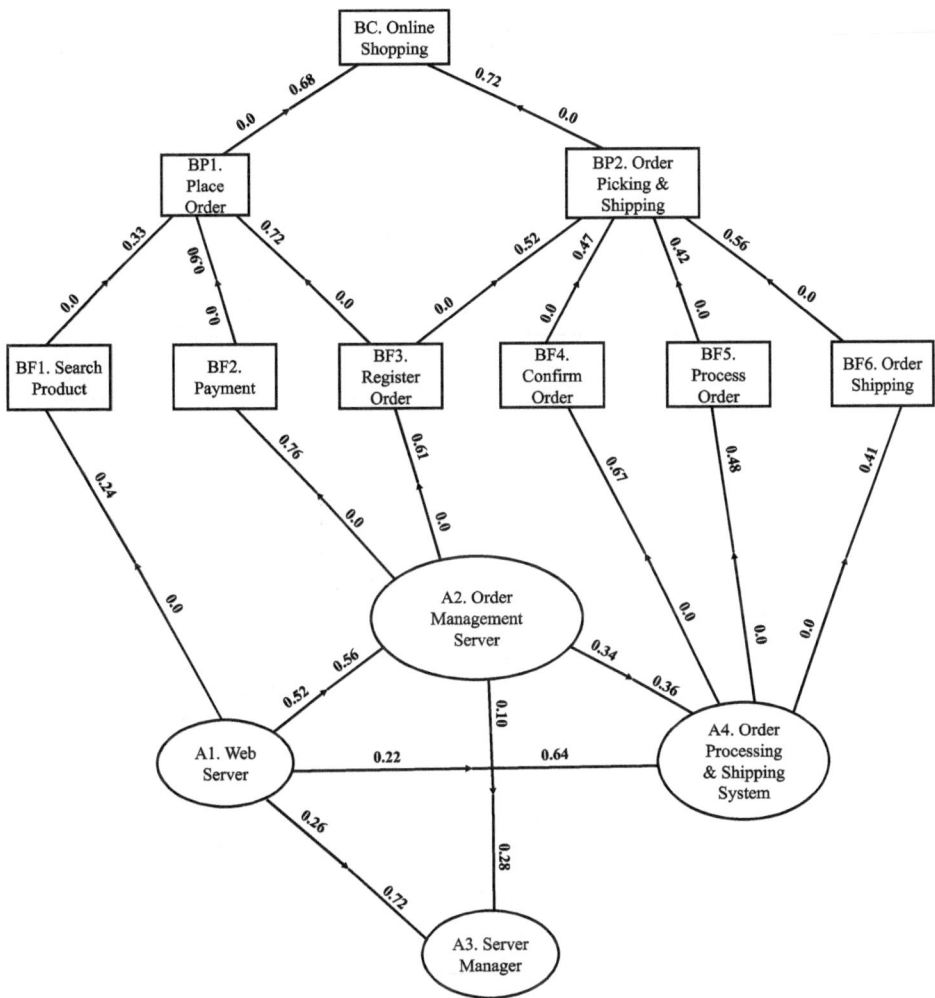

Fig. 3. Business Logic Model with conditional impact probabilities. This model represents the probability graph of the business logic model and it is generated from the resource dependency model and the mission dependency model. It includes assets, business functions and processes, business company, as well as dependencies between nodes, and contains the impact probabilities.

business logic model. The adjacency matrix generated from Tables 3 and 4 is summarized in Table 5.

Figure 3 represents the probability graph of the Business Logic Model built from the adjacency matrix associated to Table 5. The graph includes assets, business functions and processes, and the business company. It also depicts the dependencies between the different nodes on the graph. Figure 3 depicts the probability graph of the business logic model. This graph aims to help security

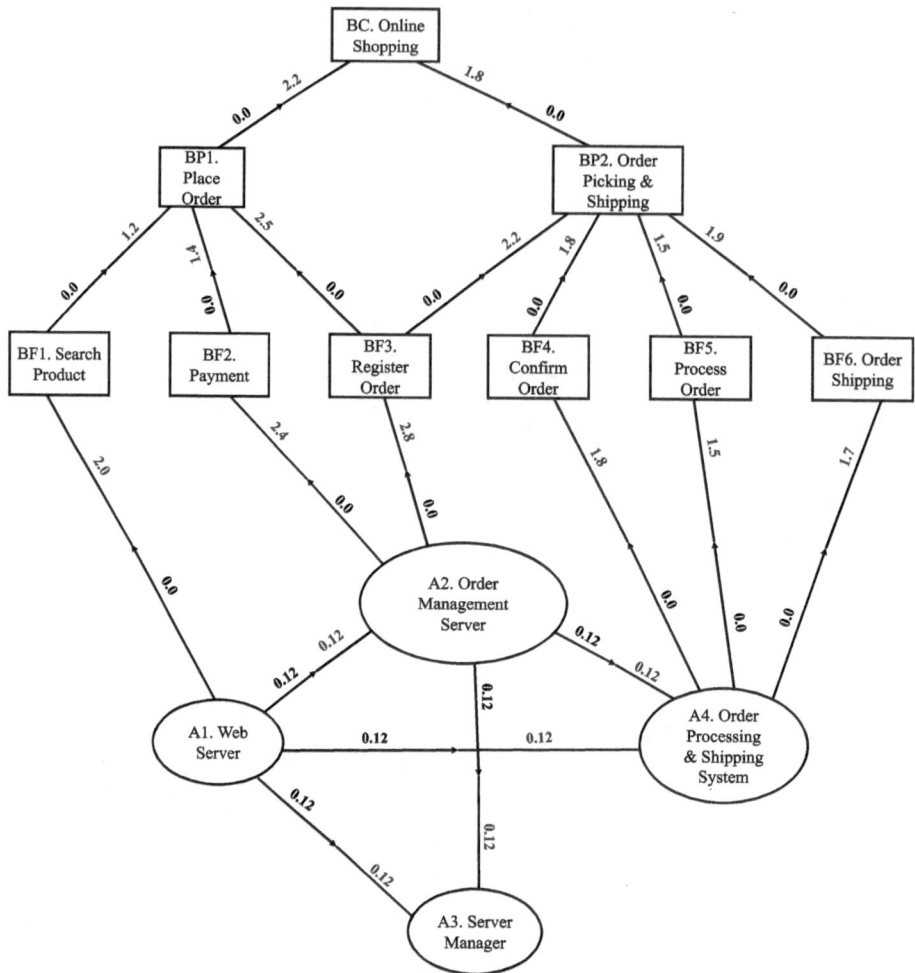

Fig. 4. Business Logic Model with downtime tolerances. This model represents the temporal graph of the business logic model and it is generated from the Business Entity downtime tolerance matrix and the Inter-asset downtime tolerance matrix. It includes assets, business functions and processes, business company, as well as dependencies between nodes, and contains the downtime tolerances.

experts to understand the impact of the attack and its propagation within the resources of an organization. Figure 4 displays the temporal graph associated to our business logic model, but highlighting the downtime tolerances instead of conditional probabilities.

As shown in Table 5, the business logic model requires an adjacency matrix presenting the interdependencies between the different nodes and containing the impact probabilities. Another adjacency matrix is required to provide the downtime tolerances between the nodes. The downtime tolerance represents an

Table 5. Complete adjacency matrix with conditional impact probabilities.

	A1	A2	A3	A4	BF1	BF2	BF3	BF4	BF5	BF6	BP1	BP2	BC
A1	0.0	0.56	0.72	0.64	0.24	0.0	0.0	0.0	0.0	0.0	0.0	0.0	0.0
A2	0.52	0.0	0.28	0.36	0.0	0.76	0.61	0.0	0.0	0.0	0.0	0.0	0.0
A3	0.26	0.1	0.0	0.0	0.0	0.0	0.0	0.0	0.0	0.0	0.0	0.0	0.0
A4	0.22	0.34	0.0	0.0	0.0	0.0	0.0	0.67	0.48	0.41	0.0	0.0	0.0
BF1	0.0	0.0	0.0	0.0	0.0	0.0	0.0	0.0	0.0	0.0	0.33	0.0	0.0
BF2	0.0	0.0	0.0	0.0	0.0	0.0	0.0	0.0	0.0	0.0	0.9	0.0	0.0
BF3	0.0	0.0	0.0	0.0	0.0	0.0	0.0	0.0	0.0	0.0	0.72	0.52	0.0
BF4	0.0	0.0	0.0	0.0	0.0	0.0	0.0	0.0	0.0	0.0	0.0	0.47	0.0
BF5	0.0	0.0	0.0	0.0	0.0	0.0	0.0	0.0	0.0	0.0	0.0	0.42	0.0
BF6	0.0	0.0	0.0	0.0	0.0	0.0	0.0	0.0	0.0	0.0	0.0	0.56	0.0
BP1	0.0	0.0	0.0	0.0	0.0	0.0	0.0	0.0	0.0	0.0	0.0	0.0	0.68
BP2	0.0	0.0	0.0	0.0	0.0	0.0	0.0	0.0	0.0	0.0	0.0	0.0	0.72
BC	0.0	0.0	0.0	0.0	0.0	0.0	0.0	0.0	0.0	0.0	0.0	0.0	0.0

estimation of how long the business functions or processes can be impacted without propagating the operational impact to other business entities. The adjacency matrix with downtime tolerances building the temporal graph depicted in Fig. 4 is generated from the Business Entity downtime tolerance matrix and the Inter-asset downtime tolerance matrix. The Business Entity downtime tolerance matrix is presented by the same graph as the mission dependency model shown in Fig. 4, but it displays the downtime tolerances instead of the impact probabilities. This model is also built manually because it requires knowledge of the company's business activities.

The inter-asset downtime tolerance matrix is presented by the same graph as the resource dependency model, shown in Fig. 3, but highlighting the inter-asset downtime tolerances instead of the conditional impact probabilities. The inter-asset downtime tolerances values are set to 10% of the minimum interdependency value between the business entities, since the impact propagation delay between the assets is considered to be much lower than the impact propagation delay between the business entities. For instance, in our examples, the minimum interdependency value between the business entities could be set to 1.2 h (or any other representative value extracted from the Business Entity downtime tolerance matrix) and the inter-asset downtime tolerances to 0.12 h.

4 Implementation of Our Approach

A proof-of-concept tool, hereinafter referred to as the *Business Impact Analyser* (BIA), is available on a companion code repository[1]. This tool implements all the concepts and models of our contribution. Next, we detail some additional functionalities that have been included in the BIA tool, as well as an evaluation of performance and scalability associated to the tool.

4.1 Business Impact Analyser Functionalities

In addition to the business logic formalism underlying our contribution, the following additional functionality is also included in our models:

- Monitoring of assets criticality.
- Impact probability of shock events affecting business functions and processes.
- Monitoring of critical time.

Assets Criticality. One of the tasks performed by the BIA tool is identifying the critical assets in the company. To identify the assets that contribute the most to the propagation of impacts on business functions and processes, our tool computes the value of the impact probability on the Business Company (BC) node when a shock event occurs on this asset with a local conditional probability equal to 1. The criticality value is between 0 and 1, in which 0 indicates that this asset has no impact on the BC, and 1 indicates that the impact of the attack is very high.

Impact Probability. The existence of an external event may have an impact on the business functions and processes. To estimate how the impact of this external event will propagate in the architecture and to assess the impact propagation of this shock event to the missions, we compute the impact probability, which represents the degree to which this shock event impacts the missions. In order to assess the impact probability of a shock event on the missions, we implemented the Monte-Carlo approximation.

We have implemented the Monte-Carlo approximation to randomly explore the business graph a certain number of times (*ntimes*), while counting the number of times each node has been impacted by the shock events, which we want to assess their impacts on business entities. The steps of one graph exploration are defined as follows:

1. Initiate a queue Q, with the input list of Shock Events.
2. Explore node n, from Q.

---

[1] A companion git repository with the code of the Business Impact Analyser (BIA) tool is available at https://gitlab.com/tsp-soccrates-components/bia.

3. For each edge of node n, try the impact probability of the edge by drawing a random number. If the drawn number is lower than the probability of impact of the tested edge, then the node at the other end of the edge is considered as impacted. If a node is impacted and was never added to Q, then add it to Q, and increase the impact counter for this new node.
4. While Q is not empty, repeat from Step 2.

The approximated impact probability is computed by taking the mean of the impact counter, i.e., *Impact Probability* gets as value *Impact Counter* divided by *ntimes*.

Critical Time. The critical time represents the time when an impact to a business process or function will not have a significant operational impact on the company. To compute the critical time, we built the business logic model with the downtime tolerances, which is an estimate of how long the business functions or processes can be impacted without propagating the operational impact to other business entities. The Dijkstra algorithm is used to find the shortest path from the shock event to the Business Company and compute the critical time, which is equal to the sum of the downtime tolerance values between the nodes in the shortest path.

4.2   Applying the Functionalities

In this section, we show how to apply All the aforementioned functionalities over our contribution, in the example shown in Sect. 3.3, Table 5. The results of applying the functionalities are described next.

– **Assets Criticality Computation**—Figure 5 shows the probability graph after calculating the criticality of assets. Three levels of criticality have been configured to help the operator identify the assets that contribute the most to the propagation of impacts on the business entities. The criticality value is displayed in orange above each *Asset node*. According to this value, the assets are colored:
  - In red: The most critical asset is *A2. Order Management Server*.
  - In orange: The medium critical assets are *A1. Web Server* and *A4. Order Processing & Shipping system*.
  - In green: The low critical asset is *A3. Server Manager*.
– **Impact Probability Computation**—Figure 6 displays the probability graph after the computation of the impact probability of a shock event on the business entities.
  The shock event is given as input in this format: (Name of the shock event, target asset, impact probability). In this example, it is presented by: (*Shock Event, Order management server*, 0.84).
  After calculating the impact probability of this shock event on the business, our tool displays the impact probability above each mission and according to the calculated values, it displays the missions in three colors:

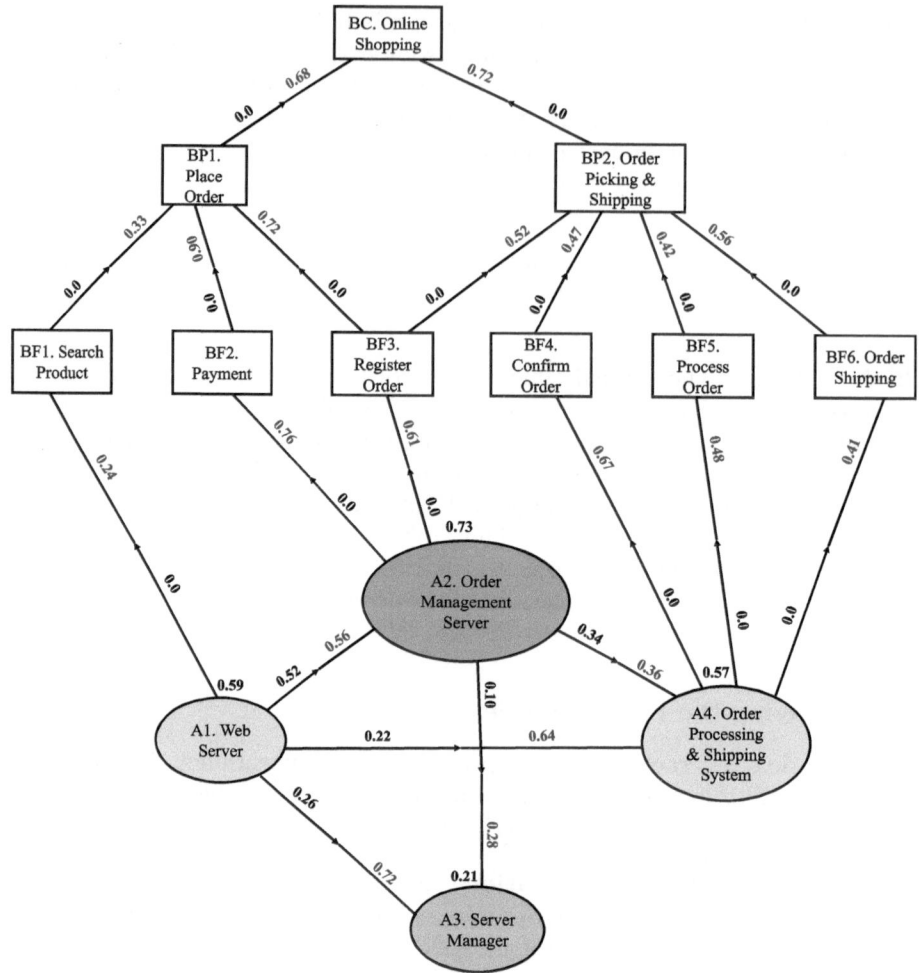

Fig. 5. Asset criticality computation. The computation of the criticality of assets aims to identify the most critical assets in the architecture. In this graph, the most critical asset, which is colored in red, is *A2. Order Management Server*. (Color figure online)

- In red: The business impact of this shock event on the business process *BP1. Place Order* is high.
- In orange: The business impact of this shock event on the business functions *BF2. Payment* and *BF3. Register Order*, the business process *BP2. Order Picking & Shipping* and the business company *BC. Online Shopping* is moderate.
- In green: The business impact of this shock event on the business functions *BF1. Search Product*, *BF4. Confirm Order*, *BF5. Process Order* and *BF6. Order Shipping* is low.

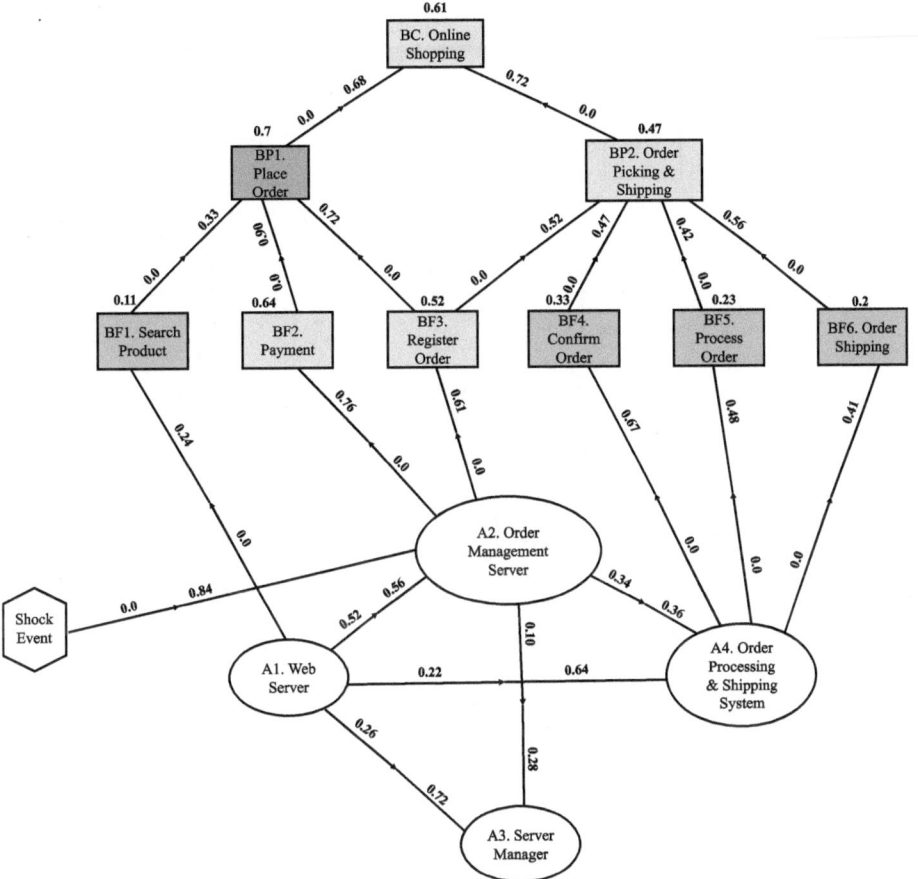

Fig. 6. Impact probability computation. The impact probability computation is performed on the probability graph using the Monte-Carlo approximation. This graph shows the impact of the Shock Event targeting *A2. Order Management Server* on the business entities. The most impacted business entity in this graph, which is colored in red, is the business process *BP1. Place order.* (Color figure online)

– **Critical Time Computation**—Figure 7 shows the temporal graph after the critical time has been calculated. Our tool calculates the critical time value and displays the shortest path as a dotted line from the shock event to the business company. In this example, the shortest path is: *Shock Event -> A2. Order Management Server -> A4. Order Processing & Shipping System -> BF5. Process Order -> BP2. Order Picking & Shipping -> BC. Online Shopping* and the critical time, which is the sum of the downtime tolerances between the nodes in the shortest path, is equal to 5.02 h.

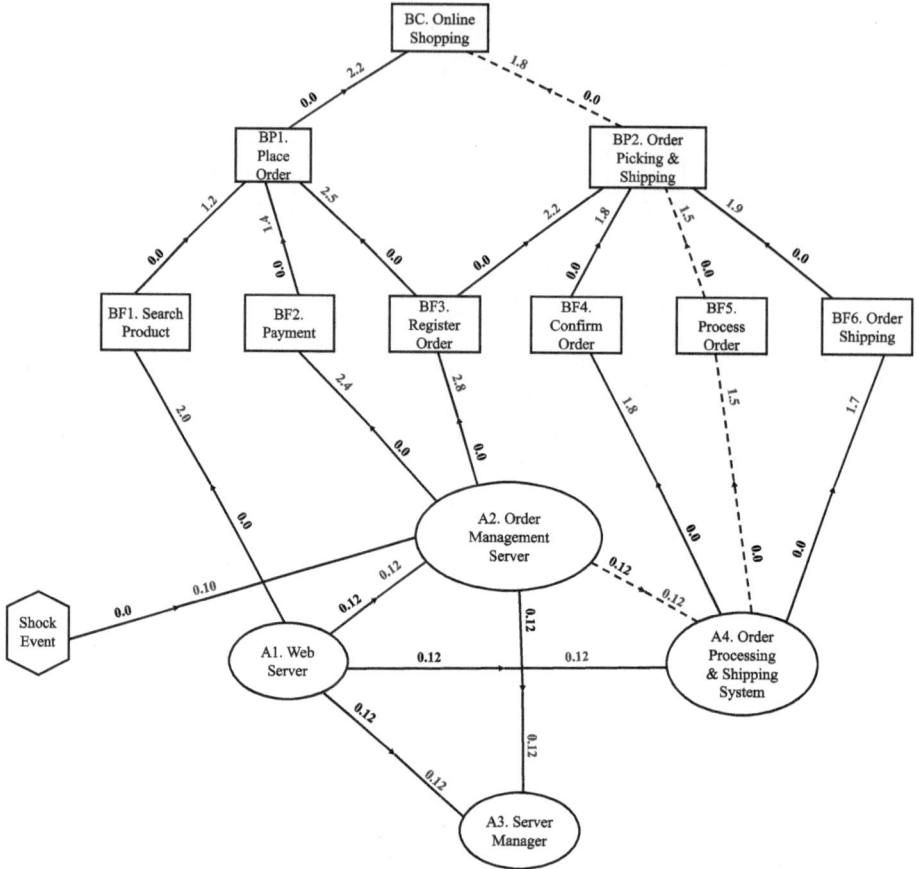

Fig. 7. Critical time computation. The critical time is calculated using Dijkstra's algorithm. The algorithm first finds the shortest path between the shock event and the Business Company *BC. Online Shopping*, and then calculates the critical time, which is equal to the length of the shortest path. The shortest path is displayed in dotted, and the critical time is equal to 5.02 h.

4.3   Scalability Evaluation

In this section, we evaluate the scalability of our proof-of-concept tool[2]. We evaluate the main computations conducted by the tool, to estimate the time required to compute the criticality of assets of a representative example. The tool is implemented using Python REST (Representational State Transfer) APIs

---

[2] Available at https://gitlab.com/tsp-soccrates-components/bia.

Table 6. Time spent to compute the criticality of assets depending on the number of assets and edges. This table shows the different tests performed by our tool, as well as the time required for each test to calculate the criticality of assets. To distinguish the edge density in the graph, three colors are used. The black rows indicate that for these tests, all assets communicate with each other, which means that all assets send and receive data. The orange rows indicate that for these tests, half of the assets communicate with each other, and the rest of assets only receive data. The green rows indicate that for these tests, only a quarter of assets communicate with each other, and the rest of assets only receive data.

# of Assets	# of Edges	Computation Time (in seconds)	Computation Time per Asset (in seconds)
50	2 450	1.96	0.039
50	1 224	1.09	0.021
50	613	0.69	0.013
250	62 250	24	0.095
250	31 124	12.2	0.048
250	15 562	6.45	0.025
1 000	999 000	197.6	0.197
1 000	499 500	101.9	0.101
1 000	249 750	52.4	0.052

and the following external libraries: NetworkX[3], NumPy[4], Uvicorn[5], FastAPI[6] and Pandas[7]. The evaluation is conducted on a 3-core CPU system, with 3 GB of memory and 20 GB of storage.

Table 6 displays the results of various tests run by our tool with various models and numbers of assets and edges. It also displays the different tests performed by our tool and the time required to calculate the criticality of assets for each test. Each line of this table presents a test, with the size of the graph and the time required to compute the criticality of assets. The first and second columns display the number of assets and edges in the graph. The third column shows the time required to calculate the criticality of all assets, and the fourth column shows the time required to calculate the criticality of one asset. For example, for a graph with 50 assets and 2450 edges, our method calculates the criticality of these 50 assets in 1.96 s. Several tests were performed with different edge densities. Three colors are used to differentiate the edge density in the graph:

- Black: all assets in the graph communicate with each other, which means that all assets send and receive data.
- Red: half of the assets in the graph communicate with each other, and the rest of the assets only receive data.

---

[3] https://networkx.org/.
[4] https://numpy.org/.
[5] https://www.uvicorn.org/.
[6] https://github.com/tiangolo/fastapi.
[7] https://pandas.pydata.org/.

– Green: a quarter of the assets in the graph communicate with each other, and the rest of the assets only receive data.

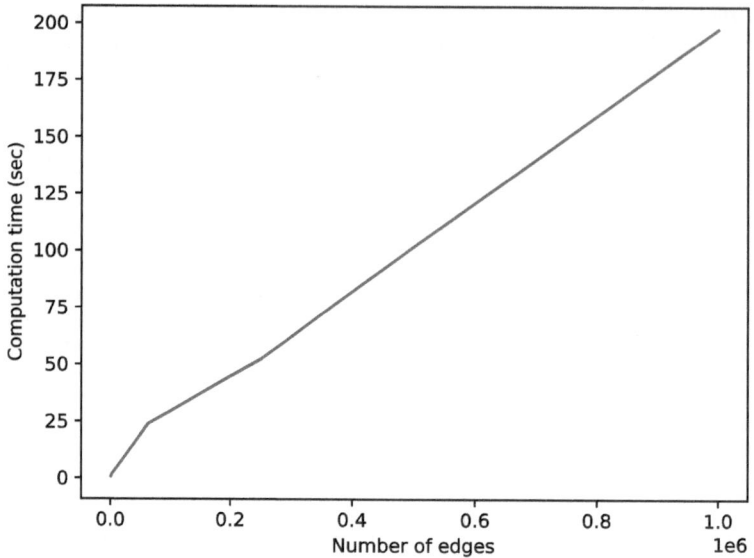

Fig. 8. Dependency between computation time and size of the graph. This plot shows the dependency between the time required to calculate asset criticality and the size of the graph. For example, for a graph with 1224 edges, our tool calculates the criticality of assets in 1.09 s.

Figure 8 displays the dependency between the computation time and the number of edges in the graph. The time required to compute the criticality of assets increases with the number of edges. F.i., the computation time for a very large graph with 1000 assets and 999000 edges is less than four minutes. These results confirm that our tool can perform computations in a reasonable time.

To sum up, we have presented in this section a practical implementation of our contributions in Sect. 3, in a proof-of-concept tool. We have validated how to quantify the operational impact of shock events on a company's business entities using a representative example. One of the limitations of our work is the automation of models generation, for instance, the mission dependency model. Future research directions to address this limitation is discussed next.

5   Future Directions for Research

The mission dependency model is built manually because there is no easy method for automating the generation of this model. The generation of the mission dependency model requires defining the list of business functions and business processes, as well as the interdependencies between them. In order to define the list of the business functions and processes of the company, a lot of work needs to be done to derive these functions and processes from the Business Process Model and Notation (BPMN) model.

In addition to the difficulty of retrieving the list of business nodes, a high level of expertise is necessary to determine the relationships and the weight of dependencies between business nodes. In other words, and in addition to using BPMN, an expert knowledge on the business activities is also required to build manually the mission dependency model.

Building the mission dependency model can make it difficult to keep the model up to date if the business activity of the company changes. Defining and implementing a standard model capable that can automate the generation of the list of business nodes from the BPMN as well as evaluating the relationships and dependencies between these nodes would improve the reliability of our model without requiring a high level of expertise and skills in business aspects.

Downtime tolerance represents an estimate of how long the business functions or processes can be impacted without propagating the operational impact to other business entities. Computing the critical time using a dynamic downtime tolerances is considered a contribution for further research, which means that the values of the dependencies between the business entities, which are the downtime tolerances, are dependent on the time of the occurrence of the attack on the technical asset. For example, if a technical asset that supports business functions is targeted by an attack during the night, the propagation of the operational impact may not be very significant since these business functions may not be very necessary during the night. Thus, the downtime tolerance should be much longer than the downtime tolerance if the attack took place during the day when these business functions are very necessary to the company's activity.

Finally, a research work could be conducted on the quantification of the operational impact of attacks on business entities when the company has a redundant technical assets, which means the redundant asset will replace the asset targeted by the attack and perform its tasks, and in this case the operational impact will be very low on the business entities of the company.

6   Conclusion

In this chapter, we have surveyed existing methods aiming to assess the impact of external events on business entities. We have started our study with a focus on two main families: financial and operational impact assessment.

We have also presented a practical contribution for assessing the operational impact of attacks targeting the infrastructure assets and affecting the execution and performance of the company's activities. We have provided detail on a novel method to assess the operational impact of shock events on a company's business entities, based on adding to previous contributions a new business logic model. The new model enables the assessment of impact probabilities associated to external events on missions and the calculation of critical time, as well as the computation of the criticality of technical assets, that helps security analysts to identify the most critical assets. Our model have been tested and validated on realistic use cases and real data provided by stakeholders in a practical tool. The code of the tool is available online. We have also evaluated and validated the scalability of the computations performed by our tool, via several tests with different business logic models and on large graphs, and all of the tests performed show that our tool is able to perform computations and provide mission impact assessment in a reasonable time.

In terms of future research, we have pointed out to further automation in the generation of the mission dependency models, as well as the necessity to further assess the operational impact of external events against missions when a company has redundant assets, and further contribution to better compute the critical time using a dynamic downtime tolerances.

Acknowledgments. Authors acknowledge support from the European Commission (Horizon Europe projects SOCCRATES and AI4CCAM, under grant agreements 833481 and 101076911).

References

1. Böhme, R., Nowey, T.: Economic security metrics. Dependability Metrics: Adv. Lect. 176–187 (2008)
2. Cao, C., Yuan, L.-P., Singhal, A., Liu, P., Sun, X., Zhu, S.: Assessing attack impact on business processes by interconnecting attack graphs and entity dependency graphs. In: Kerschbaum, F., Paraboschi, S. (eds.) DBSec 2018. LNCS, vol. 10980, pp. 330–348. Springer, Cham (2018). https://doi.org/10.1007/978-3-319-95729-6_21
3. Cashell, B., Jackson, W.D., Jickling, M., Webel, B.: The economic impact of cyber-attacks. Congressional research service documents, CRS RL32331 (Washington DC), vol. 2 (2004)
4. Claise, B.: RFC 3954: CISCO systems netflow services export version 9 (2004)
5. D'Amico, A., Buchanan, L., Goodall, J., Walczak, P.: Mission impact of cyber events: scenarios and ontology to express the relationships between cyber assets, missions and users. In: 5th International Conference on Information Warfare and Security, pp. 1–11 (2010)
6. de Barros Barreto, A., da Costa, P.C.G., Yano, E.T.: Using a semantic approach to cyber impact assessment. In: STIDS, pp. 101–108 (2013)
7. Dongre, S., Mishra, S., Romanowski, C., Buddhadev, M.: Quantifying the costs of data breaches. In: Staggs, J., Shenoi, S. (eds.) ICCIP 2019. IFIP Advances in Information and Communication Technology, vol. 570, pp. 3–16. Springer, Cham (2019). https://doi.org/10.1007/978-3-030-34647-8_1
8. Freund, J., Jones, J.: Measuring and Managing Information risk: a FAIR Approach. Butterworth-Heinemann (2014)
9. Gonzalez-Granadillo, G., Dubus, S., Motzek, A., Garcia-Alfaro, J., Alvarez, E., Merialdo, M., Papillon, S., Debar, H.: Dynamic risk management response system to handle cyber threats. Futur. Gener. Comput. Syst. **83**, 535–552 (2018)
10. Jakobson, G.: Mission cyber security situation assessment using impact dependency graphs. In: 14th International Conference on Information Fusion, pp. 1–8. IEEE (2011)

11. Liu, C., Singhal, A., Wijesekera, D.: A layered graphical model for mission attack impact analysis. In: 2017 IEEE Conference on Communications and Network Security (CNS), pp. 602–609. IEEE (2017)
12. Matheus, C.J., et al.: SAWA: an assistant for higher-level fusion and situation awareness. In: Multisensor, Multisource Information Fusion: Architectures, Algorithms, and Applications 2005, vol. 5813, pp. 75–85. SPIE (2005)
13. Motzek, A., Gonzalez-Granadillo, G., Debar, H., Garcia-Alfaro, J., Möller, R.: Selection of pareto-efficient response plans based on financial and operational assessments. EURASIP J. Inf. Secur. **2017**(1), 1–22 (2017)
14. Motzek, A., Möller, R.: Context-and bias-free probabilistic mission impact assessment. Comput. Secur. **65**, 166–186 (2017)
15. Motzek, A., Möller, R., Lange, M., Dubus, S.: Probabilistic mission impact assessment based on widespread local events. In: NATO IST-128 Workshop: Assessing Mission Impact of Cyberattacks, NATO IST-128 Workshop, Istanbul, Turkey, pp. 16–22 (2015)
16. Mukherjee, P., Mazumdar, C.: "Security concern" as a metric for enterprise business processes. IEEE Syst. J. **13**(4), 4015–4026 (2019)
17. Musman, S., Tanner, M., Temin, A., Elsaesser, E., Loren, L.: Computing the impact of cyber attacks on complex missions. In: 2011 IEEE International Systems Conference, pp. 46–51. IEEE (2011)
18. Orlando, A.: Cyber risk quantification: Investigating the role of cyber value at risk. Risks **9**(10), 184 (2021)
19. Sun, X., Singhal, A., Liu, P.: Towards actionable mission impact assessment in the context of cloud computing. In: Livraga, G., Zhu, S. (eds.) DBSec 2017. LNCS, vol. 10359, pp. 259–274. Springer, Cham (2017). https://doi.org/10.1007/978-3-319-61176-1_14

Technologies And Challenges

Enabling the Human-as-a-Security-Sensor Paradigm in the Internet of Things

Dennis Ivory$^{(\boxtimes)}$, George Loukas, and Diane Gan

Centre for Sustainable Cyber Security, University of Greenwich, London, UK
{dennis.ivory,g.loukas,d.gan}@greenwich.ac.uk

Abstract. Over the last two decades, there has been growing realisation that the user is not the weakest link in cybersecurity. Involving the user in a human-in-the-loop fashion in the process of security can have benefits in several aspects, including in cyber intrusion detection. The human-as-a-security-sensor paradigm has shown that it is possible to involve the user as a valuable source of data for detection, and in fact with a predictable level of accuracy. However, this paradigm has currently only been applied in conventional computer systems, such as desktop computers. Our aim here is to extend it for Internet of Things (IoT) environments too, specifically in detecting command injection attacks against IoT devices, whereby a user can be informed automatically about a new IoT device activity that has been detected on the network and can reason as to whether this is legitimate or not. The activity detection is based on a time series forecasting approach, where the assumption is that an abrupt change in the trend of network traffic rate, is an indication of a new activity having been triggered. Our evaluation of two time series forecasting approaches for different training window sizes, as well as of activity detection based on the best-performing of the two approaches, has shown that this is a realistic method for notifying the user.

Keywords: Internet of Things · IoT Security · HaaSS · Intrusion detection

1 Introduction

Cyber security threats are considered a prime adoption barrier for Internet of Things (IoT) technologies. Indicatively, according to F-Secure's 2019 attack landscape report[1], the number of IoT attack events measured from January through June was twelve times higher when compared with the same period in 2018. Similarly, Symantec's 2019 Internet Security Threat report[2] has concluded that targeted attack groups increasingly focus on IoT as a soft entry point. And more

---

[1] https://blog.f-secure.com/attack-landscape-h1-2019-iot-smb-traffic-abound/.
[2] https://www.symantec.com/security-center/threat-report.

© The Author(s), under exclusive license to Springer Nature Switzerland AG 2025
N. Pitropakis and S. Katsikas (Eds.): *Security and Privacy in Smart Environments*, LNCS 14800, pp. 75–97, 2025.
https://doi.org/10.1007/978-3-031-66708-4_4

recently reported by Marlin Communications[3] report that in the first half of 2023 there has been a surge of IoT attacks totalling to 77.9 million attacks, well on track to beat the previous record of 112.3 million attacks in 2022, and in certain countries such as India seeing a 311% increase in the amount of IoT malware attacks they are seeing.

1.1 Background and Motivation

A first requirement for protecting against such threats is to be able to detect them in an accurate and timely manner. Traditional cyber intrusion detection measures are system-specific and based entirely on technical mechanisms, typically training a machine learning based system to recognise known attack patterns. Although highly successful for a wide range of security threats especially on the Internet, this approach can miss completely previously unknown threats or those that are based on human deception. Previous research has shown that it can be highly beneficial to involve the user in the process of detection, in what is referred to as a human-as-a-security-sensor (HaaSS) role. According to Heartfield and Loukas [15], who have developed the concept, "HaaSS is the paradigm of leveraging the ability of human users to act as sensors that can detect and report information security threats".

The HaaSS paradigm has been evaluated successfully in conventional desktop computing environments, for instance to defend against email phishing and other semantic social engineering threats [14]. In a conventional computing environment, such as a desktop computer used to browse the web or social media, there are several cues that can make a user suspicious of a deception attempt, from unusual uniform resource locator (URL) and source email addresses to low followers-to-following ratio in social media, unusual choices of text or requests, etc. In an Internet of Things environment, these cues may be absent, as such devices follow a seamless design principle which provides the user with very little information about their internal operation. The HaaSS paradigm is already being extended by different research groups, with a feedback loop in project ACCEPT[4], integrated in an operational Security Information and Event Management solution in [45] and adapted for AI threats in EPSRC project CHAI[5]. In all existing cases, the focus has been on conventional computing environments, such as corporate desktop computers running Microsoft Windows. The applicability in an IoT environment of networked embedded systems, such as those found in a smart home, has not been evaluated to date.

This work has one primary contribution, which is to demonstrate the feasibility of applying the HaaSS approach to the paradigm of IoT by offering a first implementation example. Towards this goal, it further contributes to network-based IoT device activity identification through a novel application of time series forecasting.

---

[3] https://marlincomms.co.uk/blog/2023-cyber-threat-landscape-5-trends.
[4] https://accept.cyber.kent.ac.uk.
[5] https://project-chai.org.

2   Related Work

This chapter is structured is a look at the previous related work in the area of Human-in-the-loop (HITL) approaches in general, as well as work related to Network traffic and IoT activity recognition, and Time series machine learning.

2.1   Literature of HITL Approaches in Cyber Security

Leveraging Social Media. While global security trend analysis is often based on network data collection by a number of organisations (e.g., *Kaspersky* [20], *FireEye* [11], *Talos* [7]), it can also benefit from human user input as recorded online, such as in blogs and social media. For instance, Tsai and Chan [42] have developed a method for automatically analysing blogs to identify relevant keywords for specific cyber security threats and set the foundation for monitoring and identifying trends. More recently, Lippmann et al. [5] have demonstrated a methodology for finding signs on social media of malicious activity or preparations for impending cyber attacks, which can be useful in a more targeted manner, for example as early warning. The information used may come from the attackers themselves sharing or looking for information on particular vulnerabilities and targets, or may be the first reports of actual attacks.

Khandpur et al. [21] have proposed a new framework for crowd sourcing on social media platforms to detect cyber attacks against the same social media platforms. The three types of attacks that they have focused on are distributed denial of service attacks, data breaches and account hijacking. Their framework uses a typed query approach that allows for the searching of any chosen query (including denial of service attacks, data breaches, and account hijacking) and has also been shown to work on large data sets with no prior training. However, this framework still needs expanding so that it supports a wider range of attacks.

Attack Detection. The vast majority of work in the area of HITL in cyber security are related to detection. Haack et al. [13] have proposed bringing the human into the loop by taking users away from the role of managing agents and instead creating a shared initiative that will allow for them to work alongside and assist the technological agents in their role. By having a supervisory influence over the proposed system, the human can function in a role that they believe to be most appropriate at any given time, thus giving them the ability to assist and improve the cyber security system at multiple levels.

Malisa et al. [28] have developed a system that is able to identify potential Android mobile application phishing attempts. The users were shown a number of different mock spoofed application and then asked to identify the characteristics of them. This data was then used as training samples to develop a machine learning model for automated detection of the app spoofing attacks in their system. This example shows that a human can be involved at design time, where their sensing and detection capabilities are leveraged to develop an automated security system which allows for more accurate detection of certain types of attacks.

Asher et al. [3] discuss and show the importance and necessity to have a human involved in the role of security triage analysis. The experiments that they run involved both expert and novices being shown a combined total of 12,200 network events and then classifying them as either malicious networks events or non-malicious, with the experts correctly identifying 55% of the events with a false detection rate of 15% and the novices correctly identifying 44% of the events with a false detection rate of 18%. By doing this, they have shown the potential of having a human-based intrusion detection system. Naturally, experts who have had more experience detecting these type of threats are able to detect them with a higher accuracy.

An investigation into the effectiveness of utilising users as defence analysts by Rajvian et al. [35] established the hypothesis that effective collaboration and team work between cyber defence analysts is almost non-existent. In response, they proposed the idea of forming teams and collaborative partnerships to help improve the effectiveness of the users as defence analysts. Experimental results showed that forming larger single teams (6 people) was far more effective at solving alerts than the multiple smaller teams (3-person team and 5-person team). However, in all but one of their experiments they found that agents who were part of a heterogeneous team were able to perform better in any group size and detected more events when compared to the homogeneous teams.

Frameworks for HITL. Heartfield and Loukas [15] have introduced the paradigm of the human-as-a-security-sensor (HaaSS), where the user in an organisation is actively involved in detecting cyber threats. They have developed both a HaaSS framework and a practical implementation in the form of a Microsoft Windows application, named Cogni-Sense, which allows the user to actively report signs of attacks that they have spotted. Cogni-Sense has been evaluated with 26 users of different profiles running the application on their personal computers for a period of 45 days, which were shown to be able to consistently outperform commercial technical security systems designed specifically for the types of attacks employed. These were deception-based attacks (referred to as semantic social engineering attacks [14]), such as application masquerading, spearphishing, rogue USB devices, etc. Importantly, Cogni-Sense benefits from the same team's previous work in developing models that can predict the reliability of a human sensor in detecting a given attack [17]. So, the reports produced by a human sensor are accompanied by what is referred to as H score, which is a prediction accuracy metric based on a machine learning model. This H score can be used by a security operations centre to prioritise the reports received by different users, as well as a continuous metric for measuring user detection efficacy at an organisational and individual level.

Tyworth et al. [43] argue for a human-in-the-loop approach to the study of situation awareness in computer defence analysis (CDA), their work highlights the need for a more human-centric approach to situation awareness in CDA. Having carried out experiments using a living lab framework that they have also defined in this work, they have concluded that CDA work is distributed between both the human actors in the system and the technological agents and operate in different domains such as intrusion detection or strategic analysis.

Kumar et al. [24] have looked at the vulnerabilities both in the technological and the human involvement in IoT cyber-physical systems, they have produced a holistic framework that aims to help mitigate security threats to an IoT based cyber-physical system. They aim to do this by taking the combined approach of using traditional detection methods through machine learning as well as human sensing, the proposed response to an attack is to enable intrusion tolerance and allow the system to retain as much usability as possible for the users.

Cranor [9] discusses the fact that the human is quite commonly an issue when they are within the loop of a cyber security incident but also states that there are certain situations that a human must be in the loop due to it not being feasible or cost-effective to replace them. For this reason, they propose that when a human needs to be in the loop within a cyber security system, then the system designers should do as much as possible to support and aid the users in their role to increase the chance of them performing their security-critical function successfully. They have proposed a framework that provides a systematic approach to help identify any potential issues that may cause human failure within the loop, by helping identify and address these issues at the system design level or can be used by system operators to help identify the root cause and security failures that have been attributed to 'human error'.

Tools for HITL. Vielberth et al. have put the concept of HaaSS to the test by integrating it in an operational Security Information and Event Management solution in [45]. In the process, they have identified the artifacts that a human sensor can report into four categories: threat sources, threat events, entities, and expected impact. Their Integrated HaaSS (IHaaSS) is able to connect human sensor data to standard data structures used in cyber threat intelligence.

Staheli et al. [39] have argued that the current tools used by security analysts lead to isolated work, where they can realistically rely only on their own record keeping. In response, the authors have developed CARINA, a collaborative investigation system, where the information presented to analysts may also include evidence files and annotation produced by other analysts.

Stembert et al. [40] have proposed an application for HITL cyber security, where they propose that the human user can be leveraged to assist with the detection of spear phishing emails in a corporate setting. This is achieved by giving the users a platform on which they can quickly report any suspicious emails. The same platform also provides the user with a short educational message, which will then further increase the user's knowledge and make it more beneficial to have a human-in-the-loop approach to cyber security. They have validated this tool with an initial experiment with promising results.

At commercial space, Cofense [8] have developed a number of applications for the detection, reporting and management of phishing attacks. The main application is Cofense Reporter which allow the users to report any phishing emails that they receive, this application of human-in-the-loop cyber security focuses on using the users as security sensors and enabling them to easily report anything they detect. This can then also be used with Cofense triage which is a phishing-specific security orchestration, automation and response (SOAR) platform that will perform analysis on the reported emails and will even leverage trusted informants to make sure that the email has been correctly reported. These applications have been built on the same ideas and research that was presented by Stembert et al. [40] and has used this to create a number of applications actually utilises the users in a human-in-the-loop situation.

Also at commercial space, Wombat security [36] offers a number of applications and services that are all human-in-the-loop based and rely on the users to report potential phishing activity, similar to the services ad applications offered by [8]. However, Wombat security also allows for the reporting of SMS and USB based phishing attacks. An other company, Koodous [23] has created an online collaborative platform that allows for the detection and identification of malicious android APKs. This has been achieved through the combination of online analysis tools with human analysts and their interactions with one another.

Some internal security teams at certain organisations have also taken it upon themselves to develop and implement human-in-the-loop cyber security platforms that allow them to leverage their user base and make use of them for the detection of potential cyber threats. The cyber emergency response team at Oxford university have developed their own platform and information security policy that employs and enables the user base to become sensors for cyber threats and actively encourages both students and staff to become involved in the cyber threat protection for the university by reporting suspected email and website cyber threats through the online platform that they have been provided with [33].

There are also some online open-source community platforms that use the human-in-the-loop paradigm to help provide better cyber security. An example of this is the commonly used website PhishTank [34], which provides a platform for anonymous reporting, for suspected phishing emails and websites and gives users the ability to verify the reports of other users. This data is crowd-sourced from the users and the verification of the reports is done by majority consensus of the other users. Another similar platform to this is Millersmiles [29] which also allows users to report phishing emails using a form but does not allow for user verification as in the case of PhishTank.

HITL Approaches for the Security of Cyber-Physical Systems. Ma et al. (2017) [26] have proposed a closed loop HITL reference model for cyber-physical systems which takes into account "semantic, interactive and iterative aspects". They define this as a "Cyber Physical Human System" which integrates the human into the traditional decision making process. Their work identified that there are different models depending on the particular type of cyber-physical application. This means that currently a smart home will not have the same model as smart transportation or a smart grid. Their model incorporates human feedback into decision making processes and conclude that more work is needed in this area, especially with regards to cloud computing and security, but does not yet tackle either.

Gontar et al. [12] proposed that the human pilot is an integral part of the correct functioning of an aeroplane and as such they should be included in a defence layer to protect the aircraft from cyber attacks. They argue that cyber attacks are an imminent threat against aircraft and the effects of such an attack may be difficult to distinguish from a technical problem. For system malfunctions within an aircraft, the pilot will use standard procedures and will also get warnings and advice from the on-board systems. However during a cyber attack, pilots may be unprepared and these same systems may give false advice jeopardising safety. Pilots undergo considerable training in order to be able to deal with every technical eventuality and have trust in the systems' reliability in predicting errors and consequent outcomes. This means that they will have difficulty dealing with unexpected and contradictory situations which could arise due to a cyber attack. To test this, Gontar et al. ran a series of experiments on volunteer pilots using a flight simulator in which there were false alarms and/or warnings given and the effect on the pilot's responses were measured. They conclude that systems should be installed in aircraft to help identify the source and possible consequences of a cyber attack to assist pilots' in making informed decisions during a flight. The pilot will always be in the loop as they need to distinguish between a technical malfunction and a cyber attack and act appropriately for the safety of the aircraft.

In [47], the focus was on protecting against Global Positioning System (GPS) spoofing attacks on unmanned aerial vehicles (UAVs), which could not be identified by cyber detection methods. Their method utilised the GPS system and the on-board camera images to assist the human operator to determine if there is a mismatch between these two by comparing the two systems visually to identify if GPS spoofing was occurring. The authors ran a series of simulated experiments to determine if these operators could identify GPS spoofing attacks by visually comparing the GPS data and the view from the camera. The operators of unmanned aerial vehicles (UAVs) often control a number of machines and there were a number of false alarms given as well. The results of these simulated exercises show that the operators were slightly better at detecting false alarms than hacking events. They found that the human operator detected over 80% of the spoofing attacks correctly and so can positively contribute to the detection of cyber-attacks.

Within the railway network system, there are a number of strategies which are used to detect a cyber attack, including automated intrusion detection and protection systems. The question to be addressed is whether the human operator can detect a dangerous event of cyber nature. In [30], human drivers were assessed using human-in-the-loop principles and a number of simulated scenarios. The threats included displaying a false speedometer reading or even the complete loss of the speedometer display, the loss of breaking systems or the jamming of the cameras used to monitor passengers at a station. The authors have argued that the human in the loop must be able to deal with cyber induced events which have not been blocked by automated defence systems.

2.2   IoT Device Activity Recognition

In this work, we explore whether users may be able to contribute to the detection phase in relation to security breaches in IoT. The challenge here is that IoT devices' seamful design can make it very difficult for users to even know that an IoT device is active, let alone whether this activity looks suspicious or not. For this reason, we summarise here the related literature in IoT device activity recognition keeping in consideration attacks such as command injection, where an IoT activity is triggered by a malicious entity without the user being aware. In such cases, the user needs to be quickly informed of a suspicious activity so as to confirm whether they have requested it or not.

Bai et al. [1] have tackled the problem of IoT device recognition based solely on the network traffic streams that are generated. They captured the traffic from the IoT environment in a standard PCAP format and extracted features that were relevant for identifying the different types of IoT devices such as numbers of packets in total and specific protocols. Specifically for the packet length feature, they have applied standard statistical functions such as Min, Max, Mean, Standard Deviation, Skewness and Kurtosis. On the generated data set, they then applied a series of different standard machine learning and deep learning classifiers. They observed a significantly higher performance when a cascade convolutional neural network combined with a long short term memory layer. This work was useful for the identification of the classifiers that are likely to perform well in this context and those that are not. Indicatively, the random forest classifier which is often a well performing approach, achieved an extremely low accuracy of 30.1% between 16 different types of smart home devices. However, the classification is happening offline on a collected data set and is limited to detecting the presence of particular types of devices and not if the device is actively being controlled by a user.

Shadid et al. [38] have performed recognition of different IoT devices through the analysis of bidirectional network traffic that is produced by the aforementioned IoT devices. The way in which this was done was through the use of a standard packet capture of the network using the PCAP format. From this data, a number of features were then selected for later analysis (size of packet sent, size of packet received, inter-arrival times of packets sent and inter-arrival times of packets received). This large dataset was then separated by each device's MAC address and then these datasets were used to train different classifiers, including Random Forest, Decision Tree, SVM, k-Nearest Neighbour, Artificial Neural Network and Gaussian Naïve Bayes. Once these classifiers had been trained, they were tested against a separate data set to measure their accuracy. The highest accuracy achieved was with the Random forest classifier with an average accuracy of 99.9% and the lowest being the Gaussian Naïve Bayes classifier with an accuracy of 91.9%. However, the study was only carried out on four different IoT devices and in a controlled offline environment without real user interaction.

Network Traffic Analysis. Sekwatlakwatla et al. [37] have developed a method for the prediction of network traffic levels in a cloud computing environment and the prediction of the flow of that traffic through the utilisation of Auto Regressive Integrated Moving Averages (ARIMA) and Artificial Neural Networks (ANN). The results that these two models produced are both very accurate, with ARIMA being the most accurate of the two on average. However, predictions were conducted based on monthly or yearly network flows, which does not allow short-term activity recognition as would be required in a cyber security use case.

Nie et al. [32] have developed a method for the prediction of traffic utilising a Deep Belief network and Spatiotemporal Compressive sensing. The way in which this was carried out was using a Discrete Wavelet Transform (DWT) to extract the low-pass approximation and the high-pass approximation from the collected network traffic, once the decomposition of the data has been completed. The authors have highlighted the importance of having small-time periods for the prediction. They use intervals of five minutes, which is too large to identify individual activities or small sessions of activity in a cyber security use case. For identifying individual activities or small sessions of activity a smaller interval time should be used such as one or two second intervals.

Time Series Data Classification. Madan et al. [27] have proposed an approach to the forecasting of network traffic using Discrete Wavelet Transform (DWT), Auto Regressive Integrated Moving Averages (ARIMA) and Recurrent Neural Networks (RNN) and they performed this using a dataset. The aim of this work is to provide an approach to predicting the future trends of network traffic for such applications such as congestion control, anomaly detection and bandwidth allocation. For this combined approach the data is first passed through the DWT to decompose it into a Nonlinear component and a Linear Component. This was then passed through ARIMA and RNN. These two forecasts were then combined to form the final forecast. They also used ARIMA and RNN separately using the same data to compare the results of the two methods. The Normalised Root Mean Squared Value (NRMSE) ranged from 0.197 to 0.009 for RNN, 0.240 to 0.010 for ARIMA and ranged from 0.191 to 0.008 for their proposed approach that combined ARIMA and RNN. However, the approach proposed here is all taking place offline on the data set that had been previously collected, and is restricted by the smallest level of granularity of the data which in this case is five minutes, so does not allow for true real time forecast of network traffic.

Vafeiadis et al. [44] have focused on accurately monitoring network traffic on high-speed networks, where they have pointed out that the volume of traffic generated requires sampling to address, but this instead affects accuracy. They instead proposed to model the network traffic with a sufficiently accurate time series model. They tested their approach using an offline dataset that was captured from a live network in ten second intervals. Then an ARIMA model was used to generate a representation of the network traffic and to signal the occurrence of packet loss in a real time environment. Their conclusion was that a sampling rate of 600 s must be used to achieve the most accurate representation and detection. In a cyber security setting, such a sampling rate would be impractical as it would mean missing entirely the maliciously triggered activities that would be of short duration.

2.3   Conclusion

The aim of looking at the related work regarding HITL approaches work related to Network traffic and IoT activity recognition, and Time series machine learning was to identify and analyse the extent and means by which human users have been involved in different aspects of protection against cyber attacks. The examples identified in the literature were diverse, ranging from systems that assist airline pilots with the identification of potential cyber attacks [12] to involving users in identifying characteristics of masquerading mobile applications [28]. By looking at this wide range of HITL systems, especially including cyber-physical systems, we have concluded that the adoption of such an approach to a system in an IoT environment is a promising direction for research. We have also identified a possible approach using time series forecasting for the identification of IoT device activity.

As a primary challenge of adopting HaaSS in IoT is the lack of context for users to be able to identify suspicious IoT behaviour, we will explore the option of providing the user with notifications of IoT activities detected. For this reason, we have surveyed the literature of activity recognition technologies used in IoT. We have observed that all related research uses the network traffic as the primary source of data for identification of IoT devices and their activity, and typically through the use of machine learning. A variety of machine learning models have been used for example to predict the levels of network traffic as well as how the traffic will flow through the network. However, as the area of application is not cyber security, in all cases the forecasting was evaluated on much longer time scales than what would be required by a real-time system to be practical for helping detect cyber attacks. In the following chapter, we will evaluate whether activity recognition can indeed be sufficiently fast and accurate to be practical for this purpose.

3   Context Generation for Smart Home HaaSS Through IoT Activity Recognition

Consider the attack model whereby an adversary has gained access to the network or the application that is used to control a smart home's network of IoT devices. This adversary then injects commands to IoT devices at will, for example requesting that electrical appliances in the kitchen are switched on and off repeatedly to cause electrical damage, that a door is unlocked, or a home security sensor (e.g., motion detector or occupancy presence detector) is switched off. These are examples of command injection attacks which lead to unauthorised IoT activities to be triggered without the legitimate user's approval or knowledge. Common injection is a very common type of attack in smart home IoT [16]. Examples that have been reported at academic level or in the wild include: Smart lock backdoor pin code injection [10], Spoofed voice activating smart lock [2], BLE automated smart lock door opening [18], Smart home power generator tripping [25], Remote access for WeMo command and control [6], Man-in-the-middle smart TV injection [19], Rogue payment via audio-triggered home assistant [31] and others.

As the outcome of command injection is that a new activity is triggered, it is expected that there will be new network traffic generated every time this is the case. For example, if Amazon Echo receives a command (via voice or an app) to connect to the smart lights and turn them off, then it needs to transmit network traffic to its cloud servers, which in turn need to communicate with the smart lights' cloud servers to return a command to the smart lights through the home network's WiFi router. Even if the user is not aware of this command having been received by the smart devices, the network traffic may indicate that a new activity has been triggered. In this chapter, we explore the potential to detect this network activity accurately. The logic is akin to the notifications that are sent for example by Gmail when it recognises that someone has logged in from a computer or geographical location that is different than usual. In that case, the user is notified that a new login activity has been detected.

At a technical level, being able to notify a user about new IoT activity involves first of all recognising automatically the generation of new activity. We can approach this problem as a type of time series forecasting. We assume that it is always the case that there is only one new activity at a time, i.e., that even if there are two activities starting in close succession, their difference in time is non-negligible. With this assumption, what an IoT activity recognition system can be based on is a time series forecasting algorithm that predicts what the next say 5 s of network activity will look like and deduce that new IoT activity has been generated if the actual traffic in these 5 s is significantly higher than what was predicted. By limiting the forecasting to each individual source IP address, we can always know which device (i.e., which IP) has generated new activity, and accordingly notify the user.

As a proof of concept, we have limited the type of data captured to the network packet rate data, and have implemented and compared two time series forecasting algorithms that are both lightweight and can easily be integrated in low-power devices, such as a raspberry PI acting as the smart home's gateway. Figure 1 summarises the proposed approach, which includes network traffic capture, time series forecasting, user notification and user feedback. The analysis in the next section is agnostic of the implementation approach, but it is worth nothing that a product based on this research would likely involve a simple notification on the user's mobile device that would direct the user to confirm whether the activity detected is expected or not, potentially with a link to an automated report with further detail.

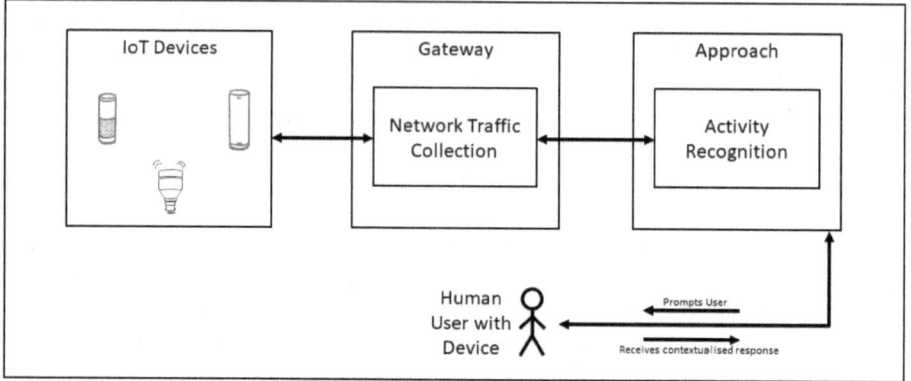

Fig. 1. The workflow for the proposed approach to activity recognition

3.1 Network Traffic Data Acquisition and Features

A key part of the proposed approach is the data collection. To have visibility to as much inbound and outbound data from all devices as possible, we have set up a Raspberry PI to act as a network bridge through which all IoT Internet traffic is routed and as such can be captured. This has been done to allow for the traffic capture to be carried out in closer proximity to the devices and to help negate any overload of traffic or noise from other parts of the network. This also ensures that it will not interfere or impact the function of any of the devices. In detail, a Raspberry Pi has been configured with two network interfaces so that it can offer a connection to our network on one interface and on the other interface offer a wireless access point that acts as a transparent bridge to our network. This is to allow for the capturing of network traffic of our chosen IoT devices, while allowing the device to connect to their required/used cloud services so that they may perform their intended activities. The actual collection of the data is then performed through the use of standard network capture tools, such as TCPDump [41]. As an example, see Fig. 2 for a data capture in the form of a PCAP file visualised in wireshark [46]. Each line in the PCAP file corresponds to one packet.

The packets captured include data such as: Order in which the packet was captured, Timestamp (the time the packet was received), Source of the packet, Destination of the packet, Protocol used, Size/Length of the packet and other extra information contained within the packet. From all of this information useful features can then be extracted. For simplicity, we take into account only the

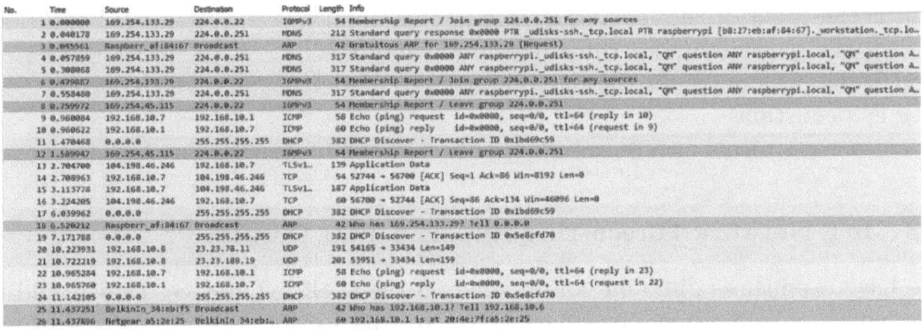

Fig. 2. An example of captured IoT network traffic in a PCAP format

timestamp and the existence of a packet, not its content, size or other characteristics. The timestamp and number of packets sent from or received by a particular IoT device's IP are used to compute the packet rate (in packets/s) corresponding to that device at any point in time. In practice, we sample the average packet rate periodically (every 1 sec in our implementation) so as to form a time series of data points that correspond to the same length of time each.

3.2 Time Series Analysis Models

The approach's core component is the analysis of the time series of packet rates provided above so as to predict how the traffic is expected to change in the very short term. If the traffic increases considerably faster than expected, then we assume that this indicates a new activity having been triggered on a particular device. For comparison, we have identified two statistical models, Auto-Regressive Integrated Moving Average (ARIMA) [4] and its simpler version, Auto-Regressive Moving Average (ARMA). Their underlying fundamental principle is that it is possible to predict/forecast a value at a future time based upon the values of past observations.

An ARIMA (p, d, q) model can be expressed as:

$$Y_t = c + \phi_1 y_{dt-1} + \phi_p y_{dt-p} + ... + \theta_1 \epsilon_{t-1} + \theta_q \epsilon_{t-q} + \epsilon_t \tag{1}$$

An ARMA (p, q) model can be expressed as:

$$X_t = c + \epsilon_t + \sum_{i=1}^{p} \varphi_i X_{t-i} + \sum_{i=1}^{q} \theta_i \epsilon_{t-i} \tag{2}$$

where:

- p is the order of the autoregressive polynomial (preceding/lagged values)
- d is number of times the data has been "differenced" to produce stationary data

- q is the order of the moving average polynomial (preceding/lagged values)
- φ is the parameters for the autoregressive model
- θ is the parameter for the moving average model
- c is a constant
- ϵ are error terms (white noise)

3.3 IoT Network Forecasting Experimental Results

We have used three different commercial off the shelf IoT devices that would be common in a smart home or smart office. In this section, our aim is to evaluate the two different forecasting methods against each other taking into account their accuracy and the time it takes to generate the forecast values based on different sizes of time windows of previously observed values. In all cases, the forecasting refers to the next 5 s. We have used this as a realistic timeframe in which a command can be injected in an IoT network and a user can determine whether they authorised a detected IoT activity or not.

The accuracy of our forecasting was measured by comparing the expected values of the network to the values that where forecast by the statistical model, for our purposes we use (pps_{actual}) as the total packets that were counted in a 5 s test window, and ($pps_{forecast}$) to represent the number of packets that had been forecast for that specific 5 s test window. We evaluate forecasting accuracy at the time where there is no new activity triggered, so ideally the forecast values should be equal to the actual ones. As a measure of forecasting accuracy, we use Residual Forecasting Error as in [22], as follows:

$$RFE = 1 - \frac{|pps_{actual} - pps_{forecast}|}{\max(pps_{actual}, pps_{forecast})} \tag{3}$$

In the case that there was no forecasting completed (in our experiments, seen as "forecasting failure" with ARMA at some low window sizes), we considered the predicted value to simply stay the same as the previous value. We also consider the case of having only one IoT device in a network or multiple (three) devices active at any point in time. In the former case, there is no background IoT traffic. In the latter case, there is significant background IoT traffic in the network. The devices involved were a LIFX smart lightbulb, a Belkin smart socket, and an Amazon Echo smart speaker. The experiments were performed based on a predefined time-scheduled script of commands injected by a user (by voice for Amazon Echo and via an app for the Belkin smart socket and LIFX smart lightbulb). They included four cases as described in the following subsections. In all cases, we repeated the experiments 10 times each. In the figures that follow, we report the mean values of these 10 iterations in each case.

We have implemented the ARIMA and ARMA models in Python. The processing was performed on a Raspberry PI 3 Model B Quad Core CPU 1.2 GHz 1 GB RAM Motherboard, so the forecast delay values reported correspond to that level of processing power. Importantly, the Raspberry PI was left in data collection mode while carrying out this processing. So, it was not processing in offline mode, but while also collecting data for future processing. This is because the aim is that no more than one device is needed to be added to an existing smart home environment for supporting a HaaSS implementation. So, the same device (here, a Raspberry PI) is used to collect the data and perform any processing required.

A.1 Amazon Echo Without Background Traffic. Here, the Amazon Echo was activated without any other devices on the network. The voice command used was "Alexa, what is the traffic like?". This setup corresponds to the best-case scenario where there is no background traffic. We have evaluated the fore-casting accuracy and the time it takes to reach a forecast for different training window sizes, ranging from 5 s to 30 s. For ARIMA, the residual error ranged from 24.98% to 28.21% (Fig. 3). As the training window size dictates the volume of data that need to be processed in generating a forecast, the larger the training window the longer the time it takes. The forecast delay started at 0.59 s for a training window of 5 s and reached 2.23 s for the window of 30 s (Fig. 4). For ARMA, the residual error was consistently extremely high. In terms of forecast delay, it ranged from 0.15 s to 0.3 s, which was considerably lower than ARIMA in all cases. An observation that can be made here is that ARIMA's residual error is consistent across different training window sizes, but at the expense of increased forecast delay.

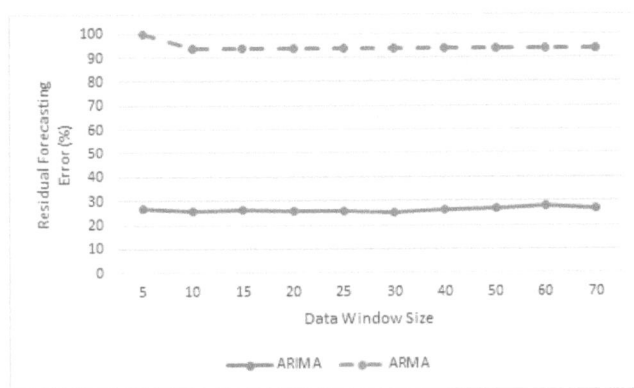

Fig. 3. A.1 - Residual forecasting error for the Amazon Alexa with no other device on the network

A.2 Amazon Echo with Background Traffic. Here, we repeat the same experiment as above, but at the time that a voice command is issued to the Amazon Echo, two other devices are on the network (Belkin Smart socket & LIFX Bulb). The voice command used was again "Alexa, what is the traffic like?".

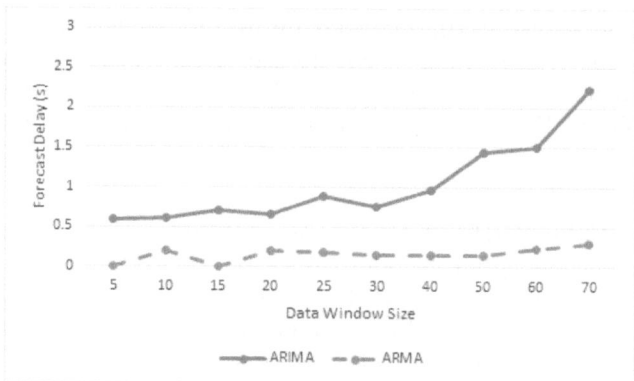

Fig. 4. A.1 - Forecast Delay for the Amazon Echo with no other device on the network

As shown in Figs. 5 and 6, the residual error is now higher for both ARIMA and ARMA, and the forecast delay increases. ARIMA's residual forecasting error ranges from 37.46% to 40%, and forecast delay from 0.93 s to 2.93 s. ARMA's residual forecasting error ranges from 18.34% to 100%, and forecast delay from 0.19 s to 0.49 s. ARIMA was much more consistent than ARMA and much more accurate for training window sizes up to 50 s, at which point ARMA outper-formed ARIMA (while also having very low forecast delay).

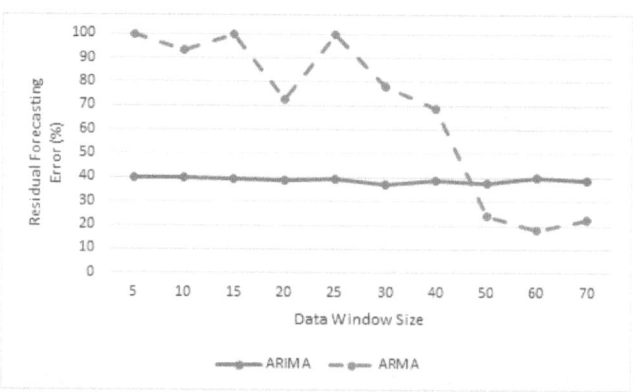

Fig. 5. A.2 - Residual forecasting error for the Amazon Echo with the Belkin Switch and LIFX Bulb also on the same network

A.3 Belkin Smart Socket with Background Traffic. Here, the Belkin Smart socket was activated with the two other devices on the network (LIFX Bulb & Amazon Echo). The command was "toggle switch". As shown in Figs. 7

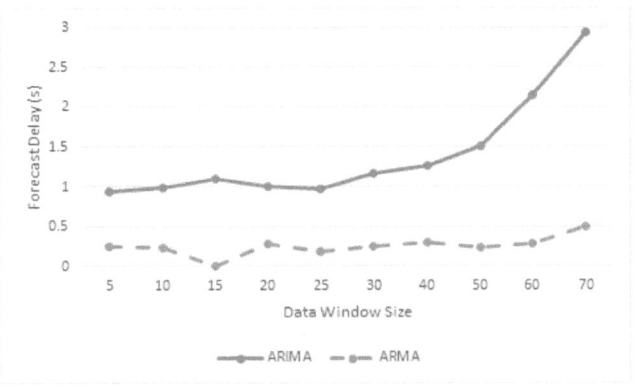

Fig. 6. A.2 - Forecast Delay for the Amazon Echo with the Belkin smart socket and LIFX Bulb also on the same network

and 8, the residual forecasting error for ARIMA ranges from 26.91% to 40.02%, and forecast delay from 0.39 s to 1.58 s. ARMA's residual forecasting error ranges from 2.42% to 51.07% residual error, and forecast delay from 0.15 s to 0.45 s. Here, ARMA was more accurate than ARIMA in small window sizes and less accurate in large window sizes, while still faster in all cases.

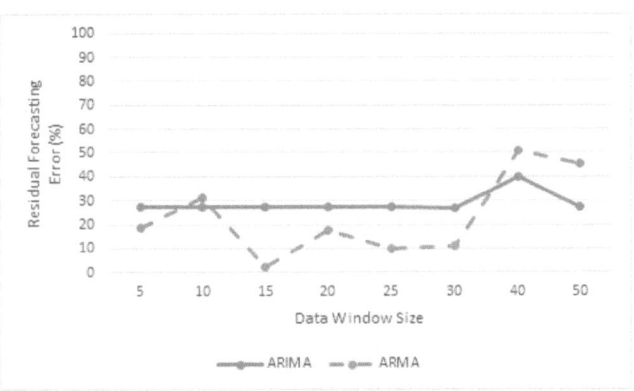

Fig. 7. A.3 - Residual forecasting error for the Belkin Smart Socket with the LIFX Bulb and Amazon Alexa also on the same network

A.4 LIFX Bulb with Background Traffic. Here, the LIFX Bulb was activated with Amazon Echo and Belkin smart socket on the network. The command was "turn on". As shown in Figs. 9 and 10, the residual forecasting error for ARIMA ranges from 20.24% to 21.57%, and forecast delay from 0.59 s to 2.48 s. ARMA's residual forecasting error ranges from 12.65% to 100%, and forecast

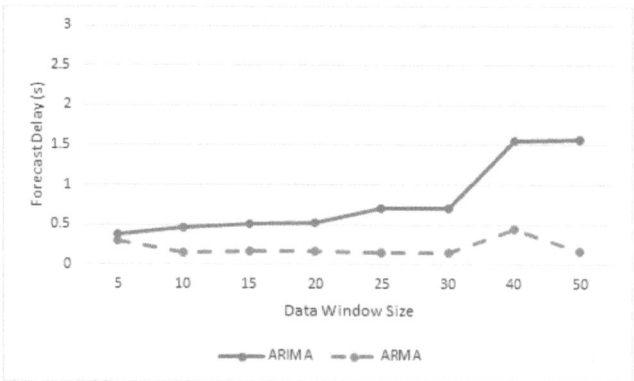

Fig. 8. A.3 - Forecast Delay for the Belkin Smart Socket with the LIFX Bulb and Amazon Echo also on the same network

delay from 0.15 s to 0.26 s. In this case, ARIMA was more accurate than ARMA in almost all cases, but again at the expense of incurring considerably higher forecast delay.

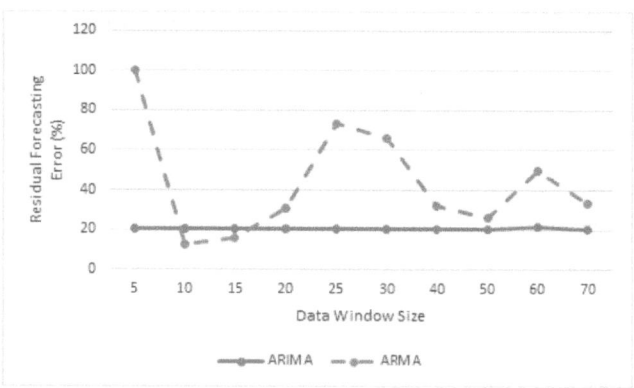

Fig. 9. A.4 - Residual forecasting error for LIFX Bulb with the Amazon Echo and Belkin Smart Socket also on the same network

4 IoT Network Activity Recognition

Here, we use a static threshold based rule system, whereby an increase in traffic is recognised as a new activity if the actual average packet rate over the last test window is higher than a given ratio of what was forecast. The ratio was computed simply as:

$$\text{Difference Ratio} = \frac{\sum pps_{actual}}{\sum pps_{forecast}} \tag{4}$$

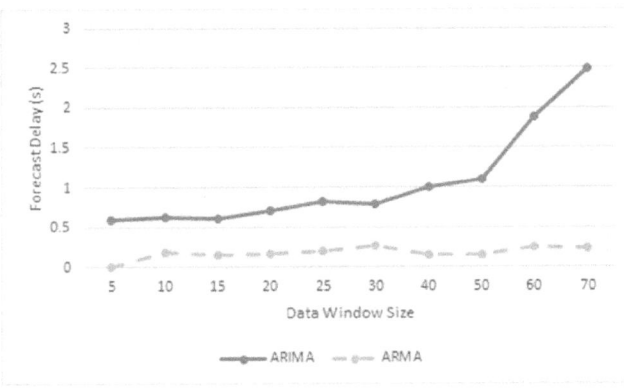

Fig. 10. A.4 - Forecast Delay for LIFX Bulb with the Amazon Echo and Belkin Smart Socket also on the same network

Indicatively, we try ratio threshold values of 2, 4, 6, 8, 10 and 50.

In a practical implementation, measures should be taken to ensure that users are not notified multiple times about the same activity. This could be done by monitoring when this activity is completed (i.e., a consistent drop in network traffic back to the previous normal levels) or to have a cool-down period of time during which we do not monitor for new activity from that particular device after a first notification is triggered. This is out of scope at this phase of the work.

To evaluate this simple threshold based system, we again take the four experimental cases presented in the previous section. As forecasting model, we chose ARIMA, as it is clear that ARIMA is much more reliable than ARMA in terms of accuracy, achieving consistently satisfactory residual forecasting error at all window sizes. The forecast delay is much higher, but within a range that is acceptable for the type of application proposed. Specifically, we use a training window size of 25 s, which is realistic given our requirement for good accuracy achieved with forecast delay that is considerably lower than the test window of 5 s. If the forecast delay increased beyond the test window, then it would generate an infinitely increasing backlog of forecast tasks.

Figures 11 and 12 summarise the results in terms of false negative rate (missing a new activity) and false positive rate (detecting new activity where there is not one). Again, we take 10 runs per case, which means a total of 40 5-s windows where there is no new activity, and 40 5-s windows where there is new activity. We observe experimentally that a ratio threshold value of 6 or 8 can achieve at the same time both a low false negative rate (25%, 30%) and a low false positive rate (5%, 2.5%). It would also be possible to implement the ratio threshold value of 2 to achieve very low rates of false negatives (2.5%) but at the cost of increasing the rate at which false positives occur (30%).

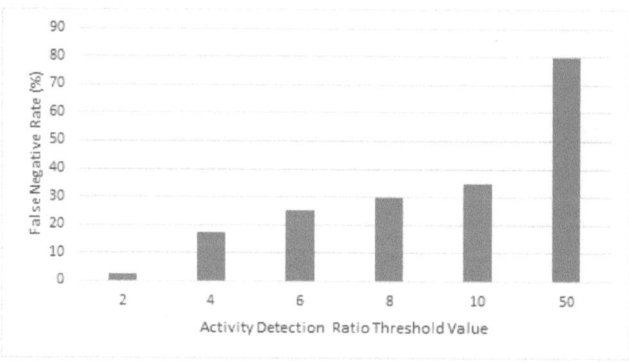

Fig. 11. Average across all devices - False Negative rate

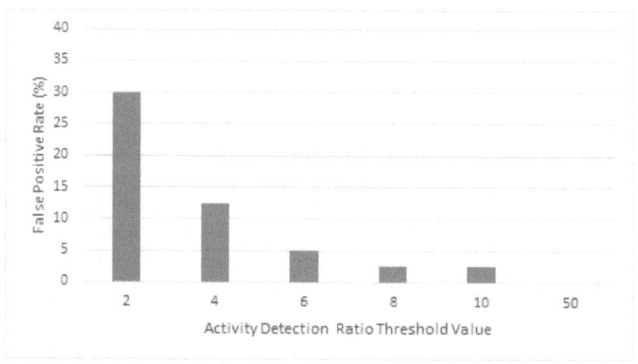

Fig. 12. Average across all devices - False positive rate

5   Conclusion

In this work, we have demonstrated that expanding the HaaSS paradigm to IoT environments is possible as an approach to helping secure IoT systems. We have reviewed related work regarding HITL approaches to cybersecurity and work related to Network traffic and IoT activity recognition, and Time series machine learning. We have also explored the practicality of using time series forecasting as a means for the automated detection of activity in an IoT network. In this context, we have aimed for a time series forecasting method that can be sufficiently accurate for different use cases (with and without background IoT traffic) without incurring a very high forecast delay. Through our experimental evaluations with different IoT environment setups, we can conclude that ARIMA provides consistency of residual forecasting error across almost all training window sizes. However, it achieves this at the expense of considerable forecast delay. Yet, this delay is still comfortably lower than the 5 s of the test window we have aimed for, even on a resource-constrained Raspberry PI. This makes it a promising approach as the basis of an activity notification provision for IoT users.

References

1. Bai, L., Yao, L., Kanhere, S.S., Wang, X., Yang, Z.: Automatic device classification from network traffic streams of Internet of Things. In: 2018 IEEE 43rd Conference on Local Computer Networks (LCN) (2018). https://doi.org/10.1109/lcn.2018.8638232
2. BBC: How hackers could use doll to open your front door (2017). https://www.bbc.co.uk/news/av/technology-38966285/how-hackers-could-use-doll-to-open-your-front-door
3. Ben-Asher, N., Gonzalez, C.: Effects of cyber security knowledge on attack detection. Comput. Hum. Behav. **48**, 51–61 (2015). https://doi.org/10.1016/j.chb.2015.01.039
4. Box, G.E.P., Jenkins, G.M.: Time Series Analysis: Forecasting and Control - First Edition. Holden-Day (1976)
5. Campbell, J., Jr., Mensch, A.C., Zeno, G., Campbell, W.M., Lippmann, R.P., Weller-Fahy, D.J.: Finding malicious cyber discussions in social media. Technical report, MIT Lincoln Laboratory Lexington United States (2015)
6. CERT: Vulnerability note vu 656302 (2014). http://www.kb.cert.org/vuls/id/656302
7. Cisco: Talos intelligence: latest cyber security threat intelligence services (2023). https://www.talosintelligence.com/
8. Cofense: Phishing prevention. https://cofense.com/
9. Cranor, L.F.: A framework for reasoning about the human in the loop. In: UPSEC (2008)
10. Fernandes, E., Jung, J., Prakash, A.: Security analysis of emerging smart home applications. In: 2016 IEEE Symposium on Security and Privacy (SP), pp. 636–654. IEEE (2016)
11. FireEye: Recent zero-day exploits. https://www.fireeye.com/current-threats/recent-zero-day-attacks.html
12. Gontar, P., Homans, H., Rostalski, M., Behrend, J., Dehais, F., Bengler, K.: Are pilots prepared for a cyber-attack? A human factors approach to the experimental evaluation of pilots behavior. J. Air Transp. Manag. **69**, 26–37 (2018). https://doi.org/10.1016/j.jairtraman.2018.01.004
13. Haack, J.N., Fink, G., Maiden, W.M., Mckinnon, D., Fulp, E.W.: Mixed-initiative cyber security: putting humans in the right loop. Mixed-Initiative Multiagent Systems (MIMS) (2012)
14. Heartfield, R., Loukas, G.: A taxonomy of attacks and a survey of defence mechanisms for semantic social engineering attacks. ACM Comput. Surv. (CSUR) **48**(3), 37 (2016)
15. Heartfield, R., Loukas, G.: Detecting semantic social engineering attacks with the weakest link: implementation and empirical evaluation of a human-as-a-security-sensor framework. Comput. Secur. **76**, 101–127 (2018)
16. Heartfield, R., et al.: A taxonomy of cyber-physical threats and impact in the smart home. Comput. Secur. (2018)
17. Heartfield, R., Loukas, G., Gan, D.: You are probably not the weakest link: towards practical prediction of susceptibility to semantic social engineering attacks. IEEE Access **4**, 6910–6928 (2016)
18. Ho, G., Leung, D., Mishra, P., Hosseini, A., Song, D., Wagner, D.: Smart locks: lessons for securing commodity internet of things devices. In: Proceedings of the 11th ACM on Asia Conference on Computer and Communications Security, pp. 461–472. ACM (2016)

19. Jacoby, D.: How I hacked my home (2014). https://www.blog.kaspersky.com/how-i-hacked-my-home/5756/
20. Kaspersky: Cyberthreat real-time map. https://cybermap.kaspersky.com/
21. Khandpur, R.P., Ji, T., Jan, S., Wang, G., Lu, C.T., Ramakrishnan, N.: Crowdsourcing cybersecurity: cyber attack detection using social media. In: Proceedings of the 2017 ACM on Conference on Information and Knowledge Management - CIKM 2017 (2017). https://doi.org/10.1145/3132847.3132866
22. Khoda, M.E., Razzaque, M.A., Almogren, A., Hassan, M.M., Alamri, A., Alelaiwi, A.: Efficient computation offloading decision in mobile cloud computing over 5G network. Mob. Netw. Appl. **21**(5), 777–792 (2016). https://doi.org/10.1007/s11036-016-0688-6
23. Koodous: Platform for the detection of malicious activity in android applications. https://koodous.com/
24. Kumar, S.A., Bhargava, B., Macedo, R., Mani, G.: Securing IoT-based cyber-physical human systems against collaborative attacks. In: 2017 IEEE International Congress on Internet of Things (ICIOT) (2017). https://doi.org/10.1109/ieee.iciot.2017.11
25. Liu, Y., Hu, S., Zomaya, A.Y.: The hierarchical smart home cyberattack detection considering power overloading and frequency disturbance. IEEE Trans. Ind. Inf. **12**(5), 1973–1983 (2016)
26. Ma, M., et al.: Data and decision intelligence for human-in-the-loop cyber-physical systems: reference model, recent progresses and challenges. J. Signal Process. Syst. **90**(8–9), 1167–1178 (2017). https://doi.org/10.1007/s11265-017-1304-0
27. Madan, R., Sarathimangipudi, P.: Predicting computer network traffic: a time series forecasting approach using DWT, ARIMA and RNN. In: 2018 Eleventh International Conference on Contemporary Computing (IC3) (2018). https://doi.org/10.1109/ic3.2018.8530608
28. Malisa, L., Kostiainen, K., Capkun, S.: Detecting mobile application spoofing attacks by leveraging user visual similarity perception. IACR Cryptology ePrint Archive (2015)
29. Millersmiles: Anti-phishing services. https://www.millersmiles.co.uk/
30. Millot, P., Mouchel, M., Paglia, C.: The human operator as the ultimate barrier to cyber attacks. In: 2018 IEEE Industrial Cyber-Physical Systems (ICPS) (2018). https://doi.org/10.1109/icphys.2018.8390774
31. Morley, K.: Amazon echo rogue payment warning after tv show causes Alexa to order dolls houses (2017). http://www.telegraph.co.uk/news/2017/01/08/amazon-echo-rogue-payment-warning-tv-show-causes-alexa-order
32. Nie, L., Wang, X., Wan, L., Yu, S., Song, H., Jiang, D.: Network traffic prediction based on deep belief network and spatiotemporal compressive sensing in wireless mesh backbone networks. Wirel. Commun. Mob. Comput. **2018** (2018)
33. University of Oxford: Information security - report an incident. https://www.infosec.ox.ac.uk/report-incident
34. PhishTank: Community-based phish verification system. https://www.phishtank.com/
35. Rajivan, P., Janssen, M.A., Cooke, N.J.: Agent-based model of a cyber security defense analyst team. Proc. Hum. Factors Ergon. Soc. Annu. Meeting **57**(1), 314–318 (2013). https://doi.org/10.1177/1541931213571069
36. Wombat Security: Wombat security, security awareness training software. http://www.wombatsecurity.com/

37. Sekwatlakwatla, P., Mphahlele, M., Zuva, T.: Traffic flow prediction in cloud computing. In: 2016 International Conference on Advances in Computing and Communication Engineering (ICACCE) (2016). https://doi.org/10.1109/icacce.2016.8073735

38. Shahid, M.R., Blanc, G., Zhang, Z., Debar, H.: IoT devices recognition through network traffic analysis. In: 2018 IEEE International Conference on Big Data (Big Data) (2018). https://doi.org/10.1109/bigdata.2018.8622243

39. Staheli, D., et al.: Collaborative data analysis and discovery for cyber security. In: WSIW@ SOUPS (2016)

40. Stembert, N., Padmos, A., Bargh, S.M., Choenni, S., Jansen, F.: A study of preventing email (spear) phishing by enabling human intelligence. In: Intelligence and Security Informatics Conference (EISIC), pp. 113–120. IEEE (2015)

41. Tcpdump: Tcpdump/libpcap public repository (2019). https://www.tcpdump.org/

42. Tsai, F.S., Chan, K.L.: Detecting cyber security threats in weblogs using probabilistic models. In: Yang, C.C., et al. (eds.) PAISI 2007. LNCS, vol. 4430, pp. 46–57. Springer, Heidelberg (2007). https://doi.org/10.1007/978-3-540-71549-8_4

43. Tyworth, M., Giacobe, N.A., Mancuso, V.F., Mcneese, M.D., Hall, D.L.: A human-in-the-loop approach to understanding situation awareness in cyber defence analysis. ICST Trans. Secur. Saf. 1(2) (2013). https://doi.org/10.4108/trans.sesa.01-06.2013.e6

44. Vafeiadis, T., Papanikolaou, A., Ilioudis, C., Charchalakis, S.: Real-time network data analysis using time series models. Simul. Model. Pract. Theory 29, 173–180 (2012)

45. Vielberth, M.: Human-as-a-security-sensor for harvesting threat intelligence. Cybersecurity 2(1), 23 (2019)

46. Wireshark: Wireshark software repository (2019). https://www.wireshark.org/

47. Zhu, H., Elfar, M., Pajic, M., Wang, Z., Cummings, M.L.: Human augmentation of UAV cyber-attack detection. In: Schmorrow, D.D., Fidopiastis, C.M. (eds.) AC 2018. LNCS (LNAI), vol. 10916, pp. 154–167. Springer, Cham (2018). https://doi.org/10.1007/978-3-319-91467-1_13

Intrusion Detection at the IoT Edge Using Federated Learning

James Pope[1], Theodoros Spyridopoulos[2]([envelope]), Vijay Kumar[2],
Francesco Raimondo[1], Sam Gunner[1], George Oikonomou[1],
Thomas Pasquier[3], Ryan McConville[1], Pietro Carnelli[4],
Adrian Sanchez-Mompo[4], Ioannis Mavromatis[5], and Aftab Khan[4]

[1] University of Bristol, Bristol, UK
{james.pope,F.Raimondo,sam.gunner,G.Oikonomou,
ryan.mcconville}@bristol.ac.uk
[2] Cardiff University, Cardiff, UK
{spyridopoulost,KumarV14}@cardiff.ac.uk
[3] University of British Columbia, Vancouver, Canada
tfjmp@cs.ubc.ca
[4] Toshiba Research Europe, Bristol, UK
{Pietro.Carnelli,Adrian.Mompo,Aftab.Khan}@toshiba-bril.com
[5] Digital Catapult, London, UK
Ioannis.Mavromatis@digicatapult.org.uk

Abstract. With the proliferation of Internet of Things (IoT) technologies in urban environments, cities are increasingly deploying Edge processing nodes for urban sensing. This large-scale integration of Edge nodes and sensing endpoints raises significant security concerns. For instance, existing Intrusion Detection methods cannot scale well and do not consider the privacy and energy consumption implications that emerge when applied to those systems. In addition, the use of containerised applications managed by container orchestration platforms in these environments, while enabling diverse applications and allowing scanning of the container images, can still introduce vulnerabilities. This Chapter addresses the challenge of effectively detecting malicious activities in large-scale resource-constrained IoT systems. We introduce an unsupervised distributed learning solution employing Federated Learning (FL) for real-time anomaly detection across the IoT infrastructure. Our approach involves analysing Linux system call data through a Federated Learning Framework, significantly reducing the need for central data processing. The Chapter presents a comprehensive architectural overview of the system, its core components, and the methodology for deploying and updating anomaly detection models. It also provides the performance evaluation of our approach. Our results demonstrate that the size of the clients' datasets and the use of pre-trained models play a significant role in the performance of FL models. The work presented in this chapter was supported by UK Research and Innovation, Innovate UK [grant number 53707].

© The Author(s), under exclusive license to Springer Nature Switzerland AG 2025
N. Pitropakis and S. Katsikas (Eds.): *Security and Privacy in Smart Environments*, LNCS 14800, pp. 98–119, 2025.
https://doi.org/10.1007/978-3-031-66708-4_5

The authors would also like to thank Bo Luo and Dan Howarth for their contribution to the project as members of SMARTIA Ltd.

Keywords: Distributed Artificial Intelligence · Infrastructure Protection · Edge Computing · Federated Learning · Anomaly Detection · IoT Security

1   Introduction

With cities experiencing rapid growth, local authorities strive to enhance essential services for residents, including waste management, water supply, and transportation. They have adopted various technologies to gather, analyse, and display data from sensors placed throughout the city. These sensors, often found in streetlights and public vehicles, monitor factors like noise levels and air quality. This data-driven approach benefits both citizens and policymakers, helping them meet regulatory requirements.

Notable projects, such as the University of Chicago's Array of Things [1,3,20] and the South Gloucester Council's UMBRELLA project [7,14], exemplify these efforts. These initiatives involve deploying numerous nodes across the city to collect data on noise and air pollution and perform video-based analysis. In the case of the UMBRELLA project, approximately 200 nodes, developed by Toshiba, are deployed over a 7 km area, monitoring air pollution and street lighting and providing a platform for IoT applications. To manage these applications, the UMBRELLA project employs containers with Kubernetes orchestration for Cloud and Edge deployments. While containers may be harmless, they can be exploited maliciously, leading to actions such as privilege escalation and Denial of Services (DoSs) attacks as presented in the authors' previous work in [18]. These actions pose risks to the host system and potentially the entire infrastructure. In addition, although administrators typically scan container images for vulnerabilities before deployment, identifying all potential weaknesses through static checks is challenging. Therefore, continuous monitoring of system and network activity, including data generated by containers, the host OS and network interactions, is essential.

Our approach utilises the Linux auditing system (auditd) to collect data on Linux system calls, which is then analysed using an autoencoder-based anomaly detection approach to detect malicious activities. We also introduce FL to update models in a privacy-preserving manner across a distributed network of edge devices.

In the latter part of this chapter, we demonstrate the practical implementation of such a system within an IoT testbed. The contributions of this work include the development of an AI-based intrusion detection system designed to operate efficiently on edge nodes in a distributed manner, capable of detecting intrusions in real-world smart city scenarios. This system is characterised by its federated model training and updating mechanisms. Additionally, we introduce a concrete use case and provide supporting datasets for identifying and detecting malicious containers deployed in edge systems.

The structure of this chapter is organised as follows: Sect. 2 delves into related work, providing insights into the existing research landscape. Section 3 outlines our approach and the architectural framework we employ. Section 4 presents the evaluation of our approach. Finally, Sect. 5 concludes our work and offers pathways for further research.

2   Related Work

Research has shown an increasing number of vulnerable Industrial IoT devices within the industry that are connected to the Internet [10]. Significant research has been conducted on securing Industrial IoT devices, predominantly focusing on Cloud-centric security solutions. However, Industrial IoT applications introduce additional constraints that render these Cloud-based methods unsuitable. These applications are time-constrained and critical to safety, and they typically require that data remain within the system to maintain privacy. Consequently, as highlighted in [4], it is essential to perform intrusion detection either directly on the device or close to it. This approach aligns with the Industrial Internet Security Framework (IISF) developed by the Industry IoT Consortium (IIC), according to which endpoint monitoring can be performed either internally to the endpoint or externally to it [21].

Current online Intrusion Detection Systems (IDSs) fall into two primary categories: signature-based and behaviour analysis-based. The operation of signature-based Intrusion Detection Systems (IDS) relies on signatures or rules to detect attacks. These are engineered based on insights gained from existing attacks including specific strings in network packet payloads and IP addresses that are linked to cyber attacks. However, such techniques are ineffective against more sophisticated attack variants and zero-day exploits that existing signatures cannot cover. These "unknown" attacks remain undetected by signature-based IDSs. Additionally, while the upkeep of a large database of signatures is feasible for conventional IT workstations, this approach is incompatible with the limited capacity inherent to IoT devices [6].

On the other hand, behaviour analysis-based methods focus on identifying abnormal system behaviour by detecting deviations from its normal/expected patterns as these are typically modelled using Artificial Intelligence (AI). The limitations of signature-based approaches can be effectively mitigated through AI-based behaviour analysis techniques [6]. A large number of AI-based IDSs has been proposed, ranging from multi-class classification to anomaly detection for the detection of novel attacks in IoT systems [2,5,9,11,13]. In particular, unsupervised anomaly detection methods enable the detection of novel attacks without requiring prior knowledge of existing threats [13] since they base their operation on accurately representing the system's normal state. Consequently, contrary to other AI-based techniques, anomaly detection-based IDSs are easier to train and are also capable of detecting suspected intrusions, zero-day attacks and device failures [27].

Nevertheless, significant challenges arise in the implementation of anomaly detection-based IDSs within an IoT system. These include the requirement to frequently update the model to mitigate concept drift, a phenomenon where a model's performance deteriorates over time due to the dynamic nature of the system which alters its normal behaviour. Furthermore, the need to collect data from IoT devices for analysis and further model training to address concept drift introduces additional challenges. Data collection from IoT and Industrial IoT devices can lead to privacy implications and increase the network load, deteriorating operational performance. In addition, even though a plethora of similar IoT devices can be deployed in different systems, privacy implications hinder the transfer of knowledge between these systems, further complicating the scenario [24].

Anomaly detection for intrusion detection in IoT devices utilising a FL architecture is an emerging field that addresses several of the aforementioned issues. Compared to traditional methods, FL employs a decentralised collaborative training approach that allows knowledge sharing and incremental training without compromising privacy. In its original form, FL comprises of an aggregation server and multiple distributed worker nodes. Each node performs incremental training, using its local data, on the model shared by the aggregator. The aggregator collects the model parameters from all workers and aggregates them to generate a new Global Model which is shared back with the workers [12].

The authors in [17] introduce a Network Intrusion Detection System (NIDS) leveraging FL for IoT environments. Security gateways within the network serve as edge workers, aggregating data from IoT devices to facilitate incremental training, while a centralised security service oversees model aggregation. The core of their anomaly detection mechanism employs a Recurrent Neural Network (RNN) with Gated Recurrent Units (GRUs), designed to analyse sequences of "normal" network traffic from IoT devices. Their system predicts subsequent time series values of packet sequences, utilising deviations from these predictions - when a predefined threshold is exceeded - as indicators of potential intrusions. However, while a NIDS might excel at identifying anomalies in network traffic, its ability to detect a system-level attack (e.g. privilege escalation) that may not produce deviations in network patterns is limited.

In their study [15], the authors implemented a Deep Neural Network (DNN) model for anomaly detection in healthcare IoT systems using a FL framework. Their approach yielded improved results compared to centralised DNN-based anomaly detection methods. Their anomaly detection approach employed several supervised machine learning approaches. However, their paper does not describe how the dataset was distributed in the conducted experiments, and there is no analysis of the Global Model initialisation. The increased performance over centralised approaches, despite the generalisation that naturally occurs due to parameter aggregation, indicates that the initial Global Model was chosen close to the final collaborative model, a detail not explained in the study. Furthermore, similarly to the approach in [17], this study focuses on network traffic for anomaly detection, which inherently limits its capability in identifying system-level intrusions that do not manifest through changes in network behaviour.

Similar outcomes have been found in a series of studies that explore the use of FL and anomaly detection models for Intrusion Detection in large-scale IoT networks [16, 26]. Even though these studies provide promising results when FL is used for intrusion detection, they focus only on network traffic analysis and do not explore the impact of Global Model initialisation.

The authors in [19] propose a Federated Deep Learning (FDL) approach for detecting zero-day botnet attacks in IoT edge devices. The method utilises a DNN architecture for network traffic classification and employs federated averaging (FedAvg) to aggregate model updates from multiple IoT-edge devices. The effectiveness of the developed model is demonstrated through simulations with the Bot-IoT and N-BaIoT datasets, showing significant improvements over centralised deep learning, localised deep learning, and distributed deep learning methods in various aspects, particularly in privacy preservation and efficient use of resources. However, due to relying on labelled data for classification, the method cannot detect novel attacks, an area in which anomaly detection-based methods excel.

Following a different approach to the learning process, [23] focuses on enhancing intrusion detection in network environments through the integration of Federated Learning and Reinforcement Learning. The system uses Deep Q-Networks (a reinforcement learning strategy) within a Federated Learning framework. Agents learn to detect intrusions aiming to maximise the reward signal based on detection accuracy and effectiveness. However, in the context of intrusion detection, the definition of rewards is based on the detection of intrusions, which in turn implies the need for labelled instances.

Contrasting with the prior research, our proposed research combines a FL architecture with unsupervised anomaly detection (specifically auto-encoders) for Host-based Intrusion Detection in IoT systems, relying on system data collected on the device (e.g. system calls). This approach enables the detection of both network and system-level cyber attacks, marking a significant advancement in the domain of FL-based IDS. To the best of our knowledge, this is the first work in the field of FL-based IDSs that can tackle both types of attacks. Additionally, we investigate the impact of Global Model initialisation on the model's performance, a factor not previously explored in this context.

3   Approach

Our work focuses on large-scale IoT systems where applications are typically implemented as containers on Edge devices. These applications are managed by container Orchestration platforms such as Kubernetes[1]. City-scale IoT systems, such as Toshiba's UMBRELLA project[2], rely on such platforms to automate deployment, scaling and management of applications and services across the

---

[1] https://kubernetes.io/.

[2] https://www.umbrellaiot.com/.

network of devices. For our experiments, we utilised a network of UMBRELLA nodes, which are a key component of the platform acting as an Edge device with interchangeable modules for various radios and sensors. The computing power in the node is provided by Raspberry Pi 3+ and Jetson Nano.

Our approach uses an AI Anomaly Detection method suitable for detecting novel attacks. The approach trains a model on the *normal* state of an edge device. The model reproduces feature vectors (denoted FV) extracted from events in audit logs generated on the Edge device. Ideally, the normal feature vectors are closely reproduced resulting in a small *reconstruction error* while *abnormal* feature vectors result in larger *reconstruction errors*. Threshold techniques are then used to classify among three states {*normal, uncertain, anomaly*}.

This methodology is also termed *semi-supervised* learning, as it primarily requires *normal* data for training the model, thereby eliminating the need for annotated data from predefined attacks. This approach is notably advantageous due to the sporadic nature of attack data occurrences, alongside the challenges and elevated costs involved in acquiring annotated attack data. Additionally, in contrast to traditional supervised learning methods, our semi-supervised model can detect novel attacks, thereby enhancing its applicability in the evolving landscape of cyber security threats.

Figure 1 depicts the approach performing intrusion detection on the edge to detect malicious activity by a container. The model is trained on normal data. When the difference between a feature vector and its reconstruction is above a threshold, an anomaly is triggered.

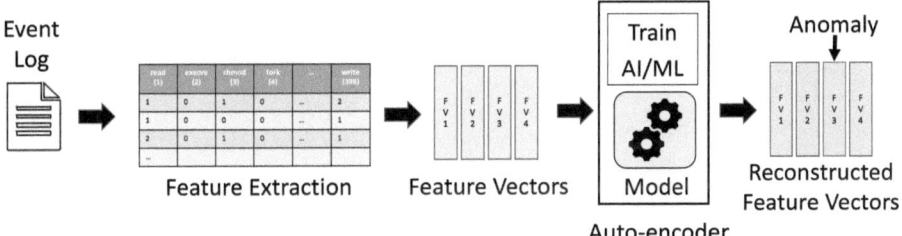

Fig. 1. Approach Overview

There exist various tools capable of logging container activities, such as Falco[3]. However, these tools do not capture host-level events, which led to our selection of the Linux Audit Daemon (*auditd*), a component of the Linux Auditing System that writes audit records to the disk. Containers running on the host system execute system calls to interact with the host OS and perform various activities - *auditd* records these system call events. Each edge device is configured to log audit events and independently train its specific model, effectively implementing a unique model for each edge device. The methodology for federating these models is detailed in Sect. 3.4.

---

[3] https://falco.org/.

3.1   Feature Extraction

Feature extraction consists of the following steps:

1. Raw auditd event logs are collected and filtered to retain events with a SYSCALL type. Most events have a SYSCALL type and are generally more informative than other event types.
2. Given a *window length* (in seconds), all events within the window are converted into counts for each SYSCALL type producing a feature vector of numbers. The feature vector length is 398, based on the number of different SYSCALL types[4].
3. The window start is advanced by the *overlap* and the process is repeated to produce another feature vector. The features are saved as a CSV file to be used later for modelling.

Figure 2 depicts how the features are determined from an event log. There are similar approaches to extracting event-based features from log files. For example, KubAnomaly [22] counts 17 SYSCALL types and 14 root access event types to produce feature vectors of length 31. Our approach uses all SYSCALL types producing feature vectors of length 398.

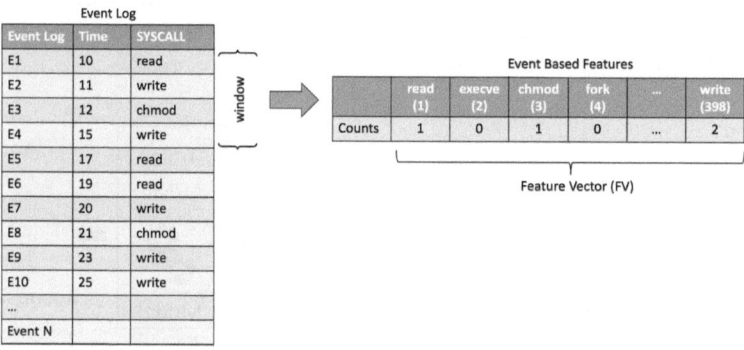

Fig. 2. The selection of event based features, where the occurrence of each event's SYSCALL type is counted.

In the *audit.log* files, the occurrence of each event type within a sliding window of specified length is counted. This is illustrated in Fig. 3. It is important to select an appropriate window length and overlap for the sliding window. If the sliding window is too short, it does not capture enough information for the model to make an accurate inference. On the other hand, if it is too long, it will respond to an attack more slowly. The overlap impacts the amount of data required to be processed and therefore has to be restricted as the target Edge devices have

---

[4] https://android.googlesource.com/platform/external/qemu/+/emu-master-dev/linux-headers/asm-arm/unistd-common.h.

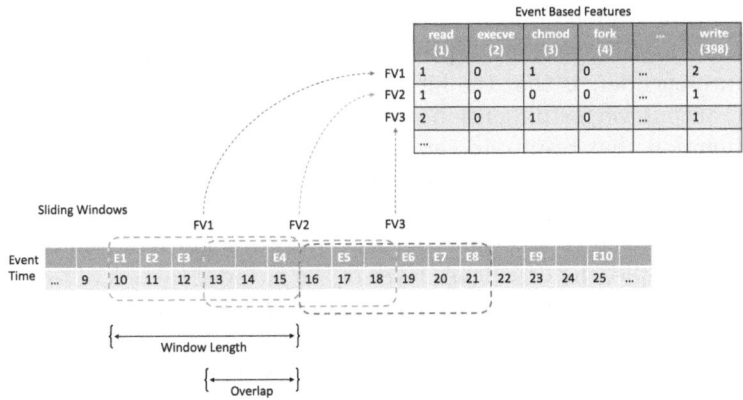

Fig. 3. The feature count is dictated in part by a sliding window passed over the events (example shows window length of 6 s with 50% overlap)

limited computing power. The overlap also partially addresses scenarios that span across adjacent windows.

For example, if the audit log covers 100 min with a window length of 2 min, 0% overlap results in 50 feature vectors, 50% overlap results in ~100 feature vectors, and 90% overlap results in ~500 feature vectors. Considering a sequence of related anomalous events (e.g. an attack) that lasts for 1 min, ideally the full sequence would be contained within one feature vector. The 0% overlap would likely result with portions of the sequence in adjacent feature vectors and thus reduce the ability of the model to detect the sequence. With 50% overlap this is unlikely and 90% very unlikely; at least one feature vector would include the majority of the sequence. One drawback of a significant overlap is that it produces a larger number of feature vectors requiring more training and inference computation. After evaluating various values with our dataset, it was found that a window length of 30 s and a 50% overlap (15 s) performs most effectively. Future work includes determining these values via an initial calibration step.

It is important to mention that the feature vectors in our analysis implicitly encode temporal information. When an anomaly is detected, the starting time of the window in which it occurred offers a rough estimation of the timing of the anomaly. While this temporal aspect is significant, it is just one facet of the broader context that a security analyst would require. Additional details such as associated users, processes, sockets, files, etc., are also crucial for a comprehensive analysis. Ideally, the output of the Artificial Intelligence (AI) model would encompass this wider range of information to improve explainability. Addressing this remains an area for future work [8].

3.2   Model Architecture

The auto-encoder model has three components: encoder, hidden representation, and decoder. Figure 4 depicts the model's architecture. The model takes an input vector and produces an output vector with a hidden vector as an intermediate, low-dimensional representation. An important design consideration for auto-encoders is that the hidden vector dimension is much smaller than the input/output vector dimension. Otherwise, the auto-encoder will learn the identity function that results in zero reconstruction error for both normal and abnormal data. Reading from left to right, each dense layer transforms a vector from the input dimensional space to the output dimensional space. Note that the intermediate vectors between the dense layers are omitted from the figure for clarity. For example, the first dense layer of the encoder takes a 398 dimensional vector and produces a 100 dimensional vector. The neurons in the layer achieve this by multiplying the input by weights, sum them, and pass the result through an activation function to produce the transformed vector.

The model learns the neurons' weights during the training process. However, there are a number of model hyper-parameters that are static during training and must be determined beforehand. The number of neurons for the encoder and decoder, the dimension of the hidden representation, the threshold values (described below), and the activation function are hyper-parameters of the model. Automatically determining model hyper-parameters is an active research area and many solutions exist [25]. A common solution is *manual tuning* through experimentation that uses a combination of modeller background knowledge and previous settings known to be effective. Though manual tuning does not result in an optimal solution, it often results in a sufficient model with minimal amount of computation when compared to automated methods. For these reasons, manual tuning was used to determine the model hyper-parameters.

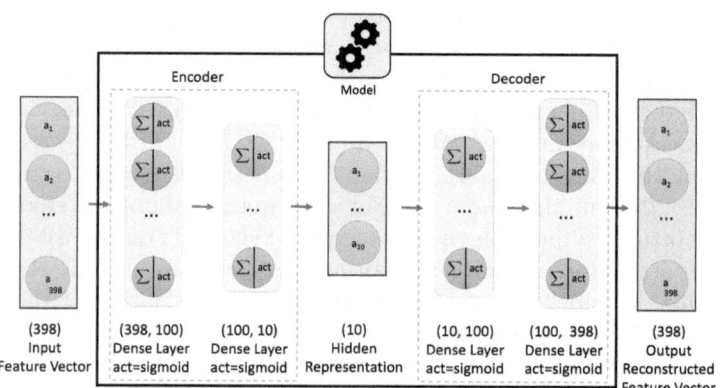

Fig. 4. Auto-encoder Architecture

Once the model has been trained, its output is used to perform inference on the validation and test data by using a classifier. The reconstruction error does not directly indicate an anomaly. It is difficult to identify the anomalies using only reconstruction errors because they keep changing when the auto-encoder is re-trained or updated. Therefore, a classifier is added on top of the reconstruction error of the auto-encoder to make the final inference as to whether the data corresponds to an anomaly. In addition, the classifier can work out the confidence of that final inference.

An upper and lower threshold is also introduced to make the final inference. It is an anomaly if the reconstruction error is above the upper threshold. If the reconstruction error is below the lower threshold, it is normal. If the reconstruction error is between the upper and lower threshold, it is uncertain whether it is an anomaly or a normal. The reconstruction error is transformed into a value ranging from 0 to 1 through a LogisticRegression model, trained on the annotated data to make a robust classification. The value indicates the confidence level of an anomaly. It is highly likely to be an anomaly when the value is close to 1 and otherwise when it is close to 0. After a series of experiments, we found that the best thresholds are 0.8 for the upper threshold and 0.5 for the lower threshold.

3.3   Dataset

Obtaining a suitable dataset with a sufficient number of annotated attacks (for evaluation, not training), including a container escape attack, proved to be difficult. Ultimately, we generated the dataset from custom experiments. Figure 5, taken from our previous work in [18] and produced here for clarity, depicts how the edge devices were configured to generate the dataset.

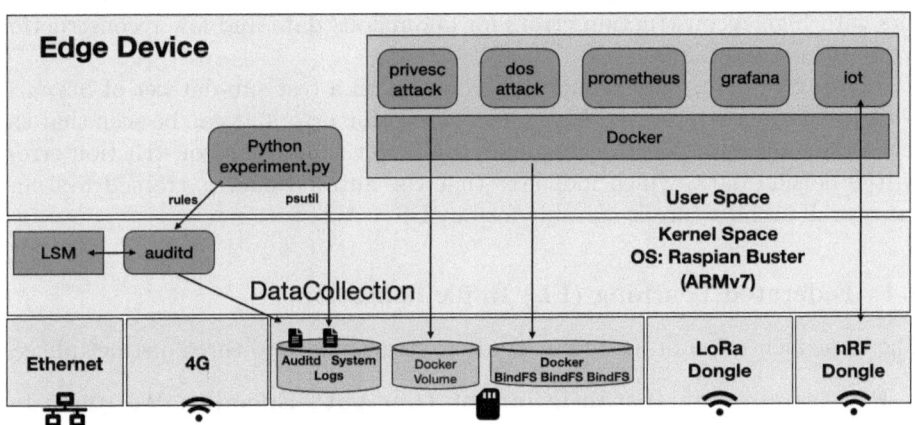

Fig. 5. Edge Configuration for Container Escape Dataset

All experiments use the same workload with a webserver and database (Grafana/Prometheus) collecting sensor data from three devices. Two of the five containers were configured to perform a container escape and launch an attack. Three scenarios are used as follows:

- *Scenario A (DoS)* where a container escape is performed using a host shell that launches a denial of service (DoS) attack. This attack involved approximately 20 system calls and many of the system calls were unusual compared to the normal behaviour. This attack would likely be easier to detect as an anomaly.
- *Scenario B (Privesc)*, where a container escape is writing to host permission file granting no password *sudo* permission to the user (i.e. launches a privilege escalation attack). This attack involved approximately 10 system calls and many of the system calls were benign compared to the normal behaviour. This attack would be more difficult to detect as an anomaly.
- *Scenario Normal*, where a benign container without any escape/attacks is used.

Each experiment runs for 15 min and all system calls associated with the container escape and attack last no more than 20 s, usually spanning across windows. There are 256 experiments. Feature extraction results in 58 normal vectors and 2 anomalous vectors per experiment. In total the dataset contains 14848 normal vectors, 128 (DoS) anomalous vectors, and 128 (Privesc) anomalous vectors.

A centralised model was trained using only normal data with anomalous and normal data used for testing. The batch size is set small to make the model more sensitive to each instance. The training parameters used were batch size: 10, patience: 30, epoch: 10000. During the training, the auto-encoder model will learn how to reproduce normal feature vectors. Since the training does not include the anomalous data, we expect the trained model to produce feature vectors with high reconstruction errors for anomalous data and low reconstruction errors for normal data.

Figure 6 presents the reconstruction errors on a test sub-dataset of *Scenario DoS*. It contains attacks and higher reconstruction errors. It can be seen that the reconstruction errors on the anomalies are larger than the reconstruction errors on the normal data, which indicates that the auto-encoder is trained well and the overall architecture is suitable for an IDS.

3.4   Federated Learning (FL) Implementation

The implementation of FL in our study is structured into three distinct phases:

1. **Pre-training or Instantiation of the Auto-Encoder:** We conducted three separate experiments in this phase. In two out of the three experiments, the auto-encoder undergoes pre-training centrally using a specific training dataset (training parameters used: batch size: 10, patience: 30, epoch: 10000). For the remaining experiment, the auto-encoder is instantiated without any pre-training.
2. **Federated Learning and Model Updating:** The model is updated following the FL methodology. Prior to the FL round, the Global model is distributed to the clients. In each FL round, the model is further trained independently on each client device during the local training phase. After

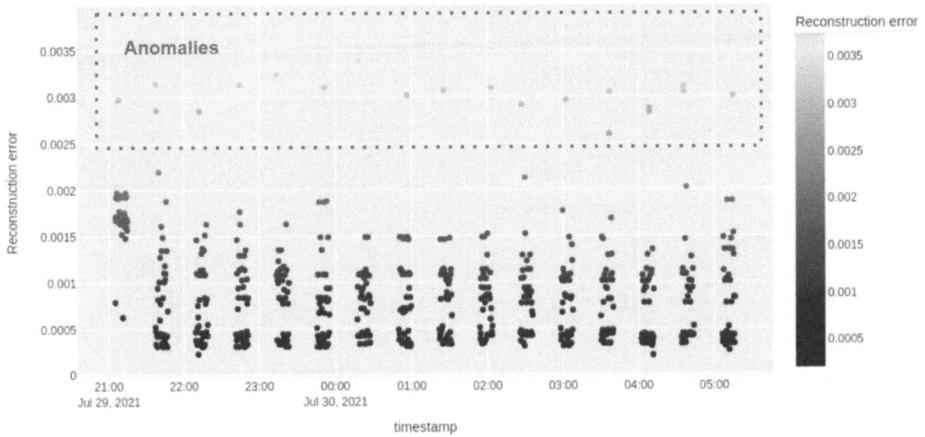

Fig. 6. Reconstruction error on test sub-dataset associated with Scenario A

local training, the updated local model parameters are sent to the server. The server then aggregates these parameters using the FedAvg algorithm to fuse the local models and update the Global model. This updated Global model is redistributed to each client, initiating a new FL round. This iterative process is repeated until the model converges.

3. **Model Assessment:** The final phase involves evaluating the model using test data.

Figure 7 depicts the FL workflow. Models at the client-side and server-side, along with training and aggregation logs, are saved in each FL round. Finally, training is performed and evaluated using the test data.

4   Evaluation

This section evaluates the approach with the dataset using a centralised and then federated models. We also describe the scores used to evaluate the models' performance.

4.1   Performance Scores

The models are evaluated using several traditional scores, provided here for convenience. The result of a prediction attempt for binary classification falls into one four cases, where we treat *anomaly* as being *positive* and *normal* as *negative.*

- TP true positive when the prediction is anomaly and is actually anomaly.
- TN true negative when the prediction is normal and is actually normal.
- FP false positive when the prediction is anomaly but is actually normal.

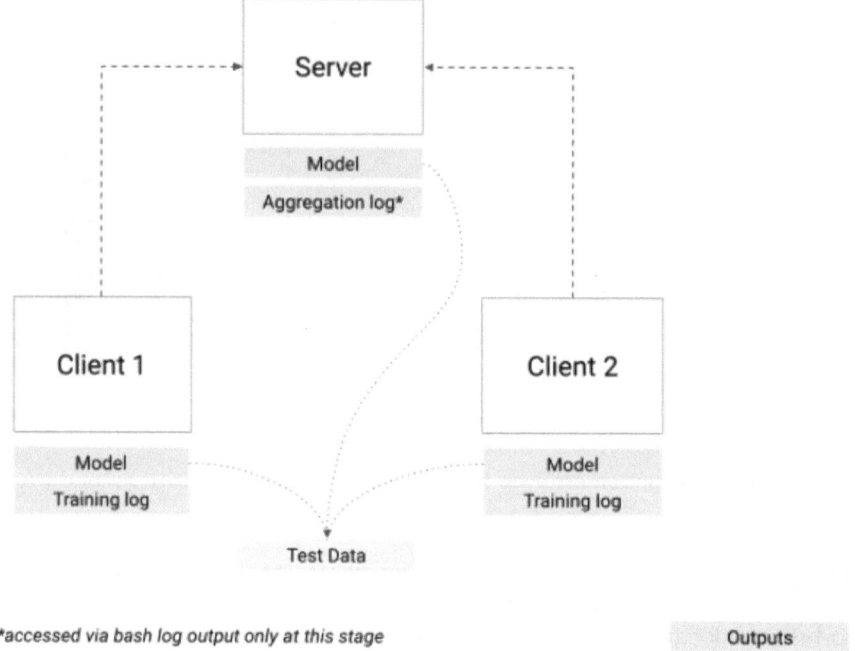

Fig. 7. Framework of the FL experiment

– FN false negative when the prediction is normal but is actually anomaly.

For a set of predictions from some experiment, these numbers are typically combined into an aggregate score as follows.

– **Accuracy:**
$$\frac{TN + TP}{TP + FP + TN + FN}$$
The accuracy is the fraction of predictions the model got right on all samples, including normals and anomalies.

– **Precision:**
$$\frac{TP}{TP + FP}$$
The precision is the proportion of positive identifications that was actually correct. This score accounts for when the model incorrectly identifies some negatives as positives (i.e. false positives).

– **Recall:**
$$\frac{TP}{TP + FN}$$
The recall is the proportion of actual positives identified correctly. This score accounts for when the model fails to identify some positives (i.e. false negatives).

– **F-score:**

$$2 \times \frac{Precision \times Recall}{Precision + Recall}$$

The F-score combines the precision and recall of the model, and it is defined as the harmonic mean of these scores.

For imbalanced datasets, where the *normal* class is much larger than the *abnormal* class, the *accuracy* is considered a poor score because it includes TNs that skew the result. A simple (but poor) model can just always predict *normal* and achieve a very high *accuracy*. Importantly, this is the situation for our anomaly detection dataset where the number of *normal* vectors is much larger than *abnormal* vectors (in our case *accuracy* = 0.983 (14848/15104)). The *precision* and *recall* do not include TNs and are more robust to the class imbalance issue. Which to prefer depends on which error *costs* more, a FP or a FN (this requires a subjective assessment). Instead of assigning costs, the F-score is biased towards the smaller of recall and precision. It is a somewhat pessimistic score of the model in this sense. The F-score does not assume a cost for errors and is also robust to the class imbalance issue. For these reasons the F-score is often preferred. Our results provide all four scores noting that the F-score is more appropriate for evaluation.

In our study, we have three classes {*normal, uncertain, anomaly*}. Though there are methods to adapt the scores to multi-class classification problems, for the purpose of evaluation, we simplify this into a binary schema based on the assumption that 'uncertain' predictions will undergo human investigation.

Consequently, in instances where the prediction is 'uncertain' but the true label is 'normal', we categorise such predictions as 'True Negative'. This categorisation is based on the rationale that human intervention would accurately identify these instances as 'normal', and this information would subsequently be utilised to refine the training dataset. On the other hand, when a prediction is 'uncertain' but the actual scenario is 'abnormal', we classify these as 'False Negative'. This is because by the time human examination occurs the anomalous activity would likely have already manifested. Moreover, such instances are not incorporated into the training set since the model relies on normal data only.

4.2 Centralised Model Results (No FL)

We train a single autoencoder model, based on the architecture provided in Sect. 3.2, using the training data and evaluate it using the unseen test data. This situation assumes all data would be available, with no Federated Learning involved. As mentioned in Sect. 3.3, the training was run for 10000 epochs, with patch size 10 and patience 30. Figure 8 presents the confusion matrix of the centralised model on the test dataset.

The model generated can identify most anomalies correctly (21 out of 23 anomalies). The model only mistakenly recognises one anomaly as normal and makes four uncertain predictions. It is assumed that the uncertain predictions will be sent to human experts for further analysis and data labelling to train

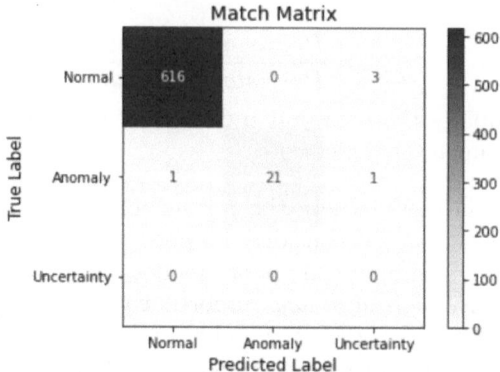

Fig. 8. Centralised Model Confusion Matrix

the model further to improve its performance. The 'human-in-the-loop' would essentially act as a separate 'asynchronous' node in the Federated Learning architecture. This scheme would improve the performance of the approach making it robust against attacks that trigger uncertain responses by the model. It should be noted that even though the model did not correctly classify one of the anomalies, it would likely not have missed the attack, as an attack is made up of several anomalies.

Table 1 shows the performance scores for the centralised model. Given all the data, the centralised model performs very well for all scores with *recall* being the lowest (21/23).

Table 1. Centralised Model Scores

Metric	Accuracy	Precision	Recall	F-score
Score	0.9922	1.0	0.913	0.9545

4.3 Federated Learning Model(s) Results

To evaluate our FL approach, we focused on three experiments: First, fully train a model using FL, utilising 100% of the training data and splitting it across the two clients. Second, pre-train a model with 50% of the training data, then update the model using the remaining 50% of the data split across two clients; and, third, pre-train a model with 90% of the training data, then update the model using the remaining 10% of the data split across two clients. The experiments fixed a number of parameters. In particular, we only used two clients and fixed the training parameters on each client to match those for centralised training. The output of the experiments was scores for the server model and both client models, based on the testing set, enabling us to compare FL performance against centralised performance.

Fully Trained. The following sections will provide a selection of results, focusing on overall server performance and individual client performance. Figure 9 shows the confusion matrices on the server side, tested against attack data in the Scenarios A (*DoS*) and B (*Privesc*) datasets.

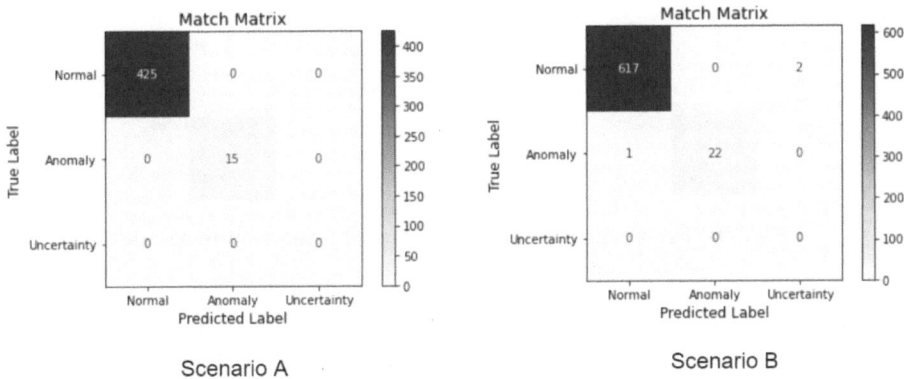

Scenario A   Scenario B

Fig. 9. Fully Trained Server Model Confusion Matrices for scenario A (DoS) and scenario B (Privesc)

Table 2 shows the scores for the fully trained server model. The results are similar to the centralised model scores with both the server and clients achieving scores above 90%, regardless of scenario. There does not appear to be any loss in performance due to the Federated Learning as compared to the centralised model.

Table 2. Fully Trained Server Model Scores

Metric	Server		Client 1		Client 2	
	Scenario A	Scenario B	Scenario A	Scenario B	Scenario A	Scenario B
Accuracy	1.0	0.9953	1.0	0.9969	0.9977	0.9938
Recall	1.0	0.9565	1.0	0.9565	1.0	0.9565
Precision	1.0	1.0	1.0	1.0	1.0	0.9565
F-score	1.0	0.9778	1.0	0.9778	1.0	0.9565

Pre-trained Model (90%). Figure 10 shows the confusion matrices for the 90% pre-trained server model. The server still performs well for Scenario A. For Scenario B the server scores are slightly reduced (with 15 total uncertain). There are 3 False Positives resulting in a *precision* score of 0.875.

Table 3 shows the scores for the 90% pre-trained server model. The results clearly show a difference between Scenario A and B for the client models. For

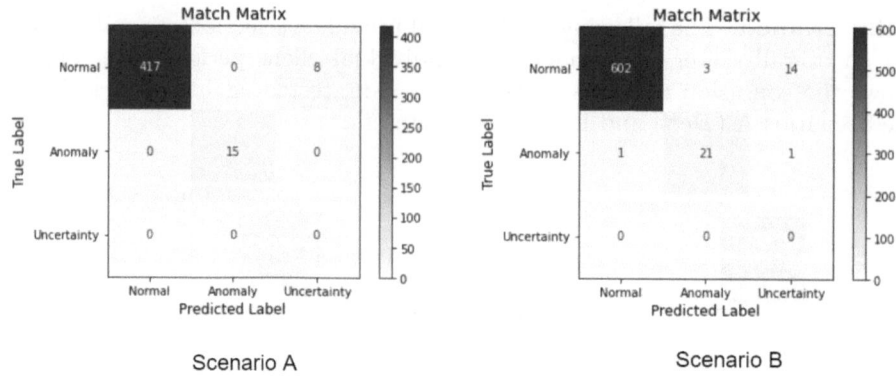

Fig. 10. Pre-trained (90%) Server Model Confusion matrices

Scenario A the clients perform very well. For Scenario B, with the exception of accuracy, the scores are notably less with F-scores of 0.75. The amount of data used for pre-training in the federated case appears to be having an impact on performance for the client models.

Table 3. Pre-trained (90%) Server and Client Model Scores

Metric	Server		Client 1		Client 2	
	Scenario A	Scenario B	Scenario A	Scenario B	Scenario A	Scenario B
Accuracy	0.9818	0.9704	0.9795	0.9595	0.9773	0.9564
Recall	1.0	0.913	1.0	0.6522	1.0	0.6522
Precision	1.0	0.875	1.0	0.8824	1.0	0.8824
F-score	1.0	0.8936	1.0	0.75	1.0	0.75

Pre-trained Model (50%). Figure 11 shows the confusion matrices for the 50% pre-trained server model. While the Scenario A results are still very good, the Scenario B confusion matrix shows a severe decline in the server models performance. There are 8 FPs and 18 FNs (taking into account the 14 predicted as *uncertain*).

Table 4 shows the scores for the 50% pre-trained server model. Due to the high number of FNs, the server model, Scenario B *recall* score is very low 0.2174% (5/23). This behaviour is mirrored in the client models with good performance for Scenario A and poor performance (particularly *recall*) for Scenario B. The F-scores are correspondingly affected with 0.4737 for Client Model 2.

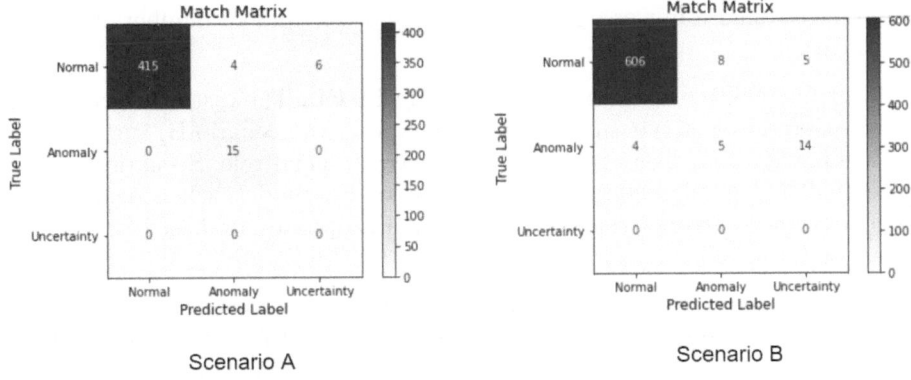

Fig. 11. Pre-trained (50%) Server Model Confusion Matrix

Table 4. Pre-trained (50%) Server and Client Model Scores

Metric	Server		Client 1		Client 2	
	Scenario A	Scenario B	Scenario A	Scenario B	Scenario A	Scenario B
Accuracy	0.9773	0.9517	0.9864	0.9642	0.9795	0.9579
Recall	1.0	0.2174	1.0	0.4783	1.0	0.3913
Precision	0.7895	0.3846	0.8333	0.7333	0.7895	0.6
F-score	0.8824	0.2778	0.9091	0.579	0.8824	0.4737

Analysis. The experiments conducted reveal significant insights into the performance dynamics of FL in intrusion detection for IoT ecosystems. A key observation is the impact of data distribution and the proportion of data among clients and servers on model efficacy.

Client Data Proportion and Model Performance: The varying performance in scenarios A and B, across different training setups (90% pre-trained vs 50% pre-trained), underscores the influence of data distribution among clients. When a substantial portion of the training data (90%) is used in the pre-training phase, the model retains a higher degree of learning, leading to better performance even after FL updates. On the other hand, with just 50% of data in the pre-training phase, the model's ability to learn and adapt deteriorates, resulting in lower accuracy, especially in more complex scenarios like B.

Fully Trained vs Partially Trained FL Models: The results indicate that fully trained FL models (i.e. no pre-training) outperform partially trained ones. However, this comes in contrast with the previous observation. A plausible explanation is that in our experiments we distributed the dataset among only two clients. As a result, the two clients had an abundance of data to train local robust models that were close to each other. Hence the FL aggregation resulted in a high-performance model. If the data were split amongst more clients, this would have resulted in worse performance due to the limited number of samples per device.

Role of Data Diversity and Complexity in FL: The distinct performance in detecting different types of attacks (Scenario A vs Scenario B) suggests that the complexity and diversity of the data play a crucial role. Scenario B, being more challenging, highlights the limitations of FL models when dealing with intricate attack patterns, especially when the FL updates are based on a smaller subset of data.

Metrics Sensitivity to Federated Learning Dynamics: The consistent accuracy across different FL setups, despite varying FPs and FNs, highlights the limitations of using accuracy as a sole metric in imbalanced datasets typical in cyber-security contexts. Precision, recall, and F-scores provide a better understanding of the model's performance, especially in FL environments where data distribution and the proportion of training data can significantly impact these metrics.

5   Conclusion

This Chapter presented an autoencoder-based Intrusion Detection System deployed using a Federated Learning framework, aimed at detecting intrusions in large-scale IoT systems. We demonstrated the effectiveness of our model in a decentralised, privacy-preserving setup. The critical insights gained from the comparison of fully trained and partially trained FL models underscore the significant impact of data distribution among clients as well as the number of clients on model performance. Our observations suggest that the fewer the number of clients participating in an FL setup, the higher the final performance due to the limited diversity among the clients. Furthermore, starting with a robust pre-trained model notably improves the overall performance. Consequently, a promising direction for future work involves the exploration of FL "islands", where a small number of clients contribute to the island's Global Model. Each island would then contribute to a higher-level FL aggregation, moving towards Hierarchical Federated Learning structures.

By training our models on normal system behaviour and utilising feature vectors from audit logs in real-world scenarios, we developed a resilient IDS capable of adapting to the evolving nature of cyber threats. The study not only reinforces the viability of FL in cyber security but also provides a foundation for future research in optimising FL deployment strategies for more complex and diverse IoT ecosystems.

In addition, with the utilisation of an "uncertain" class, we note the inability of automated solutions to address all scenarios and emphasise the importance of a "human/expert-in-the-loop". Audit logs that trigger the uncertain class can be forwarded to human experts for in-depth analysis. Human expertise complements the automated detection process, thus enhancing the overall cyber security resilience of large-scale IoT systems.

In conclusion, the research contributes to the growing body of knowledge in AI-driven cyber security, offering a practical and effective solution for intrusion detection in large-scale IoT systems. Our work opens avenues for further exploration in the field, particularly in understanding the nuances of Federated Learning in diverse and dynamic cyber security environments.

References

1. Catlett, C., et al.: Hands-on computer science: the array of things experimental urban instrument. Comput. Sci. Eng. **24**(1), 57–63 (2022). https://doi.org/10.1109/MCSE.2021.3139405

2. Chapaneri, R., Shah, S.: A comprehensive survey of machine learning-based network intrusion detection. In: Satapathy, S.C., Bhateja, V., Das, S. (eds.) Smart Intelligent Computing and Applications. SIST, vol. 104, pp. 345–356. Springer, Singapore (2019). https://doi.org/10.1007/978-981-13-1921-1_35

3. Collis, S., et al.: Introducing sage: cyberinfrastructure for sensing at the edge. In: EGU General Assembly Conference Abstracts, p. 12320. EGU General Assembly Conference Abstracts (2020). https://doi.org/10.5194/egusphere-egu2020-12320

4. Easwaran, A., Chattopadhyay, A., Bhasin, S.: A systematic security analysis of real-time cyber-physical systems. In: 2017 22nd Asia and South Pacific Design Automation Conference (ASP-DAC), pp. 206–213 (2017). https://doi.org/10.1109/ASPDAC.2017.7858321

5. Elrawy, M.F., Awad, A.I., Hamed, H.F.A.: Intrusion detection systems for IoT-based smart environments: a survey. J. Cloud Comput. **7**(1), 21 (2018). https://doi.org/10.1186/s13677-018-0123-6

6. Eskandari, M., Janjua, Z.H., Vecchio, M., Antonelli, F.: Passban IDS: an intelligent anomaly-based intrusion detection system for IoT edge devices. IEEE Internet Things J. **7**(8), 6882–6897 (2020). https://doi.org/10.1109/JIOT.2020.2970501

7. Farnham, T., et al.: Umbrella collaborative robotics testbed and IoT platform. In: 2021 IEEE 18th Annual Consumer Communications Networking Conference (CCNC), pp. 1–7 (2021). https://doi.org/10.1109/CCNC49032.2021.9369615

8. Guo, Y., Pope, J.: Spatial-temporal graph neural network for the detection of container escape events. In: Proceedings of the 16th International Conference on Agents and Artificial Intelligence - Volume 3: ICAART, pp. 326–333. INSTICC, SciTePress (2024). https://doi.org/10.5220/0012347800003636

9. Hanif, S., Ilyas, T., Zeeshan, M.: Intrusion detection in IoT using artificial neural networks on UNSW-15 dataset. In: 2019 IEEE 16th International Conference on Smart Cities: Improving Quality of Life Using ICT & IoT and AI (HONET-ICT), pp. 152–156 (2019). https://doi.org/10.1109/HONET.2019.8908122

10. Jiang, X., Lora, M., Chattopadhyay, S.: An experimental analysis of security vulnerabilities in industrial IoT devices. ACM Trans. Internet Technol. **20**(2) (2020). https://doi.org/10.1145/3379542

11. Khan, M.A., et al.: A deep learning-based intrusion detection system for MQTT enabled IoT. Sensors **21**(21) (2021). https://doi.org/10.3390/s21217016

12. Konečný, J., McMahan, H.B., Yu, F.X., Richtarik, P., Suresh, A.T., Bacon, D.: Federated learning: strategies for improving communication efficiency. In: NIPS Workshop on Private Multi-Party Machine Learning (2016). https://arxiv.org/abs/1610.05492

13. Liu, Z., Thapa, N., Shaver, A., Roy, K., Yuan, X., Khorsandroo, S.: Anomaly detection on IoT network intrusion using machine learning. In: 2020 International Conference on Artificial Intelligence, Big Data, Computing and Data Communication Systems (icABCD), pp. 1–5 (2020). https://doi.org/10.1109/icABCD49160.2020.9183842

14. B.T.E. Ltd.: UMBRELLA node (2021). https://www.umbrellaiot.com/what-is-umbrella/umbrella-node/. Accessed 06 Sept 2021

15. Mosaiyebzadeh, F., Pouriyeh, S., Parizi, R.M., Han, M., Batista, D.M.: Intrusion detection system for IoHT devices using federated learning. In: IEEE INFOCOM 2023 - IEEE Conference on Computer Communications Workshops (INFOCOM WKSHPS), pp. 1–6 (2023). https://doi.org/10.1109/INFOCOMWKSHPS57453.2023.10225932

16. Mothukuri, V., Khare, P., Parizi, R.M., Pouriyeh, S., Dehghantanha, A., Srivastava, G.: Federated-learning-based anomaly detection for IoT security attacks. IEEE Internet Things J. **9**(4), 2545–2554 (2022). https://doi.org/10.1109/JIOT.2021.3077803

17. Nguyen, T.D., Marchal, S., Miettinen, M., Fereidooni, H., Asokan, N., Sadeghi, A.R.: DÏOT: a federated self-learning anomaly detection system for IoT. In: 2019 IEEE 39th International Conference on Distributed Computing Systems (ICDCS), pp. 756–767 (2019). https://doi.org/10.1109/ICDCS.2019.00080

18. Pope, J., et al.: Container escape detection for edge devices. In: Proceedings of the 19th ACM Conference on Embedded Networked Sensor Systems, SenSys 2021, pp. 532–536. Association for Computing Machinery, New York (2021). https://doi.org/10.1145/3485730.3494114

19. Popoola, S.I., Ande, R., Adebisi, B., Gui, G., Hammoudeh, M., Jogunola, O.: Federated deep learning for zero-day botnet attack detection in IoT-edge devices. IEEE Internet Things J. **9**(5), 3930–3944 (2022). https://doi.org/10.1109/JIOT.2021.3100755

20. Potosnak, M.J., et al.: Array of Things: a high-density, urban deployment of low-cost air quality sensors. In: AGU Fall Meeting Abstracts, vol. 2019, pp. A24G–04 (2019)

21. Soroush, H., LeBlanc, J., Hirsch, F., Zhang, H., Martin, R.: Industrial internet security framework (IISF) (1.0). https://hub.iiconsortium.org/iisf. Accessed 04 Jan 2024

22. Tien, C.W., Huang, T.Y., Tien, C.W., Huang, T.C., Kuo, S.Y.: KubAnomaly: anomaly detection for the docker orchestration platform with neural network approaches. Eng. Rep. **1**(5), e12080 (2019). https://doi.org/10.1002/eng2.12080. https://onlinelibrary.wiley.com/doi/abs/10.1002/eng2.12080

23. Vadigi, S., Sethi, K., Mohanty, D., Das, S.P., Bera, P.: Federated reinforcement learning based intrusion detection system using dynamic attention mechanism. J. Inf. Secur. Appl. **78**, 103608 (2023). https://doi.org/10.1016/j.jisa.2023.103608. https://www.sciencedirect.com/science/article/pii/S2214212623001928

24. Wang, X., Wang, Y., Javaheri, Z., Almutairi, L., Moghadamnejad, N., Younes, O.S.: Federated deep learning for anomaly detection in the Internet of Things. Comput. Electr. Eng. **108**, 108651 (2023). https://doi.org/10.1016/j.compeleceng.2023.108651. https://www.sciencedirect.com/science/article/pii/S0045790623000769

25. Yang, L., Shami, A.: On hyperparameter optimization of machine learning algorithms: theory and practice. Neurocomputing **415**, 295–316 (2020). https://doi.org/10.1016/j.neucom.2020.07.061. https://www.sciencedirect.com/science/article/pii/S0925231220311693

26. Yao, W., Shi, H., Zhao, H.: Scalable anomaly-based intrusion detection for secure Internet of Things using generative adversarial networks in fog environment. J. Netw. Comput. Appl. **214**, 103622 (2023). https://doi.org/10.1016/j.jnca.2023.103622. https://www.sciencedirect.com/science/article/pii/S1084804523000413
27. Zoppi, T., Ceccarelli, A., Bondavalli, A.: Into the unknown: unsupervised machine learning algorithms for anomaly-based intrusion detection. In: 2020 50th Annual IEEE-IFIP International Conference on Dependable Systems and Networks-Supplemental Volume (DSN-S), p. 81 (2020). https://doi.org/10.1109/DSN-S50200.2020.00044

Bits and Bytes Betrayal: Unravelling the Dark Threads of Cybercrime in the Metaverse

Pankaj Pandey[✉]

Center for Cyber and Information Security, Department of Information Security and Communication Technology, Norwegian University of Science and Technology, Gjøvik, Norway
pankaj.pandey@ntnu.no

Abstract. Conventional crimes, marked by structural rigidity, have evolved as criminal organisations embrace digital possibilities facilitated by the Internet and expanding electronic commerce. Disruption, however, doesn't solely arise from technological progress; it stems from the convergence of disruptive technologies, challenging established legal frameworks. This evolution is evident in cybercrime within the metaverse, reflecting the sophistication of criminal behaviour and the emergence of new forms of illegal conduct. Our exploration has highlighted challenges within the metaverse, transcending conventional cyber threats. From the potential criminal liability of avatars to the complex interplay between legal identities and virtual personas, we've addressed issues like defamation, privacy breaches, and identity theft. The metaverse, envisioned as a utopia, harbours shadows of malevolent actors exploiting its vast potential. We discovered profound implications for personal identity as we delved into data protection and privacy breaches in this virtual space. The clash between interaction freedom and the darker side of individuals challenges us to balance liberty and security, especially with internet trolls finding refuge in the metaverse. Contemplating the metaverse's future requires addressing potential abuses, such as nefarious avatars or anonymity exploitation for harm. Robust regulations and ethical frameworks ensure the metaverse thrives without succumbing to malevolent forces undermining its transformative potential.

Keywords: Metaverse · Cyber Crime · Metaverse Crime

1 Introduction

The term "Metaverse" was introduced in Neal Stephenson's 1992 science-fiction book *Snow Crash*, depicting a scenario where individuals, acting as customisable avatars, engage with one another and software entities within a three-dimensional virtual environment that draws parallels to the physical world [1].

In the acknowledgement section at the end of Snow Crash, Stephenson penned: *The words "avatar" (in the sense it is used here) and "Metaverse" are*

© The Author(s), under exclusive license to Springer Nature Switzerland AG 2025
N. Pitropakis and S. Katsikas (Eds.): *Security and Privacy in Smart Environments*, LNCS 14800, pp. 120–148, 2025.
https://doi.org/10.1007/978-3-031-66708-4_6

my inventions, which I came up with when I decided that existing words (such as "virtual reality") were simply too awkward to use [2]. In Neal Stephenson's metaverse, users perceive an urban setting structured around a 100-meter-wide thoroughfare known as the Street. This digital landscape encompasses 65,536 km (2^{16} km) circumference of a seamless, black, spherical planet. The virtual property is owned by the Global Multimedia Protocol Group, a fictitious entity within the actual Association for Computing Machinery. Users have the option to purchase this virtual real estate and construct buildings on it. Metaverse users enter the virtual space using personal terminals, which project a top-notch virtual reality display onto the goggles they wear. Alternatively, they can access it through less advanced public terminals in booths. The experience is presented from a first-person viewpoint. Stephenson details a subgroup of individuals who opt to stay perpetually linked to the metaverse, earning them the nickname "gargoyles" owing to their unconventional and distorted appearance [2]. In the metaverse, users manifest as avatars of diverse shapes, the only limitation being their height, set *"to prevent people from walking around a mile high"* [2]. Movement within the metaverse mimics real-world options involving travel on foot or using vehicles like the monorail that traverses the entire length of the Street. The monorail makes stops at 256 Express Ports, evenly spaced at intervals of 256 km, as well as Local Ports situated one kilometre apart [2].

In today's world, the advancement of technology is rapidly facilitating the development of the metaverse, utilising tools like Virtual Reality (VR) headsets, haptic gloves, Augmented Reality (AR), and Extended Reality (XR). These innovations allow users to engage in highly interactive and immersive experiences. Businesses are beginning to explore the possibilities of the metaverse and evaluate how it could be seamlessly integrated into their established operational frameworks. The exploration of assets and ownership concepts in the metaverse, along with the design and utilisation of avatars as digital counterparts for individuals in metaverse environments, is still in its early stages. Nevertheless, the metaverse doesn't just open up possibilities for novel business models and recreational activities; it also gives rise to new types of cybercrime. This includes addressing the consequential human rights and legal and ethical considerations. Despite the metaverse being in its early phases, platforms are already finding ways to generate revenue from users keen on investing in this emerging virtual realm.

McKinsey & Company believes that the potential influence of the metaverse would vary across industries yet impact all sectors. For instance, McKinsey & Company's projections suggest the metaverse could generate a market impact ranging from $2 trillion to $2.6 trillion on e-commerce by 2030, depending on whether a base or optimistic scenario unfolds. Similarly, they anticipate an impact of $180 billion to $270 billion on the academic virtual learning market, a $144 billion to $206 billion impact on the advertising market and a $108 billion to $125 billion impact on the gaming market [3].

The Vienna University of Economics and Business has introduced a postgraduate program that can be undertaken in the metaverse [4]. Collaborating with the edtech (education technology) start-up Tomorrow University of Applied Sciences, they have created the Professional Master in Sustainability, Entrepreneurship, and Technology program. This reflects a growing trend as more business schools are venturing into the metaverse [5]. According to The New York Times, the metaverse real estate market is projected to expand by $5.37 billion by 2026 [6]. The growth is attributed largely to major global brands like Adidas, Atari, and Warner Music Group, which have acquired parcels for the purpose of selling goods or establishing virtual entertainment venues. Notably, land sales in The Sandbox, among the widely favoured metaverse platforms, reached approximately $167 million last year [6]. Majid al Futtaim, a conglomerate in the Middle East specialising in lifestyle and leisure, has inaugurated a metaverse-based shopping mall known as the Mall of the Metaverse. Located on the Decentraland platform, it is designed to provide enhanced digital experiences encompassing retail, entertainment, and leisure offerings [7]. The company asserts this initiative will enable more effective monitoring and comprehension of shopper behaviour. Major fashion labels worldwide are exploring the metaverse to engage a younger audience. Gucci, Burberry, Prada, and Balenciaga are among the participants, experimenting with various approaches. These include incorporating augmented reality for users to virtually try on 3D renditions of clothing before making actual purchases and providing Non-Fungible Tokens (NFT) owners with opportunities to acquire personalised jewellery [8].

With the above-mentioned advancements, criminals too have started to exploit the Metaverse; thus, posing new challenges. The World Economic Forum, in collaboration with INTERPOL, Meta, Microsoft, and others in a Metaverse governance initiative, has cautioned about potential issues such as social engineering scams, violent extremism, and misinformation [9]. With the increasing number of Metaverse users and technological advancements, the range of potential crimes is expected to broaden. This expansion may encompass crimes against children, data theft, money laundering, financial fraud, counterfeiting, ransomware, phishing, sexual assault, and harassment. Law enforcement faces significant challenges, as not all actions considered crimes in the physical world hold the same legal status in the virtual realm. In a surprise session at the 90th INTERPOL General Assembly in New Delhi, INTERPOL launched the first-ever Metaverse tailored specifically for law enforcement worldwide [10]. The INTERPOL Metaverse is fully operational, allowing registered users to explore a virtual replica of the INTERPOL General Secretariat headquarters in Lyon, France, without geographical or physical limitations. Users can interact with other officers via avatars and participate in immersive training courses covering forensic investigation and other policing capabilities.

In light of the above discussion, this chapter explores the benefits and concerns related to crimes that could occur in the metaverse.

2   Metaverse and Its Applications

2.1   Defining Metaverse

Interpol asserts that the Metaverse is regarded as the subsequent phase in the evolution of the Internet [10]. The Metaverse incorporates various technologies, including VR, AR, and edge computing. Its objective is to facilitate global access to shared three-dimensional (3D) virtual environments, allowing individuals to enter these spaces through avatars and fostering a feeling of "virtual presence". This is made possible by connecting to the internet and utilising specialised hardware like VR headsets or haptic suits. Europol provides a comparable description but is less focused on technology, incorporating the notion of a "digital twin" as a visually similar portrayal of the user within the simulated environment [11]. Europol's definition corresponds closely to the findings of Ritterbusch and Teichmann's examination of metaverse literature.

Ritterbusch & Teichmann conducted a systematic literature review of the academic literature from 1997 to 2022 to find a common definition of Metaverse [12]. Ritterbusch & Teichmann reviewed the literature listed in Table 1, and proposed to define Metaverse as [12]: *Metaverse, a crossword of "meta" (meaning transcendency) and "universe", describes a (decentralized) three-dimensional online environment that is persistent and immersive, in which users represented by avatars can participate socially and economically with each other in a creative and collaborative manner in virtual spaces decoupled from the real physical world.*

Table 1. Definitions of Metaverse in Different Domains [12].

Domain	Definitions	Author
Arts and Humanities	What makes the metaversal worlds different from the other online environments and graphically similar - even better - online games is the idea of the user-created content. The inhabitants are not preseneted with a ready-made world, but they can design and possess the real world copyrights of their designs; they can construct their identities from scratch, study or work to earn real money, and socialise by participating in any kind of group activities. Thus, they create collaboratively not only the graphical and interactive content but also the economical and social structures in these 3-D worlds.	Tasa and Görgülü [13]
Business Management and Accounting	In this positioning paper we will focus on the business activities and commercial applications that virtual worlds can host, and examine the wider implications of these virtual environments, often referred to as "metaverse".	Papagiannidis et al. [15]
	There were four emerging technologies that make up the so-called Metaverse - a digital domain equivalent to the atom based domain of our physical lives. These technologies are Mirror worlds (digital representations of our own atom based world), Virtual worlds (digital representations of any space, imagined or real), Lifelogging (the digital capture of information about people and objects in the real or digital worlds) and Augmented reality (sensory overlays of digital information on the real or even virtual world).	Boulos and Burden [14]

(*continued*)

Table 1. (*continued*)

Domain	Definitions	Author
Computer Science	The objective of our Metaverse is to provide users with an open, untethered, immersive environment that fools their visual senses into believing that the traditional barriers of time and space have been removed. Users access this meta-world through an interface called a Metaverse Display Portal that is (1) visually immersive, (2) self-configuring and monitoring, (3) interactive, and (4) collaborative.	Jaynes et al. [16]
	Second Life is a three-dimensional metaverse that is visualized graphically, where individuals are represented by avatars, and interact with other avatars and their environment.	Wasko et al. [17]
	Beyond the entertainment and game-play features, virtual worlds are evolving toward Stephenson's concept of a metaverse in which social and economic interactions are the main drivers. Currently, one of the best examples of this evolution is Second Life, a social virtual world in which people (called residents) can communicate, collaborate, and buy and sell not only virtual goods and services (such as clothes and real estate) but also real products through their customized virtual spaces and avatars.	Hendaoui et al. [18]
	Three-dimensional virtual worlds can be broadly classified into online games and metaverses. Meta-universes, or metaverses, are fully immersive virtual spaces that significantly differ from online games in several ways. [Key characteristics are] Seamless persistent world, User-generated content and Massive and dynamic content.	Kumar et al. [19]
	These virtual worlds or metaverses are in fact true social networks and they are useful for interaction between people in different locations.	Arroyo et al. [20]
	Until now such theories had to primarily deal with two spaces, the physical 'offline' one and the online one. This has been true even in the information age as the world's institutional and legal structures are largely still geographically based. However, new technologies made it possible to add new virtual spaces and environments, often referred to as metaverses, within which economic and social activities can take place.	Bourlakis et al. [21]
	Metaverses are immersive three-dimensional virtual worlds (VWs) in which people interact as avatars with each other and with software agents, using the metaphor of the real world but without its physical limitations. This broad concept of a metaverse builds on and generalizes from existing definitions of VWs Metaverses provide virtual team members with new ways of managing and overcoming geographic and other barriers to collaboration. These environments have potential for rich and engaging collaboration, but their capabilities have yet to be examined in depth.	Davis et al. [22]
	A virtual world (VW) is an instantiation of a metaverse a fully immersive 3D virtual space in which people interact with one another through avatars and software agents.	Owens et al. [23]
	These virtual worlds, or metaverses, are in fact true social networks, and they are useful for interaction between people in different locations.	Arroyo et al. [24]
	Metaverses are immersive three-dimensional virtual worlds (VWs) where people interact with each other and their environment, using the metaphor of the real world but without its physical limitations.	Owens et al. [25]
	The word Metaverse is a portmanteau of the prefix "meta" (meaning "beyond") and the suffix "verse" (shorthand for "universe"). Thus it literally means a universe beyond the physical world. More specifically this "universe beyond" refers to a computer-generated world, distinguishing it from metaphysical or spiritual conceptions of domains beyond the physical realm. In addition, the Metaverse refers to a fully immersive threedimensional digital environment in contrast to the more inclusive concept of cyberspace that reflects the totality of shared online space across all dimensions of representation. The progression of development culminates in a complete Metaverse that involves multiple MetaGalaxies and MetaWorld systems. A standardized protocol and set of abilities would allow users to move between virtual worlds in a seamless manner regardless of the controlling entity for any particular virtual region.	Dionisio et al. [26]

(*continued*)

Table 1. (*continued*)

Domain	Definitions	Author
	Metaverse, combination of the prefix "meta" (implying transcending) with the word "universe", describes a hypothetical synthetic environment linked to the physical world.	Lee et al. [27]
	Metaverse is a combination of "meta" (meaning beyond) and the stem "verse" from "universe", denoting the next-generation Internet in which the users, as avatars, can interact with each other and software applications in a threedimensional (3D) virtual space.	Duan et al. [28]
	A future topology for multiple virtual worlds 'metagalaxies' or the "Metaverse". The key difference between the Internet and the Metaverse, is that the Metaverse would support Rt [Real time: not turn-based and not time-based, where agents must wait for other agents to complete their actions before a new round can begin]. Because the Internet is already mixed reality (e.g., with video conferencing, web cameras depicting a live video feed of cities in the physical world, tele-operations, and projections from the net onto buildings), it is possible to conclude that the Metaverse will be necessarily mixed as well.	Nevelsteen [29]
	Metaverse is a compound word of transcendence meta and universe and refers to a three-dimensional virtual world where avatars engage in political, economic, social, and cultural activities.	Park & Kim [30]
	Metaverses use the metaphor of the real world but without its physical limitations. A virtual world is a specific instantiation of a metaverse, also referred to as a virtual space or virtual world environment. Virtual worlds provide virtual team members with new ways of managing and overcoming geographic and other barriers to collaboration. These types of environments allow for rich and engaging collaboration among team members.	Mitchell and Deepak [31]
	When the metaverse is brought to life as it was designed, it will be possible to perform many daily activities such as working, traveling, shopping, going to school, having fun by creating a 3d avatar in a digital universe. Any change users make in the metaverse will be permanently visible to almost everyone, thus providing users with greater identity and continuity of experience.	Gökçe Narin [32]
Engineering	However, though it provides a 3D view of the home, it is less realistic than a 3D virtual world, such as a 'metaverse' in which a user can walk around realistically through an avatar.	Han et al. [33]
Environmental Science	There is no single, unified entity called the metaverse. Rather, there are multiple mutually reinforcing ways in which virtualisation and 3-D web tools and objects are being embedded everywhere in our environment and becoming persistent features of our lives.	Li et al. [34]
Medicine	The most representative definition of Metaverse is that the Metaverse is a virtual world parallel to and independent of the real world. Metaverse is also considered as an online virtual world that mirrors the real world.	Chen and Zhang [35]
Social Science	"The metaverse," a kind of cyberspace world that could be considered a glorified chat room with total-body surround made possible by a sophisticated system of earphones and goggles that allowed individuals to live and act in a cyberspace peopled by iconic representations known as "avatars". These avatars could be crude artifacts with little reality, rented by the hour. In appearance these down-market avatars are some what wooden icons like those we use today. They could also run all the way up to dramatically realistic or specially constructed representations created by talented hackers either for their own use or for sale to wealthy clients.	Friedman [36]
	Technological advances in three-dimensional graphics, network connectivity, and bandwidth have just begun to enable online spaces that embody the Metaverse concepts of user creation and broad use.	Ondrejka [37]

(*continued*)

Table 1. (*continued*)

Domain	Definitions	Author
	The Second Life, SL, system by Linden Lab is a persistent 3D world, or "metaverse". Users access the online system with a proprietary client and interact with content and other "residents." Unique features include simple tools for constructing 3D objects and scripting tools for interactive content - including connectivity with external web-pages and internet resources.	Kemp and Livingstone [38]
	In particular, we consider the role of a Metaverse, understood as a globally accessible 3D virtual space and computing infrastructure - and today still a conceptual vision - as a mediator between technology trends and societal and business applications.	Rehm et al. [39]
	In this world [Metaverse] individuals interact through a perceived three dimensional landscape by creating avatars (artistically created virtual representations of individual users) that need a limited connection to the appearances of the people they represent. Each avatar is visible to all other users. and avatars interact with each other in this communal virtual space through software-specified rules.	Taylor [40]

2.2   Layers of Metaverse

Ernst & Young envisions the metaverse as a multi-layered ecosystem essential for its operation [41]. Ernst & Young's multi-layered metaverse is depicted in Fig. 1. This structure is built upon a range of technologies arranged in layers stacked on top of each other. Together, these layers form the essential ecosystem for the metaverse to operate. Fundamental components encompass core infrastructure elements like bandwidth, networks, and cloud services [41]. Additionally, blockchain technology is integrated for cryptography, identification, and ownership alongside access technologies like Augmented Reality, Virtual Reality, Apps, and Websites. These fundamental layers enable essential functionalities such as commerce, marketplaces, co-creation, and currencies. These, in turn, support a range of products such as physical/digital twins, digital assets, and NFTs, as well as a variety of experiences like hybrid shopping, live events, and immersive content [41].

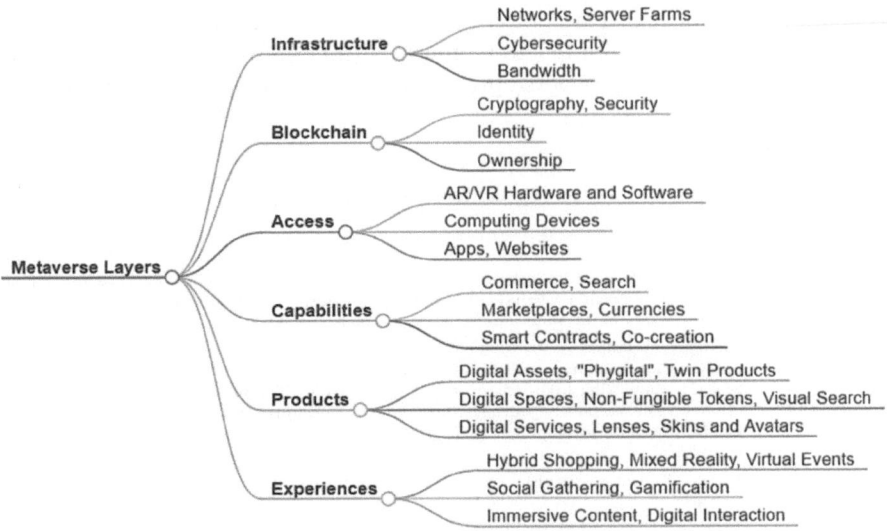

Fig. 1. Multi-layered Metaverse (adapted from [41])

The integration of technology choices significantly influences the overall user experience, emphasising the need for collaboration between marketing and technology. The technologies in play are undergoing rapid transformations. 5G and fibre optics will furnish the network infrastructure needed to accommodate extensive upload and download demands. Ongoing developments in AR and VR headset releases will enhance affordability, thereby expanding accessibility. Additionally, applications will evolve to maximise the potential of these tools, encompassing functionalities like product visualisation and visual search tools in the short term and progressing toward fully immersive experiences in the long term.

2.3 Applications

The metaverse has rapidly evolved from a concept in science fiction to a groundbreaking technological reality. Its applications span diverse fields, offering innovative solutions and immersive experiences. In entertainment and gaming, the metaverse provides a transformative platform for users to engage in virtual environments, socialise, and explore new dimensions of storytelling. Beyond entertainment, the metaverse is making significant strides in education, offering immersive learning experiences and collaborative environments that transcend traditional classroom boundaries. In the business sector, the metaverse redefines remote work, collaboration, and communication, providing a spatial computing platform enabling teams to collaborate seamlessly across distances. Moreover, the metaverse is increasingly recognised for its potential in healthcare, therapy, and training simulations. As this dynamic virtual space continues to expand, its applications are poised to revolutionise various aspects of our daily lives, ushering in a new era of interconnected, immersive experiences. A list of selected applications of metaverse presented in the academic literature is shown in Table 2.

Table 2. List of Selected Metaverse Applications.

Domain	Applications	Author
Creative Industry	Computer-rendered imagery, such as virtual photography and cinema, 3D digital portraits; Virtual calligraphy using Artificial Intelligence (AI); Production of audio and music material using AI	Lee et al., [42]
Education and Training	Engaging in hands-on learning experiences, such as virtual construction, exploring culturally significant locations (e.g., Taj Mahal), simulating high-risk situations (e.g., fire drills and medical procedures), immersive journeys into historical periods, and utilizing gaming for skill development (e.g., problem-solving and critical thinking)	Kye et al., [43]
Entertainment	Virtual Concerts organised on immersive platforms, for instance, Roblox	Park & Kim [44]
Gaming	An immersive virtual world called Sandbox, where 3D games can be created and monetised, is utilised using blockchain.	Christodoulou et al., [45]
Health and Wellbeing	Utilising augmented reality for surgical procedures; transforming services through socialization and gamification; and dynamically monitoring health and sports training.	Thomason [46]
	Testing machinery, systems, and procedures through digital twin technology; offering real-time guidance using augmented reality to surgeons during surgery; employing AI for personalized medical decision-making; conducting surgical simulations; utilising a move-to-earn strategy for rehabilitation, such as engaging patients in metaverse games as part of physiotherapy.	Chen & Zhang [47]
Hospitality and Tourism	Simulated flights in a virtual environment; employing VR for outdoor adventure experiences (e.g., kayaking in a distant location); offering location information through AR to tourists; virtual tours and hotels (enabling clients to preview before booking); experiencing destinations both in person and virtually through digital twin technology	Gursoy et al., [48]
Manufacturing and Logistics	Evaluating products and refining production processes, as exemplified by BMW utilising Ominverse to synchronise car production across its factories	Alkazzi & Rizk [49], Chang et al., [50]
Retail and Advertising	Immersive purchasing experiences and brand merchandising in virtual settings, such as those employed by Nike, Puma, and Sketchers	Kim [51]
	Brands developing digital representations, such as Gucci unveiling an augmented reality-enabled virtual sneaker; luxury brand collectibles transformed into NFTs; digital fashion, exemplified by companies like Dress X dedicated to designing exclusively virtual garments	Joy et al., [52]
	Multiple brands, including Johnson & Johnson, L'Oreal, Chuck E. Cheese, and McDonald's, have applied for trademarks related to selling virtual goods and establishing metaverse environments	Gonzalez [53]
Social Media	Virtual environments where individuals can gather and engage, like VR Chat, and VR experiences crafted by influencers for their followers	Huq et al., [54]
Work and Collaboration	Conducting meetings and establishing office spaces, as demonstrated by platforms like Branch, Gather, and Teamflow	Park & Kim [44]
	Organising and attending conferences	Thomason [46]

3   Crimes in Metaverse

Over the last few decades, significant transformations in society have been propelled by technological advancements. The introduction of what are commonly referred to as 'disruptive technologies'-those that fundamentally reshape our lifestyles, occupations, and interpersonal connections-holds considerable implications for our security landscape [93]. Criminals have consistently recognised the potential of disruptive technologies. The adaptation of traditional criminal activities like drug trafficking, terrorism, money laundering, and extortion has become evident through the integration of digital technologies. While conventional crimes may exhibit structural rigidity, criminal organisations are constantly exploring and exploiting possibilities for unlawful ventures facilitated by the Internet and the ongoing expansion of electronic commerce. As technology advances, it becomes apparent that disruption doesn't solely arise from technological progress. Instead, it emerges from the convergence of different disruptive technologies, posing challenges to established legal or regulatory frameworks through previously unexplored applications. Consequently, cybercrime within the metaverse signifies the advancing sophistication of established criminal behaviour and the rise of novel forms of illegal conduct.

A shared vision for the metaverse involves an expansive virtual space resembling the real world [94] but not constrained by time and space [95], wherein various virtual worlds coexist. This metaverse aims to replicate the physical, economic, cultural, and legal aspects of the real world and introduces novel ways for users to engage with these elements. Users could mimic real-world crimes in familiar or innovative ways, impacting the physical world directly (such as stealing property with actual monetary value) or indirectly causing harm (e.g., virtual sexual assault). Additionally, new forms of crimes may emerge through metaverse-related technologies [96]. An example of such a crime is the simultaneous virtual sexual assault of avatars, achieved through a subprogram compelling them to engage in sexual acts with each other, either graphically or textually. These types of offences have been documented in early text-based virtual worlds like LambdaMOO [97], serving as small-scale prototypes for the metaverse. Existing virtual worlds like SecondLife [98] and Horizon Worlds [99] have also witnessed similar cases, suggesting that the metaverse, by connecting these worlds, may inherit and grapple with their associated issues.

As the vision for the metaverse emphasises its boundary-free nature, a singular crime could have repercussions across multiple nations, introducing challenges in terms of technical and standard-related issues during investigations [100]. Concentrating all regulatory authority in the hands of a select few would contradict the metaverse's ideal of decentralised, democratic governance [101,102]. The concept of the metaverse should be perceived as an idea, not confined to a particular platform or tool like VR [124]. Similar to the internet, it could have a regional aspect, with countries like China or Russia having their own metaverse [125], while others utilize a universal platform [126]. It may be jurisdictional, decentralized, or less centralized, especially if built on Web3 [127]. The metaverse's structure could range from being dominated by a few tech giants, resembling today's internet, to having multiple platforms or even a single service provider. Regardless of its control or design, the metaverse will likely bring significant changes and potentially disrupt how we currently communicate and interact with the world [128].

In contrast to science fiction portrayals, the actual metaverse is unlikely to supplant all human activities [129] completely. Many individuals may still opt for traditional travel experiences, even if the metaverse can replicate such journeys' psychological and physiological responses. People continue to visit cinemas, sometimes to rewatch films they've already seen or could easily view at home, simply for the experience [called as experience goods in economics jargon [130]]. Not everything can be replaced to that degree, at least not in the foreseeable future. Certain aspects of society, or at least certain members, may choose to retain certain activities in the physical realm [131]. However, as the competition intensifies and virtual worlds become more widely accessible and commercialised, some individuals will utilise them, leading to legal implications that require further examination. Even if it eventually becomes a niche, the metaverse will still have a significant impact on many individuals [127]. From a legal perspective, the metaverse introduces numerous legal inquiries in areas such as property law [132,133], intellectual property law [134,135], contract law [136], privacy law [137,138], tax law [139,140], and tort law [141], among others [142,143]. Criminal law is also not exempt from these considerations.

The metaverse is certain to attract criminal behavior, raising significant questions for policymakers and law enforcement regarding how to address legal misconduct within this virtual realm. While scholars have previously explored the laws and regulations of virtual worlds [144,145], and some have begun examining criminal law issues in online gaming and virtual environments [146], current academic research remains limited in its classification of immersive criminal activities [147–151] and the conceptualisation of new virtual offences [152]. In prior examinations of the criminal aspects of video games and virtual worlds, scholars have often referred to the "magic circle" metaphor, suggesting that actions within these realms remain confined to that space [30]. Others argue that criminal activities in the metaverse should be treated similarly to traditional or cyber-related crimes, depending on the nature of the conduct [31]. According to the "law of the horse" argument [155,156], actions within the metaverse do not present novel legal challenges, and policymakers can apply existing legal frameworks to regulate them [151]. In essence, if a criminal act in a virtual world causes harm to a user in the real world, it should be prosecuted under criminal statutes. However, to our knowledge, while individual countries have developed policies related to the metaverse, there is currently no universally accepted framework. Given the surge in metaverse technology, an international legal framework is urgently needed to address crimes within the metaverse [103]. Such a framework would encourage collaboration between nations, streamline crime investigations, and uphold the principles of democratic governance.

From a law enforcement and crime perspective, there are two conflicting interpretations of the metaverse, each carrying distinct implications for investigation and law enforcement. According to Interpol/Europol, metaverse technology signifies a revolutionary approach to human-machine interaction, marking a pivotal moment that radically transforms the utilisation and experience of online services. This perspective emphasises technology's presentational and experiential aspects, often overlooking the foundational principles underpinning it. Moreover, it implies that the technology itself generates novel forms of interaction rather than serving as a mediator that presents existing modes of interaction differently. This perspective may lead to the hypothesis that the mere existence of new technology is adequate to create conditions for entirely new forms of crime, challenging the assertion that there are no new crimes, only *"old wine in new bottles"* [104]. In the alternative interpretation, the immersive concept of the metaverse is seen predominantly as an extension of the array of ways in which presentations and interactions occur. This expansion provides additional possibilities for typical physical, sensory experiences to be encountered, albeit in a more dispersed or remote manner (meaning that those participating in the experience don't have to be in the same physical location) [105]. Marshall and Clarkson summarised this aspect as *"The crimes committed are not novel-though the modalities may be. Prior study has concluded that technology may facilitate or even broaden the scope of a given criminal act, but the use of technology is an extension rather than creation of a criminal class"* [104].

Given the discussion above, we are of the opinion that the classification of cybercrime put forth by Maskun et al. [106] and presented in Fig. 2 would be largely applicable in the context of criminal activities within the metaverse.

4   Crime Scenarios

The metaverse, while offering unprecedented opportunities for connectivity and immersive experiences, also introduces a novel frontier for potential criminal activities. In this expansive virtual realm, crime scenarios take on diverse forms, ranging from replicated real-world crimes with tangible consequences, such as theft of virtual property with real monetary value, to more abstract offences like virtual harassment or sexual assault. As users navigate this digital space, new crimes emerge, often leveraging metaverse-related technologies to orchestrate offences that challenge traditional notions of criminality. With the metaverse's integration of various virtual worlds, the potential for innovative and interconnected crime scenarios raises important challenges that necessitate the development of robust legal frameworks and security measures in this evolving landscape. Therefore, relying on the cybercrime classification proposed by Maksun et al. [106], a selected list of crimes in the metaverse and related crime scenarios with the academic literature discussing the same is presented in Table 3, 4 and 5 below.

Table 3. Type I Crimes in Metaverse (adapted from [81]).

Crime	Crime Scenario	Author
Blockchain Attacks	Vulnerabilities in blockchain technology could be manipulated to pilfer digital assets and/or currency from users	Huq et al., [54], Annison [55]
Denial of Essential Services	Hostile entities could block access to crucial services in the metaverse, such as healthcare and education, for numerous users	Consensus among Law Enforcement Agencies and techno-legal experts

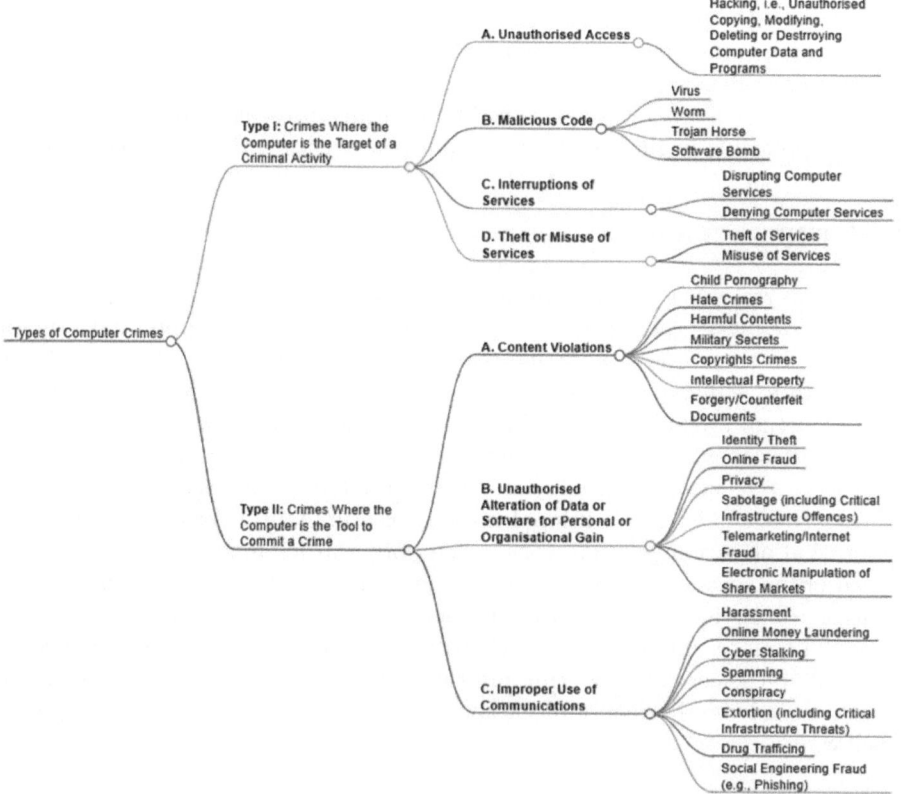

Fig. 2. Types of Computer Crimes (adapted from [106])

Table 4. Type II Crimes in Metaverse (adapted from [81]).

Crime	Crime Scenario	Author
Child Grooming	Avatars of adults (criminals) approaching the avatars of children to engage them in sexual activities	Crawford & Smith [56], Li & Lalani [57], Reed & Joseff [58], Rice [59], Russia Business News [60], Sum of Us [61]
Child Labour and Modern Slavery to Develop Metaverse Content	The desire for digital goods, assets, and services may present an opportunity to gain a competitive edge by employing child labor and engaging in modern slavery	Consensus among Law Enforcement Agencies and techno-legal experts
Child Sexual Abuse Material	Offenders and victims in remote locations could engage in paid immersive streaming of child sexual abuse material. The use of haptic suits and other immersive equipment could exacerbate the resulting harms	Consensus among Law Enforcement Agencies and techno-legal experts
Conspiring	Criminals may utilize highly detailed virtual spaces, akin to digital twins, to strategize and practice criminal activities intended for the physical world	Huq et al., [54], Allen & McIntosh [62], Wang et al., [63]
Copyright Infringement	Audio, software, visual, and graphic content, along with other copyrightable works created explicitly for the metaverse, may be repurposed and minimally altered for use in user spaces, leading to copyright infringement	Zhao et al., [65], Goossens et al., [64]
Counterfeiting	Unscrupulous individuals may produce fake digital goods, including NFTs, presenting them as legitimate products from brands (e.g., counterfeit digital Gucci bags)	Huq et al., [54], Cheong [116], Goossens et al., [64], Zhao et al., [65]
Doxing	Hostile individuals may misuse the extensive data gathered from users (such as biometric data and eye tracking) to blackmail or disgrace them	Buck & McDonnell [66], Vladimirov et al., [67]
Harassment	A user might encounter avatars that approach them for the purpose of harassment, potentially leading to pursuit across various metaverse platforms	Allen & McIntosh [62], Buck & McDonnell [66], Cheong [116], Combs [73], Di Pietro & Cresci [84], Howell [85], Identity Management Institute [86], Reed & Joseff [58], Shanker & Zytko [85], Sum of Us [61], Zhao et al., [65]
Hate Crime	A user might encounter avatars with the intent of engaging in hate crimes	Allen & McIntosh [62], Li & Lalani [57], Rice [59], Sum of Us [61], Zhao et al., [65]
Impersonating a Law Enforcement Agency	In the metaverse, criminals have the capability to impersonate law enforcement authorities for various reasons, including gathering intelligence	Bell [68], Cunha Barbosa [69], Huq et al., [54], Pinnock [70]
Impersonation Scam	Criminals have the ability to pose as service providers such as doctors and provide fraudulent medical guidance to patients in exchange for payment	Bell [68], Cunha Barbosa [69], Huq et al., [54], Pinnock [70]
Incitement to Self-harm	Multiple users could gather in a virtual environment and encourage a vulnerable user to engage in self-harm. AI-created avatars could be programmed to display increased empathy and potentially even encourage extensive self-harm	Consensus among Law Enforcement Agencies and techno-legal experts

(continued)

Table 4. (*continued*)

Crime	Crime Scenario	Author
Investment Scams	Criminals may take advantage of the excitement and buzz around the metaverse, coupled with the limited understanding of security measures, to perpetrate various scams. These may include giveaway scams, fraudulent metaverse schemes, deceptive wearable minting practices, technical support scams, spurious land expansions, rug pulls, and pump-and-dump schemes	Annison [55], Banaeian Far & Imani Rad [71], CITIC Telecom International [72], Combs [73], Dataquest [74], Huq et al., [54], Kadar, [75], Mackenzie [76], PCQuest [77], Shen [78], Smaili & de Rancourt-Raymond [79], Targeted News Service, [80]
Money Laundering	Criminals might utilise assets in the metaverse (such as cryptocurrency, virtual land, and wearables) for the purpose of laundering illegal funds	Annison [55], Banaeian Far & Imani Rad [71], Huq et al., [54] Pinnock [70]
Non-Consensual Sexual Image Offences	Unscrupulous individuals might manipulate personal, sensitive, and explicit content exchanged between users to engage in non-consensual virtual reality sexual activities. This could also include the use of deepfake technology	Annison [55], Li & Lalani [57]
Preying on Addicted Users for Extortion, Coercion or Incitement Purposes	Loan sharks and criminal organisations could target vulnerable individuals, exploiting them financially or coercing them into committing crimes	Consensus among Law Enforcement Agencies and techno-legal experts
Radicalisation	Empathetic AI-designed avatars and multiuser spaces could be employed to radicalise susceptible users, such as underage individuals	Abdulsattar Jaber [87], Buck & McDonnell [66], Howell [85], Reed & Joseff [58]
Sexual Assault	Within a virtual environment, users may face indecent and forceful interactions from avatars controlled by malicious actors, intending to commit sexual assault	Allen & McIntosh [62], Cheong [116], Clayton [82], Huq et al., [54], Li & Lalani [57], Reed & Joseff [58], Rice [59], Shanker & Zytko [83]
Stalking	A malicious actor could track a user across various metaverse platforms without being physically present in the same location. They could also employ invisible avatars to evade detection	Di Pietro & Cresci [84]; Huq et al., [54], Wang et al. [63], Zhao et al., [65]
Tax Evasion	A company solely existing in the metaverse may lack a clear jurisdiction and could potentially evade the payment of income taxes	Huq et al., [54]
Unauthorised Use of Training Materials	Criminals might misuse virtual scenarios intended for training and readiness for significant events (such as organised crime) to learn how to circumvent law enforcement measures	Consensus among Law Enforcement Agencies and techno-legal experts
Virtual Trafficking of People for Sexual Exploitation	Vulnerable users' avatars may face repeated sexual exploitation within the virtual environment without the necessity of crossing borders or disappearing	Consensus among Law Enforcement Agencies and techno-legal experts

5   Legal Aspects

In exploring the intricate dimensions of crime within the metaverse, a comprehensive understanding requires an examination of relevant cases that have emerged in response to the evolving challenges of this virtual landscape. As the metaverse continues to expand and integrate with our daily lives, legal frameworks must adapt to address the unique complexities associated with digital offences, virtual property rights, and the blurred boundaries between physical

Table 5. Type I & II Crimes in Metaverse (adapted from [81]).

Crime	Crime Scenario	Author
Broker Imposter Scam	Criminals might impersonate digital asset brokers, transferring assets between different metaverse platforms (such as Decentraland and Roblox) with the intent of stealing or defrauding owners	Huq et al., [54]
Cyber-Physical Burgalry	Malicious users might misuse VR, AR, and other intelligent sensing materials to acquire information (such as location, access, valuables) about properties and attempt physical burglaries	Huq et al., [54], Nichols [88], Wang et al., [63]
Cyber-Physical Person Attacks	VR, AR, haptic suits, and other wearables could be exploited by malicious actors to inflict harm on users, such as by manipulating the physical activity limits programmed into the devices	Huq et al., [54], Nichols [88], PCQuest [77]; Wang et al. [63]
Cyber-Physical Infrastructure Attacks	Digital twins and the integration of infrastructure with the metaverse through IoT and other technologies could be manipulated by malicious actors to strategise and execute attacks on infrastructure	Huq et al., [54]
Identity Theft for Financial Gain	Criminals may employ avatars to masquerade as fraudulent financial figures (e.g., a virtual bank teller) to gain access to users' financial information for personal financial benefit	Abdulsattar Jaber [87]; Bell [68], Cunha Barbosa [69], Dey [89], Howell [85], Huq et al., [54], Identity Management Institute [86], Khitrov [90]; Li & Lalani [57], Pinnock [70], Rosenberg [91], Smaili & de Rancourt-Raymond [79], Williams [92]

and virtual realities. This section delves into the notable legal landscape surrounding crime in the metaverse, shedding light on pivotal criminal law aspects and their implications for governance, jurisdiction, and the protection of individuals within this increasingly interconnected digital domain.

5.1 Crime

A potential situation involving criminal activity in the metaverse, with ramifications extending into the real world, might occur when an individual utilises their avatar to engage in unlawful actions within the virtual realm. Within the metaverse, various acts and failures to act may be deemed reprehensible from a criminal law standpoint. Examples include deceiving others to pilfer belongings from their avatars or persistently sending explicit images or videos to harass another avatar [107]. In a future metaverse scenario, this could escalate to inflicting harm on another avatar, causing psychological distress (or even physical pain) for the real person controlling that avatar, facilitated by the neural link connecting their brain to the virtual representation in the metaverse.

Consider, for example, a scenario in which an artificial intelligence-driven avatar independently engages with another avatar in the metaverse to purchase illegal substances like heroin, arranging for their delivery at a specific real-world location and time akin to transactions on the dark web. The actions of the avatar, influenced by potentially unforeseen habits acquired through deep machine learning from other avatars in the metaverse, raise the question of criminal liability for the natural person [108]. The avatar and the corporation lack the status of natural persons, and the challenge lies in determining the "mental state" for imposing liability [109]. Attributing the avatar's misdeeds to its creator might appear unfair, considering the unpredictable responses of the avatar, particularly when deep learning enables actions beyond the programmer's anticipation [110]. On the flip side, consistently assigning responsibility to the individual overseeing the avatar in all situations could undermine the distinct legal identity of the avatar, potentially hindering the advancement of the metaverse and the broad acceptance of the technology [111].

However, if avatars were to bear criminal responsibility, Eidenmuller's proposals for artificial intelligence could be expanded to encompass avatars. Granting legal personality to avatars might entail sanctions such as *"revoking the legal capacity of the [avatar], detaining it for some time... or destroying it"* [112]. The possibility of eliminating an avatar, restricting it to a virtual prison for a designated period, or barring the real person from the metaverse could be a solution, provided that the individual does not exploit any potential loopholes to create a new avatar [113]. There are also situations where the criminal actions of an avatar may not be attributable to the person controlling it. For instance, should an avatar contract a software virus in the metaverse, resulting in it committing a crime, the metaverse community would have to develop a range of legal defences to handle different factual situations [114].

Furthermore, the issue of sexual harassment is widespread in the digital domain, mirroring its presence in the physical world. Numerous documented cases of sexual harassment within the metaverse already exist. An illustrative incident involves a beta tester for Horizon Worlds, a virtual reality platform owned by Meta, who encountered sexual harassment when the individual's avatar was inappropriately touched by a stranger [115]. Meta's internal inquiry into the incident concluded that the beta tester could have utilised a feature called 'Safe

Zone.' This tool establishes a protective bubble around avatars, inhibiting interactions with other avatars until they leave the specified area [115].

5.2   Data Protection and Privacy

In the realms of data protection and privacy, the metaverse is poised to introduce new categories of personal data for processing. Virtual reality platforms
seem invasive, enabling companies to observe facial expressions, physiological
reactions, and biometric data. As a result, the advancement of metaverse platforms raises several unanswered queries, including issues of accountability for
data processing, responsibility for lost or compromised data, and the acquisi-
tion of consent for data processing [116]. The applicability of the EU's General
Data Protection Regulation (GDPR) to the metaverse is a subject of consideration. Additionally, the utilisation of Light Detection And Ranging sensors
(LIDAR), three-dimensional (3D) cameras and microphones in virtual reality
glasses implies the processing of diverse data related to users' private real-world
environments, such as their homes or family members. This information might
also include sensitive categories, as detailed in Article 9 of the European Union's
General Data Protection Regulation (GDPR), requiring cautious treatment during processing. In particular, the principles of minimising data and limiting purposes play a pivotal role in managing personal data within the metaverse, serving
not only as a compliance matter but also as a cornerstone for building trust [117].

User acceptance of the metaverse hinges on their ability and willingness to
engage. Moreover, the unrestricted nature of the metaverse introduces complexity; while assuming the application of the GDPR is desirable, further clarification
may be required for clauses addressing the transfer and processing of data outside the EU. The applicability of the GDPR depends on the individual's location
at the time of data processing, irrespective of their home country or citizenship.
Therefore, it might be essential to determine the location based on either the
operator of the avatar or the avatar itself, as it is the avatar's data undergoing
processing.

5.3   Defamation

As mentioned earlier, the metaverse would eliminate all constraints on freedom
of interaction. One drawback of this, as observed on social networking sites, is
that it grants individuals (commonly referred to as 'internet trolls') unrestricted
freedom to post unwarranted and false statements about a person (or entity),
potentially harming their reputation and goodwill. In such a scenario, it is suggested that the veil be lifted by drawing an analogy between *"the avatar and the
user [...] to that between a non-living business entity and a sole shareholder,
where the entity is essentially an alter ego of the controller, and thus an action
[for defamation] may be sustainable on that basis"* [118]. Suppose we embrace
the perspective that avatars lack a distinct consciousness and merely serve as a
conduit for individuals to engage in activities in the metaverse. In that case, this

proposition becomes acceptable, treating avatars as akin to a company's 'alter ego' [116].

For example, if an avatar representing a real-world CEO of a multinational company is defamed by another avatar in the metaverse, leading to repercussions on the CEO's reputation, and the metaverse community acknowledges that the avatar is associated with the real-world CEO, this could give rise to a defamation claim in the real world, considering the avatar as essentially an extension of the real-world person. The key challenge would be determining whether *"despite the differences between the physical characteristics of the avatar and herself, she and the avatar are one and the same for the purposes of a defamation inquiry"* [118]. Another factor to take into account is whether any actual harm occurred to an individual or entity in the physical world due to a defamatory statement made about an avatar in the metaverse [119].

5.4 Identity Theft

In the metaverse, the significance of identity is heightened. For instance, in cases of account takeover, an individual could utilise someone else's avatar, masquerading as that person [120]. This jeopardises one's reputation and may result in legal responsibility for the individual in the physical world. These difficulties are intensified by an increasing sense of impunity among malicious actors who think they can utilise the anonymity offered by the metaverse to avoid repercussions. Registering or formally establishing one's avatar could offer a potential solution to ensure avatars are held responsible for their actions in the metaverse.

Another solution is to regulate user behaviour contractually [118]. Massively Multiplayer Online Role-Playing Games implement terms of service that claim to contractually regulate user behaviour, offering remedies for infractions such as platform bans and the seizure of in-game assets. The confiscation of in-game assets can be significant since these assets may be exchanged among players, with the potential for players to convert them into real-world currency. For instance, popular virtual worlds like Fortnite and Roblox mandate users to agree to terms of service before accessing the game [121].

5.5 Intellectual Property Rights

In the evolving landscape of the metaverse, intellectual property law faces unique challenges and considerations, particularly concerning digital assets such as art and virtual real estate. Unlike the traditional art world, where ownership encompasses both the physical artwork and, potentially, the associated intellectual property, the metaverse introduces a paradigm shift. Here, ownership may be construed as a form of licensing or service provision rather than outright possession. This distinction implies that true ownership may still reside with the creator or original owner despite acquiring a digital asset, limiting the buyer's rights [122]. The extension of land law principles to virtual real estate in the metaverse further complicates matters, raising questions about the applicability of real-world legislation to issues like trespassing on private digital land or securing mortgages on virtual properties. Platforms and services need to integrate technical safeguards and contractual mechanisms, including comprehensive terms of use, to protect users and navigate these legal complexities. The case of *'MetaBirkins'*, involving a digital artist and a luxury brand, underscores the need for content owners to monitor and enforce their rights in the metaverse vigilantly, emphasising the necessity for a proactive approach in protecting intellectual property amidst the expanding horizons of this digital realm [123].

6   Conclusion

In the intricate tapestry of the metaverse, where the convergence of technology and virtual reality unfolds, this chapter has delved into the complex landscape of cybercrime within this emerging digital realm. Our exploration has illuminated the novel challenges and unprecedented dimensions that cybercrime assumes within the metaverse, transcending the boundaries of conventional cyber threats.

From the potential criminal liability of avatars to the intricate dance between legal identities and virtual personas, we've navigated the nuanced issues surrounding defamation, privacy breaches, and the dark corners of identity theft. The metaverse, designed to be a utopia of interconnected experiences, carries within it the shadows of malevolent actors seeking to exploit its vast possibilities.

As our investigation unfolded, we encountered the profound implications of data protection and privacy breaches in this virtual expanse, where the very essence of personal identity is at stake. The clash between the freedom of interaction and the darker inclinations of individuals, the "internet trolls" who find refuge in the metaverse, challenges us to strike a delicate balance between liberty and security.

In contemplating the future of the metaverse, we must grapple with the potential for abuse, whether through the proliferation of avatars for nefarious purposes or the exploitation of anonymity to perpetrate harm. The need for robust regulations and ethical frameworks becomes evident, ensuring that the metaverse can flourish without succumbing to the malevolent forces that seek to undermine its transformative potential.

In conclusion, *"Bits and Bytes Betrayal"* serves as a critical exploration into the evolving landscape of cybercrime in the metaverse. As we unravel the dark threads woven within this digital realm, it becomes apparent that safeguarding the metaverse necessitates a concerted effort from policymakers, technologists, and users alike. By understanding the intricacies of cyber threats in the metaverse, we can strive to forge a path toward a secure and thriving digital future where the potential for betrayal is eclipsed by the promise of innovation and interconnectedness.

References

1. Grimshaw, M.: The Oxford Handbook of Virtuality. Oxford University Press, New York (2014). ISBN 9780199826162
2. Stephenson, N.: Snow Crash, Penguin Books Ltd., Bangalore (2011). ISBN-13: 978-0241953181
3. McKinsey & Company, Value creation in the metaverse: the real business of the virtual world, June 2022
4. Moules, J.: Courses in the metaverse struggle to compete with real world: Fulfilment of initial promise made for the technology remains elusive, Financial Times, 12 March 2023. https://www.ft.com/content/4ea0dccb-aad7-4bac-95de-c6e3f122d015
5. Li, C.: 4 things you need to know about the metaverse this month, World Economic Forum, 21 December 2022. https://www.weforum.org/agenda/2022/12/metaverse-december-2022-what-to-know/
6. Kamin, D.: The Next Hot Housing Market Is Out of This World. It's in the Metaverse. The New York Times, 19 February 2023. https://www.nytimes.com/2023/02/19/realestate/metaverse-vr-housing-market.html
7. Majid Al Futtaim, Majid Al Futtaim launches Mall of the Metaverse, 15 February 2023. https://www.majidalfuttaim.com/en/media-centre/press-releases/detail/2023/02/majid-al-futtaim-launches-mall-of-the-metaverse
8. What Gucci and others learnt from the metaverse, Financial Times, 23 February 2023. https://www.ft.com/content/d4c3d51f-4568-400e-8ca9-7706539d9cae
9. Interpol, INTERPOL launches first global police Metaverse, Press Release, 20 October 2022
10. Interpol, Interpol Technology Assessment Report on Metaverse, Interpol, October 2022
11. Europol. (2023). Policing the metaverse: What law enforcement needs to know, an observatory report from the Europol Innovation Lab. Publications Office of the European Union. https://www.europol.europa.eu/publications-events/publications/policing-in-metaverse-what-law-enforcement-needs-to-know
12. Ritterbusch, G.D., Teichmann, M.: Defining the metaverse: a systematic literature review. IEEE Access **11**, 12368–12377 (2023). https://doi.org/10.1109/ACCESS.2023.3241809
13. Tasa, U.B., Görgülü, T.: Meta-art: art of the 3-D user-created virtual worlds. Digit. Creat. **21**(2), 100–111 (2010). https://doi.org/10.1080/14626261003786251
14. Boulos, M., Burden, D.: Web GIS in practice V: 3-D interactive and real-time mapping in second life. Int. J. Health Geograph. **6**(1), 51 (2007). https://doi.org/10.1186/1476-072X-6-51
15. Papagiannidis, S., Bourlakis, M., Li, F.: Making real money in virtual worlds: MMORPGs and emerging business opportunities, challenges and ethical implications in metaverses. Technol. Forecast. Soc. Change **75**(5), 610–622 (2008). https://doi.org/10.1016/j.techfore.2007.04.007
16. Jaynes, C., Seales, W.B., Calvert, K., Fei, Z., Griffioen, J.: The metaverse: a networked collection of inexpensive, self-configuring, immersive environments. In: Proceedings of the Workshop Virtual Environments, Zurich, Switzerland, May 2003, pp. 115–124, May 2003. https://doi.org/10.1145/769953.769967
17. Wasko, M., Teigland, R., Donnellan, B.: Creating innovation systems through virtual communities. In: Proceedings of the AMCIS, 2007, pp. 1–8 (2007). https://aisel.aisnet.org/amcis2007/213

18. Hendaoui, A., Limayem, M., Thompson, C.W.: 3D social virtual worlds: research issues and challenges. IEEE Internet Comput. **12**(1), 88–92 (2008). https://doi.org/10.1109/MIC.2008.1

19. Kumar, S., et al.: Second life and the new generation of virtual worlds. Computer **41**(9), 46–53 (2008). https://doi.org/10.1109/MC.2008.398

20. Arroyo, A., Serradilla, F., Calvo, O.: Multimodal agents in second life and the new agents of virtual 3D environments. In: Mira, J., Ferrández, J.M., Álvarez, J.R., de la Paz, F., Toledo, F.J. (eds.) IWINAC 2009. LNCS, vol. 5601, pp. 506–516. Springer, Heidelberg (2009). https://doi.org/10.1007/978-3-642-02264-7_52

21. Bourlakis, M., Papagiannidis, S., Li, F.: Retail spatial evolution: paving the way from traditional to metaverse retailing. Electron. Commer. Res. **9**(1–2), 135–148 (2009). https://doi.org/10.1007/s10660-009-9030-8

22. Davis, A., Murphy, J., Owens, D., Khazanchi, D., Zigurs, I.: Avatars, people, and virtual worlds: foundations for research in metaverses. J. Assoc. Inf. Syst. **10**(2), 90–117 (2009). https://doi.org/10.17705/1jais.00183

23. Owens, D., Davis, A., Murphy, J.D., Khazanchi, D., Zigurs, I.: Real-world opportunities for Virtual- world project management. IT Prof. **11**(2), 34–41 (2009). https://doi.org/10.1109/MITP.2009.35

24. Arroyo, A., Serradilla, F., Calvo, O.: Adaptive fuzzy knowledge-based systems for control metabots' mobility on virtual environments. Exp. Syst. **28**(4), 339–352 (2011). https://doi.org/10.1111/j.1468-0394.2011.00595.x

25. Owens, D., Mitchell, A., Khazanchi, D., Zigurs, I.: An empirical investigation of virtual world projects and metaverse technology capabilities. ACM SIGMIS Database DATABASE Adv. Inf. Syst. **42**(1), 74–101 (2011). https://doi.org/10.1145/1952712.1952717

26. Dionisio, J.D.N., Iii, W.G.B., Gilbert, R.: 3D virtual worlds and the metaverse: current status and future possibilities. ACM Comput. Surv. **45**(3), 1–38 (2013). https://doi.org/10.1145/2480741.2480751

27. Lee, L.H., et al.: All one needs to know about metaverse: a complete survey on technological singularity, virtual ecosystem, and research Agenda, 2021. arXiv:2110.05352

28. Duan, H., Li, J., Fan, S., Lin, Z., Wu, X., Cai, W.: Metaverse for social good: a university campus prototype. In: Proceedings of the 29th ACM International Conference on Multimedia, pp. 153–161, October 2021. https://doi.org/10.1145/3474085.3479238

29. Nevelsteen, K.J.L.: Virtual world, defined from a technological perspective and applied to video games, mixed reality, and the metaverse: virtual world, defined. Comput. Animat. Virtual Worlds **29**(1), e1752 (2018). https://doi.org/10.1002/cav.1752

30. Park, S.M., Kim, Y.G.: A metaverse: taxonomy, components, applications, and open challenges. IEEE Access **10**, 4209–4251 (2022). https://doi.org/10.1109/ACCESS.2021.3140175

31. Mitchell, A., Deepak, K.: Ethical considerations for virtual worlds. In: Proceedings of the AMCIS, 2012, pp. 1–5 (2012). https://aisel.aisnet.org/amcis2012/proceedings/PerspectivesIS/12

32. Gökçe Narin, N.: A content analysis of the metaverse articles. J. Metaverse **1**(1), 17–24 (2021)

33. Han, J., Yun, J., Jang, J., Park, K.-R.: User-friendly home automation based on 3D virtual world. IEEE Trans. Consum. Electron. **56**(3), 1843–1847 (2010). https://doi.org/10.1109/TCE.2010.5606335

34. Li, F., Papagiannidis, S., Bourlakis, M.: Living in 'multiple spaces': extending our socioeconomic environment through virtual worlds. Environ. Plan. D Soc. Space **28**(3), 425–446 (2010). https://doi.org/10.1068/d14708

35. Chen, D., Zhang, R.: Exploring research trends of emerging technologies in health metaverse: a bibliometric analysis. SSRN Electron. J. 1–32 (2022). https://doi.org/10.2139/ssrn.3998068

36. Friedman, K.: Building cyberspace: information, place and policy. Built Environ. **24**(2–3), 83–103 (1998)

37. Ondrejka, C.: Escaping the gilded cage: user created content and building the metaverse. N. Y. Law Sch. Law Rev. **49**(1), 81–101 (2004)

38. Kemp, J., Livingstone, D.: Putting a second life metaverse skin on learning management systems. In: Proceedings of the Second Life Education Workshop at the Second Life Community Convention, San Francisco, CA, USA, pp. 13–18, August 2006. https://doi.org/10.1145/1235511.1235517

39. Rehm, S.-V., Goel, L., Crespi, M.: The metaverse as mediator between technology, trends, and the digital transformation of society and business. JVWR **8**(2) (2015). https://doi.org/10.4101/jvwr.v8i2.7149

40. Taylor, J.: The emerging geographies of virtual worlds. Geogr. Rev. **87**(2), 172–192 (1997). https://doi.org/10.1111/j.1931-0846.1997.tb00070.x

41. Bonelli, F., MacSweeney, R.: How meeting customers in the metaverse can unlock lasting value, Ernst & Young, 20 October 2022. https://www.ey.com/en_no/consumer-products-retail/meet-customers-in-the-metaverse-to-unlock-lasting-value

42. Lee, L.-H., et al.: When creators meet the metaverse: a survey on computational arts (2021). arXiv preprint arXiv:2111.13486

43. Kye, B., Han, N., Kim, E., Park, Y., Jo, S.: Educational applications of metaverse: possibilities and limitations. J. Educ. Eval. Health Prof. **18** (2021)

44. Park, S.-M., Kim, Y.-G.: A metaverse: taxonomy, components, applications, and open challenges. IEEE Access **10**, 4209–4251 (2022)

45. Christodoulou, K., Katelaris, L., Themistocleous, M., Christodoulou, P., Iosif, E.: NFTs and the metaverse revolution: research perspectives and open challenges. Blockchains and the Token Economy: Theory and Practice, pp. 139–178 (2022)

46. Thomason, J.: MetaHealth-how will the metaverse change health care? J. Metaverse **1**(1), 13–16 (2021)

47. Chen, D., Zhang, R.: Exploring Research Trends of Emerging Technologies in Health Metaverse: A Bibliometric Analysis (2022). SSRN 3998068

48. Gursoy, D., Malodia, S., Dhir, A.: The metaverse in the hospitality and tourism industry: an overview of current trends and future research directions. J. Hosp. Mark. Manag. 1-8 (2022)

49. Alkazzi, J.-M., Rizk, A.: Leveraging NVIDIA's technology for the ultimate industrial autonomous transport robot. In: GPU Technology Conference (2020)

50. Chang, L., et al.: 6G-enabled Edge AI for Metaverse: Challenges, Methods, and Future Research Directions (2022)

51. Kim, J.: Advertising in the metaverse: research agenda. J. Interact. Advert. **21**(3), 141–144 (2021)

52. Joy, A., Zhu, Y., Peña, C., Brouard, M.: Digital future of luxury brands: metaverse, digital fashion, and non-fungible tokens. Strateg. Chang. **31**(3), 337–343 (2022)

53. Gonzalez, Y.: These brands have filed metaverse trademarks–and what it all means; following Meta's and Nike's lead, several food, entertainment and retail companies have filed trademarks to sell virtual goods. AdAge **93**(4), 0013 (2022)

54. Huq, N., Reyes, R., Lin, P., Swimmer, M.: Metaverse or metaworse? Cybersecurity Threats Against the Internet of Experiences. Trend Micro Research (2022)
55. Annison, T.: Elliptic Metaverse Report 2022 - The Future of Financial Crime in the Metaverse: Fighting Crypto-crime in Web3.0. Elliptic (2022)
56. Crawford, A., Smith, T.: Metaverse app allows kids into virtual stripclubs, 23 February 2022. https://www.bbc.co.uk/news/technology-60415317
57. Li, C., Lalani, F.: How to address digital safety in the metaverse. World Economic Forum, 14 January 2022. https://www.weforum.org/agenda/2022/01/metaverse-risks-challenges-digital-safety/
58. Reed, N., Joseff, K.: Kids and the Metaverse: What Parents, Policymakers, and Companies Need to Know (2022). https://www.commonsensemedia.org/sites/default/files/featured-content/files/metaverse-white-paper.pdf
59. Rice, K.: In K. Rice, Inside the Metaverse Are You Safe? Dispatches. Channel 4, 24 April 2022. https://www.channel4.com/programmes/inside-the-metaverse-are-you-safe-dispatches
60. Russia Business News. Technologies for protecting children on the Internet: Rostelecom identified 10 cyber risks of future, 16 June 2022. https://www.proquest.com/magazines/technologies-protecting-children-on-internet/docview/2677618257/se-2
61. Sum of Us. (2022). Metaverse: another cesspool of toxic content. https://www.sumofus.org/images/Metaverse_report_May_2022.pdf
62. Allen, C., McIntosh, V.: Safeguarding the metaverse: a guide to existing and future harms in virtual reality (VR) and the metaverse to support UK immersive technology policymaking (2022). https://www.theiet.org/impact-society/factfiles/information-technology-factfiles/safeguarding-the-metaverse/
63. Wang, Y., et al.: A survey on metaverse: fundamentals, security, and privacy (2022). arxiv preprint arXiv:2203.02662
64. Goossens, S., Morgan, C., Kuru, C., Ji, F., Cespedes, D.J.: Protecting intellectual property in the metaverse. Intellect. Prop. Technol. Law J. **33**(9), 11–16 (2021)
65. Zhao, R., Zhang, Y., Zhu, Y., Lan, R., Hua, Z.: Metaverse: Security and Privacy Concerns (2022). arXiv preprint arXiv:2203.03854
66. Buck, L., McDonnell, R.: Security and privacy in the metaverse: the threat of the digital human. In: ACM CHI Conference on Human Factors in Computing Systems. Session: SSPXR - Novel Challenges of Safety, Security and Privacy in Extended Reality, Online (2022)
67. Vladimirov, I., Nenova, M., Nikolova, D., Terneva, Z.: Security and privacy protection obstacles with 3d reconstructed models of people in applications and the metaverse: a survey. In: 57th International Scientific Conference on Information, Communication and Energy Systems and Technologies (ICEST), Ohrid, North Macedonia (2022)
68. Bell, C.: The metaverse is coming. Here are the cornerstones for securing it. Official Microsoft Blog, 28 March 2022. https://blogs.microsoft.com/blog/2022/03/28/the-metaverse-is-coming-here-are-the-cornerstones-for-securing-it/
69. Cunha Barbosa, D.: What security risks could be hidden in the Metaverse? We live security by ESET, 5 April 2022. https://www.welivesecurity.com/la-es/2022/04/05/riesgos-seguridad-puede-esconder-metaverso/
70. Pinnock, B.: The metaverse will not be immune to cyber threats. The Mail & Guardian, 26 July 2022. https://mg.co.za/opinion/2022-07-26-the-metaverse-will-not-be-immune-to-cyber-threats/
71. Banaeian Far, S., Imani Rad, A.: Applying digital twins in metaverse: user interface, security and privacy challenges. J. Metaverse **2**(1), 8–16 (2022)

72. CITIC Telecom International. CITIC telecom international: (metaverse business opportunities) changing consumption patterns with immersive experience & deconstructing blind spots of blockchain security applications, 5 July 2022. https://www.proquest.com/magazines/citic-telecom-international-metaverse-business/docview/2686181145/se-2

73. Combs, V.: Metaverse security: how to learn from Internet 2.0 mistakes and build safe virtual worlds. Tech Republic (2022). https://www.techrepublic.com/article/metaverse-security-learn-lessons-from-internet-2-0-mistakes-to-build-safe-virtual-worlds/

74. Dataquest. What are the security risks and privacy challenges in Metaverse. Dataquest, 15 March 2022

75. Kadar, T.: The Metaverse Fraud Question: What Are the Risks? SEON (2022). https://seon.io/resources/metaverse-fraud/

76. Mackenzie, S.: Criminology towards the metaverse: cryptocurrency scams, grey economy and the technosocial. Br. J. Criminol. (2022)

77. PCQuest. Security risks that lurk deep inside the Metaverse, 30 March 2022. https://www.proquest.com/magazines/security-risks-that-lurk-deep-inside-metaverse/docview/2645890572/se-2

78. Shen, X.: NFTs and metaverse top tech risks, officials say: government watchdog warns criminals could steal sensitive user data or access accounts to hijack money as value of cryptocurrency keeps rising. South China Morning Post, 15 February 2022. https://www.proquest.com/newspapers/nfts-metaverse-top-tech-risks-officials-say/docview/2628333575/se-2

79. Smaili, N., de Rancourt-Raymond, A.: Metaverse: welcome to the new fraud marketplace. J. Financ. Crime (2022)

80. Targeted News Service. Ala. Securities Commission: Five States File Enforcement Actions to Stop Russian Scammers Perpetrating Metaverse Investment Fraud. Targeted News Service, 12 May 2022. https://www.proquest.com/wire-feeds/ala-securities-commission-five-states-file/docview/2662617069/se-2

81. Gómez-Quintero, J., Johnson, S., Borrion, H., Lundrigan, S.: A scoping study of crime facilitated by the metaverse, 29 April 2023. https://doi.org/10.31235/osf.io/x9vbn

82. Clayton, M.: Mother, 43, has her avatar groped by three male characters in the online Metaverse. Daily Mail Online, 30 January 2022. https://www.dailymail.co.uk/news/article-10455417/Mother-43-avatar-groped-three-male-characters-online-Metaverse.html

83. Shanker, S.S., Zytko, D.: The...Tinderverse?: Opportunities and Challenges for User Safety in Extended Reality (XR) Dating Apps. Cornell University Library (2022). arXiv.org

84. Di Pietro, R., Cresci, S.: Metaverse: security and privacy issues. In: 2021 Third IEEE International Conference on Trust, Privacy and Security in Intelligent Systems and Applications (TPS-ISA) (2021)

85. Howell, J.: 3 Metaverse Security Issues that you must know. 101 Blockchains, 23 February 2022. https://101blockchains.com/metaverse-security-issues

86. Identity Management Institute. (2022). Top 10 Metaverse Risks. Identity Management Institute. https://identitymanagementinstitute.org/top-10-metaverse-risks/

87. Abdulsattar Jaber, T.: Security risks of the metaverse world. Int. J. Interact. Mob. Technol. (iJIM) **16**(13), 4–14 (2022). https://doi.org/10.3991/ijim.v16i13.33187

88. Nichols, S.: Metaverse rollout brings new security risks, challenges. Tech Target, 7 February 2022. https://www.techtarget.com/searchsecurity/news/252513072/Metaverse-rollout-brings-new-security-risks-challenges

89. Dey, V.: Data Privacy In Metaverse Is An Evolving Concern. Martech Vibe, 2 June 2022. https://martechvibe.com/martech/data-privacy-in-metaverse-is-an-evolving-concern/

90. Khitrov, A.: What will it take to stop fraud in the metaverse? Information Age, 29 March 2022. https://www.information-age.com/what-will-it-take-to-stop-fraud-in-metaverse-19707/

91. Rosenberg, L.: Evil twins and digital elves: how the metaverse will create new forms of fraud and deception. The Future, 25 April 2022. https://bigthink.com/the-future/metaverse-fraud-digital-twins/

92. Williams, C.: Facebook's Metaverse a dangerous breeding ground for crime and mental health issues, experts says: facebook wants us to move away from our phones and into a virtual reality. The scandal-plagued platform has a new - Meta - and will launch a new platform, which will be accessed through a headset and not a phone. Tech experts have raised concerns about how crime will be policed in this new universe. Especially if Facebook can't quite get a hold of the issues it has already Sydney, Australian Broadcasting Corporation, 29 October 2021. https://www.abc.net.au/radio/programs/pm/facebooks-metaverse-a-dangerous-breeding-ground/13609832

93. Europol, Do Criminals Dream of Electric Sheep? How technology shapes the future of crime and law enforcement, European Union Agency for Law Enforcement Cooperation, 2019. https://www.europol.europa.eu/sites/default/files/documents/report_do_criminals_dream_of_electric_sheep.pdf

94. David Ingram, Facebook goes Meta: Zuckerberg announces new corporate name, NBC News, 28 October 2021. https://www.nbcnews.com/tech/tech-news/facebook-goes-meta-zuckerberg-announces-major-restructuring-rcna3605

95. Huddleston Jr., T.: Microsoft's metaverse plans are getting clearer with its $68.7 billion Activision acquisition, CNBC LLC., 19 January 2022. https://www.cnbc.com/2022/01/19/microsoft-activision-what-satya-nadella-has-said-about-the-metaverse.html

96. Weimann, G., Dimant, R.: The Metaverse and Terrorism: Threats and Challenges, Perspectives on Terrorism, Issue XVII, Volume 2, June 2023

97. Dibbell, J.: A Rape in Cyberspace, The Village Voice, 18 October 2005. https://www.villagevoice.com/a-rape-in-cyberspace/

98. https://secondlife.com/

99. https://www.meta.com/en-gb/experiences/2532035600194083/

100. GlobalData Thematic Intelligence, Metaverse regulations are underway, but theme remains nascent, Verdict Media Limited, 27 September 2023. https://www.verdict.co.uk/metaverse-regulation-requires-consistent-standards/

101. Madiega, T., Car, P., Niestadt, M., Van de Pol, L.: Metaverse Opportunities, risks and policy implications, European Parliamentary Research Service, European Union, PE 733.557, June 2022

102. https://www2.deloitte.com/us/en/insights/industry/technology/emerging-regulations-in-the-metaverse.html

103. Pellegrini, C.: Conflict of Laws and the Metaverse, EAPIL-The European Association of Private International Law, 13 June 2023. https://eapil.org/2023/06/13/conflict-of-laws-and-the-metaverse/

104. Marshall, A., Clarkson, A.: Future crimes and detection methods in cyberspace. Meas. Control **41**(8), 248–251 (2008). https://doi.org/10.1177/002029400804100803

105. Angus McKenzie Marshall and Brian Charles Tompsett: The metaverse-Not a new frontier for crime. WIREs Forensic Sci. (2023). https://doi.org/10.1002/wfs2.1505

106. Maskun, A., Naswar, H. A., Syafira, A., Napang, M., Hendrapati, M.: qualifying cyber crime as a crime of aggression in international law. J. East Asia Int. Law **13**(2), 397–418 (2020). https://doi.org/10.14330/jeail.2020.13.2.08

107. Powell, A., et al.: Digital harassment and abuse: experiences of sexuality and gender minority adults. Eur. J. Criminol. **17**(2), 199–223 (2018)

108. Caldwell, M., et al.: AI-enabled future crime. Crime Sci. **9**, 14 (2020)

109. Lowry, J., Reisberg, A.: Pettet's Company Law: Company Law & Corporate Finance, 4th edn. Pearson Education Limited, London (2012)

110. Sarch, A., Abbott, R.: Punishing Artificial Intelligence: Legal Fiction or Science Fiction (2019). 53 UC Davis Law Review 323, 326

111. Lucchetti, S.: Why Artificial Intelligence Will Need a Legal Personality, LawCross-Border, 22 May 2017. https://lawcrossborder.com/2017/05/22/why-robots-need-a-legal-personality/

112. Eidenmueller, H.: "The Rise of Robots and the Law of Humans" (26 March 2017) Oxford Legal Studies Research Paper No. 27/2017. https://ssrn.com/abstract=2941001

113. Carlson, J.: Me, Myself and My Multiple Avatars, CoinDesk, 23 September 2020. https://www.coindesk.com/markets/2020/09/23/me-myself-and-my-multiple-avatars/

114. Guinchard, A.: Crime in virtual worlds: the limits of criminal law. Int. Rev. Law Comput. Technol. **24**(2), 175–182 (2010)

115. Basu, T.: The metaverse has a groping problem already, MIT Technology Review, 16 December 2021. https://www.technologyreview.com/2021/12/16/1042516/the-metaverse-has-a-groping-problem/

116. Cheong, B.C.: Avatars in the metaverse: potential legal issues and remedies. Int. Cybersecur. Law Rev. **3**, 467–494 (2022). https://doi.org/10.1365/s43439-022-00056-9

117. Biega, A.J., Finck, M.: Reviving Purpose Limitation and Data Minimisation in Data-Driven Systems, Technology and Regulation, 2021

118. Chin, B.: Regulating your second life: defamation in virtual worlds. Brooklyn Law Rev. **72**(4) 1303, 1334 (2007)

119. Lavoie, R., et al.: Virtual experience, real consequences: the potential negative emotional consequences of virtual reality gameplay. Virtual Reality **25**, 69–81 (2021)

120. Atallah, A.: What Does the Metaverse Mean for Your Digital Identity, Forbes, 18 January 2022. https://www.forbes.com/sites/forbestechcouncil/2022/01/18/what-does-the-metaverse-mean-for-your-digital-identity/?sh=438d2a697ba6

121. Ara, T.K., Radcliffe, M.F., Fluhr, M., Imp, K.: Exploring the metaverse: what laws will apply? DLA Piper, 22 February 2022. https://www.dlapiper.com/en/us/insights/publications/2022/02/exploring-the-metaverse/

122. Cheong, B.C.: Application of Blockchain-enabled Technology: Regulating Non-fungible Tokens (NFTs) in Singapore, Singapore Law Gazette, January 2022. https://ssrn.com/abstract=4009972

123. Garno, D.: Trademarks meet NFTs: Hermès sues NFT creator over MetaBirkins, ReedSmith, 26 January 2022. https://www.adlawbyrequest.com/2022/01/

articles/in-the-courts/trademarks-meet-nfts-hermes-sues-nft-creator-over-metabirkins

124. Casey Newton, Mark in the Metaverse, THE VERGE, 22 July 2021. https://www.theverge.com/22588022/mark-zuckerberg-facebook-ceo-metaverse-interview

125. Hui, M.: China is Eyeing the Metaverse as the Next Internet Battleground, QUARTZ, 17 November 2021. https://qz.com/2089316/china-sees-themetaverse-as-the-next-internet-battleground

126. Garon, J.M.: When AI goes to war: corporate accountability for virtual mass disinformation, algorithmic atrocities, and synthetic propaganda N. KY. L. REV. **49**, 181 (2022). 209–10

127. Robertson, A., Peters, J.: What Is the Metaverse and Do I Have to Care?, THE VERGE, 4 October 2021. https://www.theverge.com/22701104/metaverse-explained-fortnite-roblox-facebook-horizon

128. Ravenscraft, E.: What Is the Metaverse, Exactly?, WIRED, 25 April 2022. https://www.wired.com/story/what-isthe-metaverse

129. Sparkes, M.: What is a metaverse, 251 NEW SCIENTIST 3348 (2021)

130. Nelson, P.: Information and consumer behavior. J. Pol. Econ. **78**, 311, 312 (1970)

131. Karpf, D.: Virtual Reality Is the Rich White Kid of Technology, WIRED, 27 July 2021. https://www.wired.com/story/virtual-reality-rich-white-kid-of-technology

132. Glushko, B.: Note, Tales of the (Virtual) City: Governing Property Disputes in Virtual Worlds, 22 BERKELEY TECH. L.J. 507 (2007)

133. White, E.E.: Comment, Massively multiplayer online fraud: why the introduction of real-world law in a virtual context is good for everyone. NW. J. Tech. Intell. Prop. **6**, 228 (2008)

134. Barfield, W.: Intellectual property rights in virtual environments: considering the rights of owners, programmers and virtual avatars. AKRON L. Rev. **39**, 649 (2006)

135. Reuveni, F.: On virtual worlds: copyright and contract law at the dawn of the virtual age. Ind. L.J. **82**, 261 (2007)

136. Sheldon, D.P.: Comment, Claiming Ownership, but getting owned: contractual limitations on asserting property interests in virtual goods. UCLA L. REV. **54**, 751 (2007)

137. Zarsky, T.Z.: Information privacy in virtual worlds: identifying unique concerns beyond the online and offline worlds. N.Y.L. Sch. L. Rev. **49**, 231 (2004)

138. Leenes, R.: Privacy in the metaverse: regulating a complex social construct in a virtual world, in PRIVACY AND IDENTITY: THE FUTURE OF IDENTITY IN THE INFORMATION SOCIETY 95 (2007)

139. Camp, B.T.: The play's the thing: a theory of taxing virtual worlds. HASTINGS L.J. **59**, 1 (2007)

140. Lederman, L.: Stranger than fiction: taxing virtual worlds. N.Y.U. L. Rev. **82**, 1620 (2007)

141. Balkin, J.M.: Law and liberty in virtual worlds. N.Y.L. Sch. L. Rev. **49**, 63 (2004)

142. Kasiyanto, S., Kilinc, M.R.: The legal conundrums of the metaverse. J.CEN. BANK. L. INST. **1** (2022). 299, 305-307

143. Hackl, C.: The Metaverse is coming and it's a very big deal, FORBES, 5 July 2020. https://www.forbes.com/sites/cathyhackl/2020/07/05/the-metaverse-is-coming--its-a-very-big-deal/?sh=3b9a9105440f

144. Lastowka, G.: Virtual justice: the new laws of online worlds (2010)

145. Mnookin, J.L.: Virtual(ly) Law: the emergence of law in LambdaMOO. J. Comput.-Mediated Comm. **2** (1996)

146. Strikwerda, L.: Theft of virtual items in online multiplayer computer games: an ontological and moral analysis. ETHICS Inf. Tech. **14**, 89 (2012)
147. Lastowka, F.G., Hunter, D.: Virtual crimes. N.Y.L. SCH. L. REV. **1**, 293 (2004). 297
148. Kerr, O.S.: Criminal law in virtual worlds. U. CHI. LEGAL F. **2008**, 415 (2008)
149. Danaher, J.: The law and ethics of virtual sexual assault. In: Barfield, W., Blitz, M. (eds.) The Law of Virtual and Augmented Reality (2018)
150. Grimmelmann, J.: Bone crusher 2.0: the fourth annual Greg Lastowka memorial lecture. RUTGERS U. L. REV. **71**, 843 (2019)
151. Mackenzie, S.: Criminology towards the metaverse: cryptocurrency scams, grey economy and the technosocial. Br. J. Criminol. **62**, 1537 (2022)
152. Strikwerda, L.: Present and past instances of virtual rape in light of three categories of legal philosophical theories of rape. Philos. Technol. **28**, 491 (2015)
153. Joshua, A.T.: Fairfield, the magic circle, 11 VAND. J. ENT. TECH. L. 823, 825 (2009)
154. Susan, W.: Brenner, is there such a thing as virtual crime'? CAL. CRIM. LAW REV. **4**(1), 12 (2001)
155. Easterbrook, F.H.: Cyberspace and the law of the horse. U. CHI. LEGAL F. 207 (1996)
156. Lessig, L.: The law of the horse: what cyberlaw might teach. Harv. L. Rev. **113**, 501 (1999)

Intricacies of Critical Infrastructures

Smart Environments: Information Flow Control in Smart Grids

Argiro Anagnostopoulou[1] , Dimitris Gritzalis[1]([envelope]) , Ioannis Mavridis[2] ,
and Panagiotis Kantas[1]

[1] Department of Informatics, Athens University of Economics and Business,
Athens, Greece
{anagnostopouloua,dgrit,pkantas}@aueb.gr
[2] Department of Applied Informatics, University of Macedonia, Thessaloniki, Greece
mavridis@uom.gr
https://www.infosec.aueb.gr/

Abstract. Electrical grid is a complex system designed to deliver electricity from generation to consumers. The constantly changing energy consumption requirements, along with the technological evolution pave the way for the upgrading of the electrical grid. Internet of Things (IoT) increases the automation and intelligence of devices, reducing the need for human interaction. Moreover, Industrial Internet of Things (IIoT) describes IoT applications in industrial systems. IIoT consists of a great number of low-cost interconnected devices, including sensors, actuators, and programmable logic controller (PLC). Such environments deal with vast amounts of data originating from a wide range of devices, applications, and services. The incorporation of IIoT to the electrical grid drives to the rise of the smart grid. In this chapter we provide an analysis of the smart grid architecture, its core components, and related technologies. Furthermore, we introduce information flow control and analyze several use cases that describe different business scenarios in smart grid environments. Finally, we discuss the current challenges regarding the security of smart grids.

Keywords: Smart Environments · Smart Grid · Access Control · Information Flow · Industrial Internet of Things · Industry 4.0

1 Introduction

Electrical grid is the physical system that transports electricity from the location where it is generated to the location where it is consumed. The three primary functions of the electrical grid are: (i) the generation, (ii) the transmission, and (iii) the distribution of electric current. Traditional power plants (e.g., coal, nuclear, hydro) generate electricity which is then delivered to distributors via high-voltage transmission lines. The electricity is converted into lower voltage and is distributed via distribution lines to several consumers, either residents

N. Pitropakis and S. Katsikas (Eds.): *Security and Privacy in Smart Environments*, LNCS 14800, pp. 151–172, 2025.
https://doi.org/10.1007/978-3-031-66708-4_7

or industries. The key difference between transmission and distribution is that the first handles high voltage (>110 KV), while the second handles medium (<50 KV) and low voltage (≤1 KV). Some features of the traditional electrical grids are large-scale infrastructures, centralized energy production, hierarchical control, little energy storage, one-way communication, and passive loads [1–4].

The rise of smart environments drives to increased automation. In traditional electrical grids, the automation took place on a dedicated network and was only on the grid operator's side. Smart grids are composed of multi entities that are connected. The entities own cables, solar panels, wind turbines, etc., and they all need IT connections to exchange data with each other. They are also capable of making decisions based on economic and technical aspects. The main difference between these two environments is that the smart ones incorporate equipment to facilitate data exchange with automated or remote-controlled devices. The remaining practices regarding manufacturing and maintenance are the same as those that were carried out on a dedicated network [5].

As the number of interconnected devices increases, the attack surface is enlarged. Thus, there is a need to handle how the information flows between the actors that participate in the business processes of a smart grid environment. In order to manage such flows, we propose a method that transforms a business process into a directed graph. The graph includes details regarding the use case, such as the data category that is transmitted (e.g. sensor data, configuration data etc.) and the type of request (read or write). This will help security experts to identify how the information flows in a large-scale infrastructure. In this chapter, we study eight real-world use cases of business processes in smart grid environments.

The chapter is structured as follows: In Sect. 2 we describe the architecture of smart grids, including the key objectives of such environments. In Sect. 3 we introduce the main actors grouped by domains they belong to. In Sect. 4 we explain the term information flow control and propose a way that can transform business processes as a graph. Also, we describe eight business use cases of smart grids as a graph. In Sect. 5 we discuss the challenges that smarts grids have to overcome. The paper ends with some concluding remarks.

2   Smart Grid Architecture

In this section we describe the architecture of smart grid environments, i.e., we introduce the key objectives, along with the conceptual model that defines a smart grid.

2.1   Key Objectives of Smart Grids

Below we introduce the key objectives of a smart grid, concerning several aspects such as power grid resilience, environmental performance, and the efficiency of the provided services.

Self-healing. Transforming a traditional power grid to a smart one enables handling of technical issues without human intervention. Smart grid autonomously recognizes malfunctions in its systems, addresses them, and takes proactive actions to prevent future disruptions. Such infrastructures are capable of responding with no delay and minimize time needed to restore their operations [6].

Increase in Power Quality. Smart grid guarantees a consistent supply of electric load, avoiding voltage variations that can cause productivity losses to industries. It increases the power quality through an effective transmission from power plants, along with an effective distribution from step-down transformers (which decrease the voltage incoming to the site by increasing the electrical current) [4,6,7].

Environmental Performance. According to the Paris agreement, all the countries should reduce their greenhouse gas emissions. The adoption of the smart grid will contribute to this goal. Improving the power grid's efficiency by 5% is equivalent to removing 53M cars from the streets, reducing the global amount of CO2 emissions [8].

Line Loss Reduction. During the transmission and distribution of electric current there is approximately 7% loss of energy production. This corresponds to ∼300 billion KWh power loss. Through the advancement of the power grid, utility companies produce less electricity, reducing also their carbon emissions [8].

Resilience to Attacks. Smart grid should guarantee a consistent supply of electric load throughout natural disasters, physical or cyber-attacks. Smart grid is equipped with sensors and intelligent devices to monitor/respond immediately to a security incident, ensuring the quality and resilience of its operations. It is also equipped with early warning systems that can detect potential threats and take preventative measures. Such environments can detect and resolve problems, without human intervention [4,6,7].

Integration of Renewable Technologies. Most of the current electricity comes from conventional fossil fuels, while less than 15% comes from sustainable energy sources such as wind, geothermal and solar. In order to address this low rate, more renewable energy sources need to be added to the electrical system. The integration of renewable energy sources demands advanced management solutions. Smart grid uses several distributed green power sources benefiting smaller installations of sustainable technologies and reducing CO2 emissions [4,8,9].

Market Demand. Adopting Demand Response (DR) on smart grid may contribute to a significant reduction of energy consumption. Consumers can actively participate in the grid performance as they can alter their energy behavior based on grid information. Timely information is provided to consumers, giving them the chance to reduce their energy expenses by managing their energy usage. This means that there will be a big shift from the current power grid: the energy load will follow the generation, rather than the generation following the load [4,9].

2.2   Smart Grid Conceptual Model

NIST introduced a conceptual model that describes the components which participate in smart grids. This model consists of seven domains. The word *domain* describes a classification of elements, including organizations, people, systems, devices, or other actors that share similar goals. An actor is any entity, e.g. person, computer, or software, that is involved in the operation of the smart grid. Actors may be involved in >1 domain. Below, we present the smart grid's domains, along with examples regarding the applications for each domain [11].

Customer Domain. In traditional grids customers were the end-users of electricity. In smart grids, customers are not limited to the consumption, but they can also generate, store or manage the use of energy. The customer domain is usually segmented into sub-domains for home, building/commercial, and industrial environments. Each sub-domain has a meter actor and includes an Energy Services Interface (ESI). The ESI may reside at an end-device, in a premise-management system, in the meter, or outside the premises. ESI may also communicate with other domains via the Advanced Metering Infrastructure (AMI) or through the Internet. The ESI provides the interface to devices and systems within the customer premises, either directly or through a Home Area Network (HAN) or a Local Area Network (LAN). The customer domain interacts with the domains of distribution, generation, markets, service provider, and operations. Indicative examples of applications in the customer domain are the home automation, the industrial automation, the storage, or the microgeneration.

Markets Domain. Markets domain is responsible for the purchase and trading of the grid resources and services, as well as for handling the entities that share market price, and supply and demand balance within the smart grid. Controlling the information flows between the market domain and the energy supply domain (such as generation, Distributed Energy Resources (DER), or customer domains) is crucial to achieving a highly efficient combination of production and consumption. Relationships among markets for wholesale and distribution, as well as communication processes among market operators and participants at each level, continue to be a challenging issue. This is because market activities and values (retail-wholesale) are governed by economic regulations and legal frameworks that change over time. Markets domain communication systems should bolster e-commerce guidelines concerning integrity and non-repudiation in order to be trustworthy, traceable, and auditable. Moreover, as the share of energy supplied by small, distributed energy resources grows, the requirements for acceptable delay in communication systems should be officially established. Indicative examples of applications in the market domain are the market management, the retailing, the DER aggregation, trading, or market operations.

Service Provider Domain. The services offered by actors in this domain are essential to producers, distributors, and consumers of the electrical grid. These services range from more complex ones, such as management of energy consumption and home energy production, to the basic ones like customer account management and billing. Electric Service Providers (ESP), current third parties, and new competitors may all provide services for this domain. However, defining standards and interfaces that work across multiple networking technologies, as well as maintaining consistent messaging semantics are the main challenges in protecting the smart grid in a dynamic, market-driven environment. When delivering a service, the service provider must ensure that all the requirements are addressed regarding the electrical power network's cybersecurity, reliability, continuity, integrity, and safety. All the domains, except the transmission one, interact with the service provider domain. While the communications with the markets and customer domains are critical for enabling economic expansion through the development of intelligent services, the communication with the operation domain is essential for maintaining situational awareness and system control. One of the benefits of the smart grid for the service provider industry is the opportunity for non-utility service providers to offer products and services to customers, utilities, and other stakeholders at more competitive prices. As customers actively participate in the electricity production process, they are able not only to reduce the cost of business services for other smart grid domains, but also enhance the amount of power generated while consuming less energy. Last but not least, adjusting consumption to enhance the performance of the smart grid and better matching power usage with operating conditions, like cost or lack of supply, are important considerations. Indicative examples of applications in the service provider domain are the customer management, installation and maintenance, home management, energy management, and billing.

Operations Domain. Actors in this domain have primary responsibility to ensure the electric grid reliability. With the adaptation of smart grid, more services will be introduced from the providers. Because the markets and service provider domains are changing, regulated utilities that own and manage the distribution system require services related to the planning and operation of the points at which customers receive their electricity. With the emerge of smart grid, the operations domain has access to various energy management systems used in the traditional power grid for reliable and efficient power system analysis and operation. Indicative examples of applications in the operations domain are the monitoring, control, analysis, and operational planning.

Generation Including DER Domain. Electricity production is the process of turning various energy sources, like wind, solar radiation, and moving water, into electrical energy. The generation domain, which serves as the primary source of power, is connected to the transmission, distribution, and customer domains. The generation domain must necessarily communicate with the transmission and distribution domains in order customers to be satisfied. Also, the generation domain should communicate the Key Performance Indicators (KPI) and Quality of Service (QoS) needs. Both in direct communication (through operations domain) and in indirect communication (through markets domain), the generation domain should respond appropriately to address loss of energy. Energy outage and generator failure are two examples of communication scenarios that could result in routing of electricity from other sources. Communication at the bulk system and distribution levels, including equipment located behind the meter, is quite important for the increasingly widespread of Distributed Energy Resources (DER). DER is related to generation, storage, and demand response provided in the distribution and customer domains, along with service provider-aggregated energy resources. Finally, generation including DER domain may impose restrictions on gas emissions, increase in the use of renewable energy sources, and provide storage to manage changes in renewable power generation. There are various types of physical actors that could play a role in the generation including DER domain, such as PLC, protection relays, RTU, equipment and fault monitors.

Distribution Domain. The distribution domain connects the transmission and customer domains, along with the meters for distributed generation, distributed storage, and consumption. DER (such as electrical storage) and medium-scale assets (such as community energy solar installations) may also be found in the distribution domain. The implementation, configuration, and interoperability of control devices as well as their interactions with entities from various domains are factors that impact the reliability of the distribution network. Advanced sensing and control functions of the smart grid are housed in the distribution domain. The adoption of faster communication systems to manage and optimize energy-related processes like power flow, generation, and consumption in real time is causing concern among all stakeholders. DER falls into the categories of distribution and transmission entities. This indicates a general need to increase the awareness and reliability of the distribution system. Supplementary sensing devices (fault circuit indicator) and domain operational functions (stabilize) are also included in the distribution domain. However, interactions between the markets and the distribution domains may have an impact on regional production and consumption, as well as the electrical and structural components of the distribution domain. In some cases, service providers interact with the customer domain through the distribution domain's infrastructure, which would alter the communication systems chosen for utilization in the distribution domain.

3   Smart Grid Actors

An actor is defined as a participant in an action or a process. An actor can be a hardware, a software, a computer system, a person, or a company. In the smart grid, an actor may be a person who is responsible for making decisions or exchanges necessary information for the successful operation of the infrastructure [11]. Tables 1, 2, 3, 4, 5, 6 and 7 introduce the classification of the actors based on the domain they belong to [12].

4   Information Flow Control in Smart Grid Environments

In the IIoT environments, as the number of interconnected devices is increased, the attack surface is enlarged. This is the reason why there is need to control how the information flows between the actors that participate in the business

Table 1. Actors of Transmission Domain.

Actor	Description
Phasor Measurement Unit (PMU)	A device that measures the phase angle, amplitude, and frequency of an electrical grid to assess the state of the system
Transmission Intelligent Electronic Device (IED)	A device that receives sensor data and issues commands
Transmission Remote Terminal Unit (RTU)	A device that sends: (1) status and measurement data from the transmission substation to the SCADA, (2) control commands from SCADA to the field equipment

Table 2. Actors of Generation Domain.

Actor	Description
Plant Control System	A local control system at a bulk generation plant, known as Distributed Control System (DCS)

processes of a smart grid. Smart grids bring significant changes to the information systems. For instance, new information flows emerging from the electricity grid, new actors such as decentralized renewable energy producers, new applications (e.g. electric vehicles and connected houses), as well as new communicating equipment (e.g. smart meters, sensors, and remote-control points) [16].

All information systems are composed of subjects and objects. An object contains data or operations to manipulate the data, such as databases or files. A subject is an entity, i.e., user or transaction, that manipulates an object. In order a subject to access data, it issues an operation to the respective object. A transaction refers to a sequence of operations on an object [10]. It is crucial for information security to define proper access control rules and track how information is propagated by computing systems during execution. Thus, information flow control aims to enhance both the confidentiality and the integrity of the information [13].

Let assume that an object o supports one of the basic operations OP (e.g., read or write). An access rule is composed of a tuple $<s, o, op>$, where s stands for subject, o for object, and op for operation. The pair $<o, op>$ is called access right. An authorizer grants an access right to a subject s. Subject s is allowed to manipulate the object o in an operation op only if s is granted an access right $<o, op>$ [10]. Now, let suppose that a subject s_i is granted with the access right $<f, read>$ on an object f, and an access right $<g, write>$ on object g. Suppose another subject s_j is granted with an access right $<g, read>$. Let assume that s_i reads data d in the object f and then writes data d to the object g. The s_j is not allowed to read data in the object f. However, the s_j can obtain data d in the object f by reading data d stored in the object g. As a result, information in the object f illegally flows into the s_j via the s_i and the object g [10].

Table 3. Actors of Distribution Domain.

Actor	Description
Field Crew Tools	A set of tools that includes mobile computing and handheld devices for field engineering and maintenance
Geographic Information System (GIS)	A system for managing remote assets that provides utilities with asset information network connectivity for advanced applications
Distribution Sensor	A device that measures a physical quantity and converts it into a signal that can be read by an observer or an instrument
Distribution Data Collector	A device that gathers data from multiple sources and modifies or transforms it
Distribution Remote Terminal Unit (RTU)	A device responsible for the remote control, measurement, and data collection in field devices
Intelligent Electronic Device (IED)	A device that receives data from sensors and power equipment and issue control commands

Table 4. Actors of Customer Domain.

Actor	Description
Customer	An entity that pays for electrical services. Customers may also provide power
Customer Energy Management System (EMS)	An application service or device that interacts with home appliances and gets information to make decisions to better control energy consumption
Customer Distributed Energy Resources (DER)	Energy generation resources for generation and storing of energy on the customer's site for energy-related activities
Customer Premise Display	A device that displays data regarding the consumption and cost to the customer on location
Home Area Network Gateway (HAN Gateway)	An interface that enables communication and sharing of real time usage information with the rest domains
Meter	A device used for the transfer of products and measuring usage from one domain or system to another. Energy Usage Metering Device (EUMD) A meter used for information-monitoring purposes

Table 5. Actors of Service Provider Domain.

Actor	Description
Aggregator/Retail Energy Provider	An entity that facilitates services on behalf of the consumer
Billing Entity	An entity that produces invoices in order to collect payments from clients
Energy Service Provider (ESP)	An entity that offers energy efficiency services and products along with retail electricity, natural gas, and clean energy options
Third Party Entity	An entity that performs a business operation

Below we propose a way to depict the business processes as a graph in order to control information flows among the devices in a smart grid network. The graphs include details such as the data category that is transmitted (e.g. sensor data, configuration data etc.) and the type of request (read or write). Below we analyze eight use cases that referred to real business processes in smart grid environments. For each use case we firstly give a brief description, and then we depict this information in a graph. Table 8 introduces the abbreviations, along with a brief description, that are used in the described scenarios.

4.1   Bulk Meter Readings

This use case describes the process of Bulk Meter Readings message collection. Every 15 min the electrical consumption is logged in the Smart Meter and every 4 h these data are sent to the AMI Head-End system. Once a day the AMI Head-End sends batches to the MDM system. MDM is responsible for the verification, estimation, editing, and checking for missing meters. The bulk meter reads are finally transmitted from the MDM to CIS and ODS. ODS send these reads to the Customer Web Portal.

In detail, a Smart Meter contains an internal board, known as the Meter Metrology Board, where its functions are set up and carried out. The Smart Meter's Metrology Board continuously logs information about electrical consumption, together with 15-min interval data blocks. The Bulk Meter Read Data, which includes meter readings and the interval data, are sent by the Meter Metrology Board every 4-h to the Smart Meter's NIC. The Smart Meter's NIC is an AMI Network Interface Component (NIC) that works with the Meter Metrology Board. The NIC then transmits the meter data to the AMI Head-End via the AMI Network, with the AMI Head-End transmitting batches of Bulk Meter Read Data to MDM for verification, estimation, and modification, as well as checking for missing meters. The bulk meter reads are finally transmitted from the MDM to CIS and ODS. ODS send these reads to the Customer Web Portal. An ODS is a subsystem that keeps track of all metering messages and events [14]. Figure 1 presents the information flows of the Bulk Meter Readings process.

Table 6. Actors of Operation Domain.

Actor	Description
Advanced Metering Infrastructure (AMI) Head-end	An infrastructure that handles the communication of data among third-party or non-head-end systems and the AMI network
Customer Information System (CIS)	A system that enables businesses to oversee attributes of their customer's relationships
Bulk Storage Management	An entity that manages energy storage systems connected to the bulk power grid
Customer Service Representative (CSR)	An entity that provides customer services
Distribution Management Systems (DMS)	A system that supports power grid operations, such as topology processor, contingency analysis, or loss analysis
Distribution Engineering	An entity that provides operations for technical planning, management, and system upgrade for the distribution system
Energy Management System (EMS)	A system that supervises, manages, and optimizes the transmission and generation performance
Distribution Supervisory Control and Data Acquisition (SCADA)	A system that collects and analyzes data and implements operational controls to manage distributed assets
ISO/RTO Operations	An entity that provides overview of load management and security evaluation for the transmission grid
Load Management Systems (LMS)	A system that sends load management commands to devices at customer premises to reduce load during peak/emergency situations
Demand Response Management System (DRMS)	A system that sends pricing or other signals to devices at customer premises to request the decrease or increase of the load
Meter Data Management System (MDMS)	A system that stores meter information, such as energy consumption, energy generation, or meter logs
Outage Management System (OMS)	A system that identifies outages and restores power
Wide Area Measurement System (WAMS)	A system that keeps track of all phase measurements and substation hardware, and offers information using visual modeling

Table 7. Actors of Markets Domain.

Actor	Description
Independent System Operator/Regional Transmission Organization	Market-participating ISO/RTO control center that does not operate the market
Energy Market Clearinghouse	Large-scale energy market operation system that provides overview of market signals for distribution companies

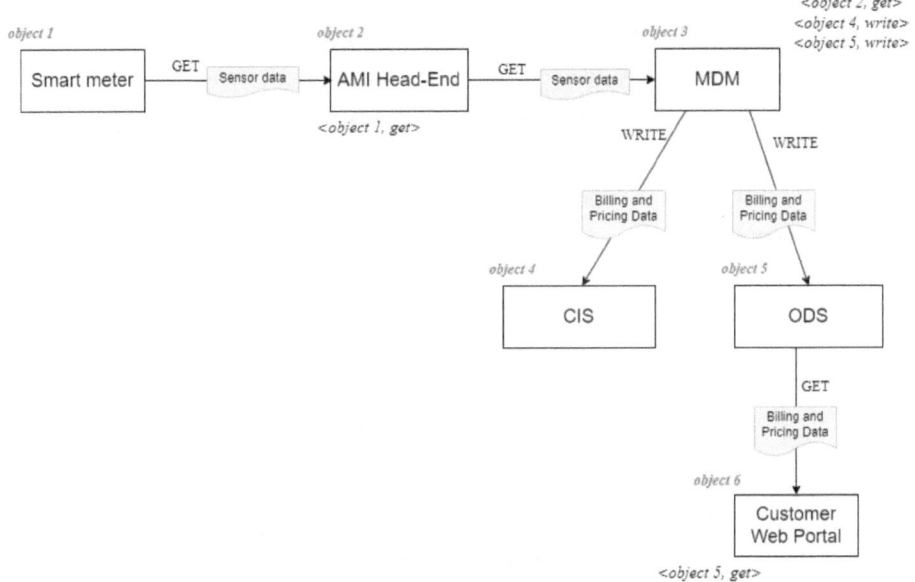

Fig. 1. Bulk Meter Readings.

4.2 DR HAN Pricing and Event Customer Opt-Out

CE exchanges Demand Response (DR) messages with the CIS and the DR Application regarding billing and pricing information. In detail, CE interacts with CIS and DR Application to transfer either current tariff CPP or DLC pricing data to the HAN Devices (such as PCT, Load Control Switch, Smart Meter, IHD). The customer chooses whether to take part in the DLC event. The customer's response is sent from the appropriate CPP or DLC HAN Device to AMI Head-End, and then to DR Application. DR Application sends to CIS the customer's choice regarding the DLC event acceptance or opt-out messages for the corresponding DLC program customer billing adjustments. Finally, DR Application sends the CPP program and DLC event information to the Customer Web Portal where that information is provided to the customers enrolled in those CPP

Table 8. Abbreviations used in use cases.

Abbreviation	Description
AMI	Advanced Metering Infrastructure
AMI Head-End	Back-office system than controls the AMI
CIS	Customer Information System that records customer's data and billing
CE	Customer Engagement provides customers with customized text messaging via the PCT
Customer Web Portal	Interactive website that enables the exchange and display of information for the customer
DR Application	A system for managing demand response and devices for load control, pricing, and messages
IHD	In-Home Display is a small portable device with a screen that presents basic information to the customer such as consumption data, price information or demand response signals
Load Control Switch	Electric switch that can be remotely commanded to open or close
MDM	Meter Data Management is a software that performs smart meter's data aggregation, validation, estimating, and editing
ODS	Operational Data Store is a data warehouse, which stores operational data, i.e. metering events and messages
PCT	Programmable Communicating Thermostat
Smart Meter	Digital meter used in measuring watts, vars, var-hours, volt-amperes, or voltage-ampere-hours
Vendor Meter Firmware	A tool to develop meter programs. It is used in the field for direct meter updates

or DLC programs [14]. In Fig. 2, we present the information flows that occur during the DR HAN Pricing and Event Customer Opt-Out process.

4.3 In-Field Programming of Smart Meter and Meter Firmware Upgrade

A Smart Meter is equipped with an optical port that can be connected and programmed with a Vendor Meter Firmware Tool. This tool is a software that can upload (apply) programming and firmware upgrades. Specifically, a Smart Meter upgrades its firmware through the Smart Meter's Optical port. The electrician can load new firmware through the Vendor Meter Firmware/Program Tool software using a laptop computer equipped with an optical probe. The Smart Meter will acknowledge the successful completion or the failure. The AMI Head-End

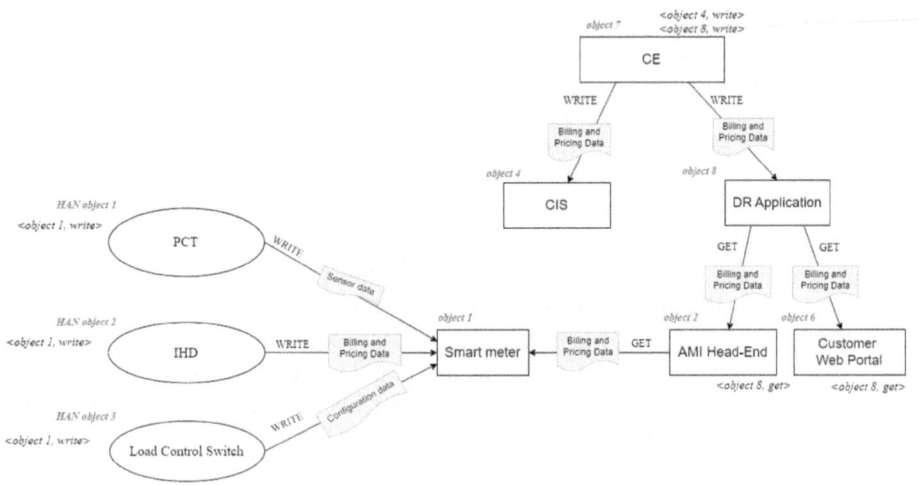

Fig. 2. DR HAN Pricing and Event Customer Opt-Out.

system will receive this acknowledgment of the program change upon completion or failure and send this information to ODS [14]. Figure 3 presents the information flows of the In-Field Programming of Smart Meter and Meter Firmware Upgrade process.

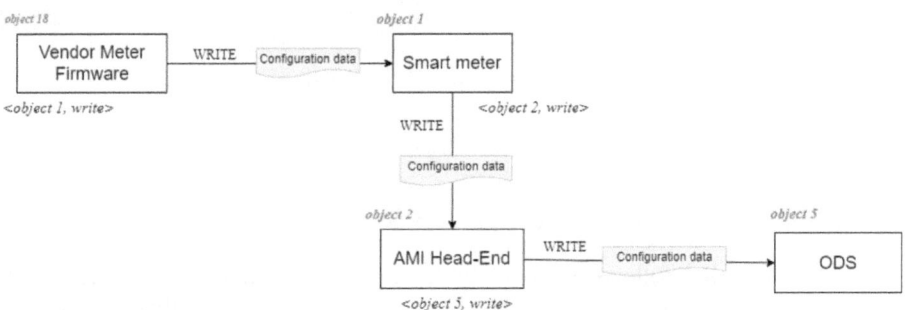

Fig. 3. In-Field Programming of Smart Meter and Meter Firmware Upgrade.

4.4 Meter Remote Connect/Disconnect

This use case addresses the messages exchanged between CIS and Smart Meter through the AMI Head-End and AMI Network when a meter connect/disconnect request is issued by CIS. In detail, due to the AMI Head-End system we can interact remotely with Smart Meters. We can remotely switch the internal meter, when necessary, for several reasons such as Remote Connect for Move-In or Disconnect for Move-Out, Remote Connect for Reinstatement on Payment or Disconnect for Non-Payment, Unsolicited Connect/Disconnect Event etc. The Customer Service Representative (CSR) is issued a remote disconnect in the CIS. The message is transmitted to the AMI Head-End and then forwarded via the AMI Network to the corresponding Smart Meter. The message is received at the NIC of the meter and then it is converted and sent to the Meter Metrology Board. The Meter Metrology Board causes the Internal Meter Switch (part of the Smart Meter) to connect or disconnect. When an action takes place, a confirmation is sent back to the Meter Metrology Board and thus to the AMI Head-End via the AMI Network. Finally, the acknowledgment is transmitted by the AMI Head-End both to the CIS and the ODS, where it is logged as a meter event [14]. In Fig. 4, we introduce the information flows that occur during the Meter Remote Connect/Disconnect process.

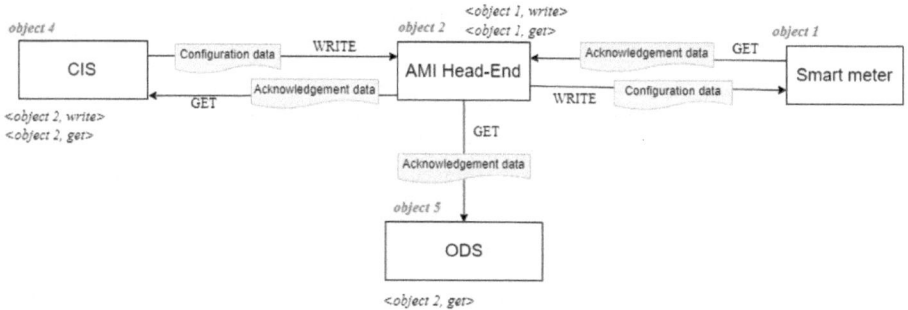

Fig. 4. Meter Remote Connect/Disconnect Use Case.

4.5   Community Energy Storage (CES) Energy Dispatch

The CES Controller follows a predetermined schedule when polling CES Units. The CES Units return data regarding the capacity and availability of energy of each unit. Through the master RTU, the CES Controller polls the circuit breaker in the substation as well. When the load exceeds a certain level discharge, the CES Controller triggers a load following event based on the values received. The CES Controller determines the level of participation of each CES Unit and issues device commands to each CES Unit. At every device command issued by the CES Controller, each participating CES Unit responds to the CES Controller with an acknowledgement. The Distribution SCADA (D-SCADA) system is constantly polling the CES Controllers and gathers data from them for display within D-SCADA and delivery to the Distribution Historian [14]. Figure 5 depicts the information flows of the CES – Energy Dispatch process.

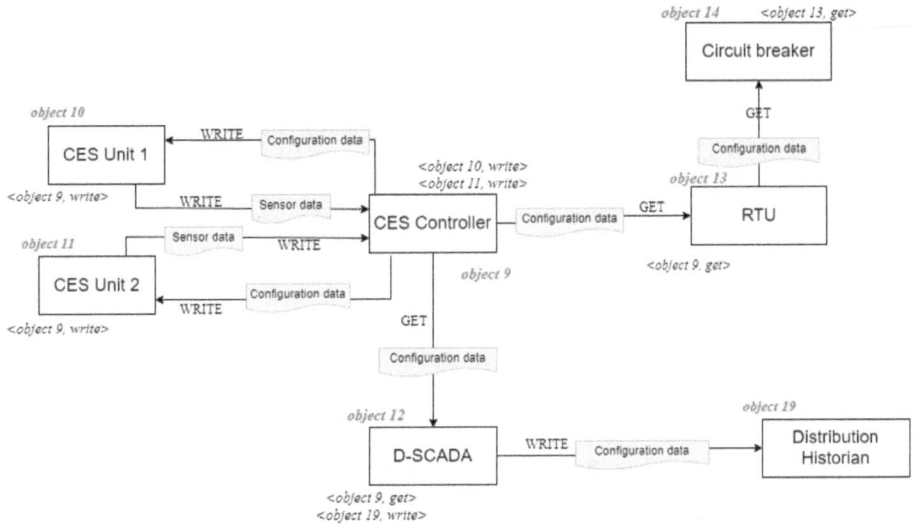

Fig. 5. Community Energy Storage (CES) - Energy Dispatch.

4.6   Direct Load Control Event

A DR solution handles direct load control programs. DR also presents interactions with consumers to communicate direct load control information using
an IHD. The IHD provides the customer with essential data and information,
such as usage data, price information, or DR signals. It manages the transmission of direct load control actions to PEV, load-control switches (type of electric
switch that can be opened or closed in response to an external command), HAN
devices, and smart appliances. A DR Application initiates a DR event, and the
AMI Head-End module sends the DR command to the Smart Meter through
the AMI Network. The Smart Meter then sends it on to the direct load control devices via the HAN. The DR profile includes a variety of commands and
messages, such as those related to temperature change, pricing information, and
informational messages. The HAN devices could be a PCT, a Load Control
Switch, or an IHD [14]. In Fig. 6, we present the information flows that occur
during Direct Load Control Event process.

4.7   Outage Management System Poll Multicast

This use case describes the Outage Management System (OMS) Poll. The OMS
initiates the meter polling request. It issues a poll of certain Smart Meters that
will enable operations personnel to determine if an outage is still valid. The OMS
Poll is a multicast which can be initiated either manually or automatically. This
is used when trying to diagnose the extent of an outage event. In detail, the
OMS issues an OMS Poll Multicast Request to the AMI Head-End. The OMS
Polling Multicast Request travels from the AMI Head-End, through the AMI

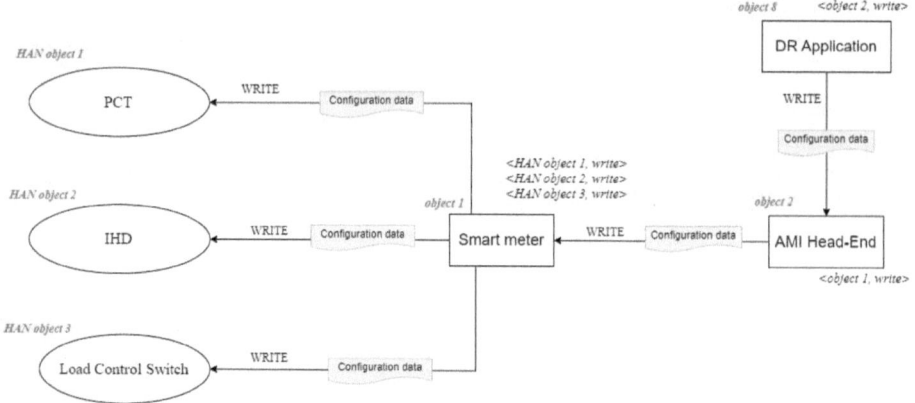

Fig. 6. Direct Load Control Event.

Network to the multicast set of the NIC of the Smart Meter. The NIC transmits the polling request to the Meter Metrology Board where the voltage is actually read and returns the value to the NIC. The NIC sends the meter reading back up through the AMI Network to the AMI Head-End. The AMI Head-End sends the meter reading to the OMS [14]. Figure 7 presents the information flows of the Outage Management System Poll - Multicast process.

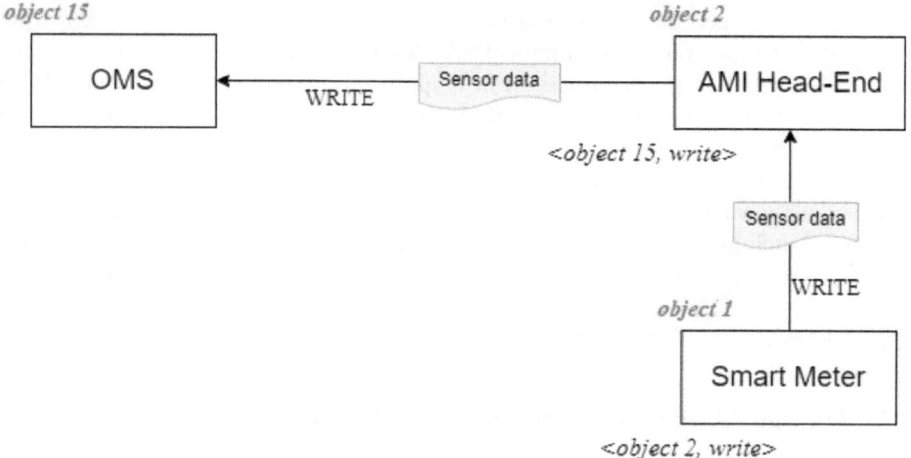

Fig. 7. Outage Management System Poll - Multicast.

4.8   Outage Notification

Many AMI systems offer endpoints with a "last gasp" message transmission capability to inform the utility that the endpoints have lost power. This last-gasp transmission serves as a surrogate for the customer's call. AMI systems using these "last gasp" capabilities work well in helping the OMS and dispatcher understand and efficiently respond to widespread outage conditions. This use case is unique as it is initiated at the NIC when it detects a zero-voltage event lasting more than a programmed period of time. The NIC will issue a last gasp message that flows back to the enterprise without any request from an enterprise system. The Last Gasp Message is routed through the AMI Network to the AMI Head-End. The AMI Head-End sends the message to the ODS and the Outage Filter. The Outage Filter sorts the message and sends it to the Trouble Ticket System (TTS). TTS sends the message to the OMS and the ODS. The OMS system updates the outage ticket and sends the update to TTS and ODS [14]. Figure 8 depicts the information flows of the Outage Notification process.

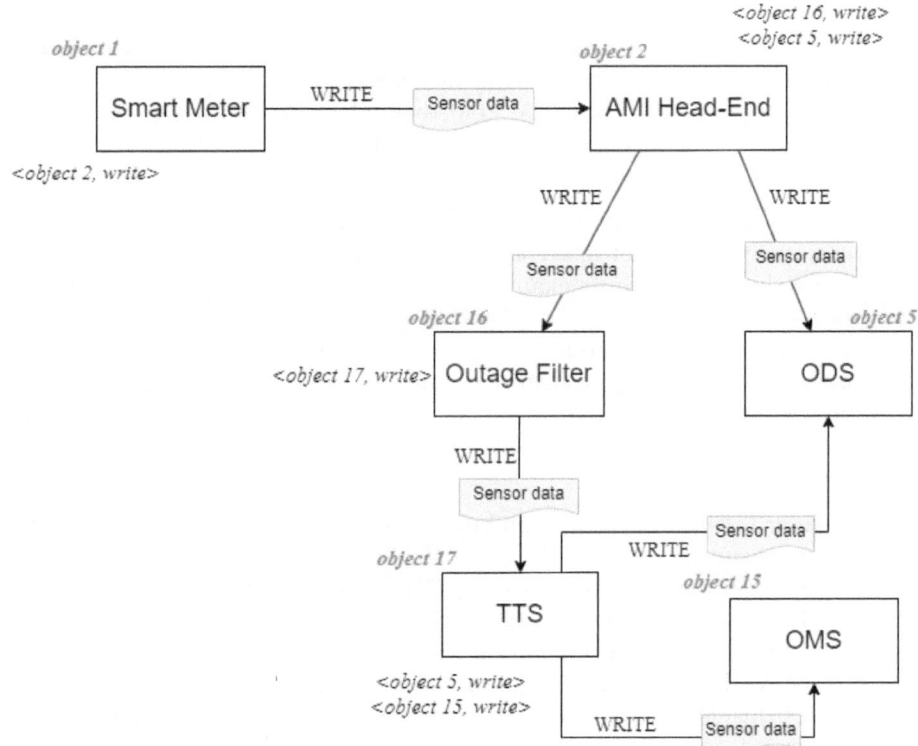

Fig. 8. Outage Notification.

5   Security Challenges of Smart Environments

The management of electricity distribution and consumption will undoubtedly be improved by adopting the concept of smart grid. This approach will benefit consumers, energy providers as well as grid operators. However, intelligent operations will raise new vulnerabilities affecting the entire electricity network, particularly in the area of information and communication system security [15]. In this section we overview the main security challenges that smart grid environments are facing.

Challenge #1. Assessment of the information security requirements. The smart grid's principal nervous system will be Information and Communications Technology (ICT), with data and information flows flooding through all domains. The security triad (confidentiality, integrity, availability) of data will be of utmost priority in the operational areas of the smart grid (generation, transmission, distribution, and consumption). Especially when dealing with customer's data, such as usage data, we should ensure confidentiality throughout the phase of storage and transmission. It is necessary for the infrastructure to define the requirements for adequate data protection. Each domain, along with the operators involved, should individually implement the appropriate security measures. The adoption of the smart grid should establish the required security controls. Finally, the infrastructure must be able to implement the proper techniques, such as data encryption, tunneling, authentication, digital certificates, etc., for the data stored in the smart grid [15].

Challenge #2. Numerous intelligent devices. A smart environment is composed of a great number of interconnected devices. Smart meters and AMI are likely to be the most notable examples of the various electronic and information processing devices that are deployed in smart grids. Moreover, a significant amount of Intelligent Electronic Devices (IEDs) and related information and communications technology will also be introduced by substation automation or by smartening of transformer facilities. Grid operators are facing significant challenges in designing, implementing, and preserving a dependable and scalable infrastructure [15].

Challenge #3. Physical security and perimeter fencing. Since smart meters are installed in consumers' homes (HAN/BAN/IAN), they are not directly located under the supervision of retail providers. Consequently, retail providers are vulnerable to firmware compromise or tampering by customers who intend to spread malware against smart grid software products (like DR), commit fraud, or gain access to the AMI network. Although most transformer control centers are physically housed inside secured buildings, this does not ensure that they are sufficiently secured against physical attacks. Unfortunately, in case of an insider threat an entity can gain physical access to these installations and thus he can get access to the communication system [15].

Challenge #4. Outdated and insecure communication protocols. All devices can accept connections from other devices attempting to communicate with them, regardless of whether that device is authorized or not. This is because most of the communication protocols currently used for the automation and management of electrical grid were never designed with security in mind. None of these protocols employ encryption or message integrity mechanisms, despite the fact that the absence of these techniques is a widely known vulnerability. Thus, communications are vulnerable to eavesdropping, and session hijacking. On the other hand, a completely new collection of communication protocols is being developed to handle new applications in smart grids, including end-to-end authentication and encryption [15].

Challenge #5. Numerous stakeholders and connections to other utilities. Due to the complexity of the smart grid's infrastructure, it is necessary for a great number of diverse stakeholders to work together. Thus, the number of actors that are involved has increased, such as customers, small power producers with DER, energy retailers, private electric vehicles (PEV), etc. It is challenging to quickly coordinate the functions of such a diverse group of stakeholders. Each group has its own organizational procedures, management goals, information communication needs, regulation standards, in order to deliver dependable, and reliable electricity. The use of smart meters by utilities, like gas and heating, to remotely read and process energy usage data in the near future creates the need for a flexible, interoperable, and well-communicated electric grid that can sustain all information sharing required between various utilities. As a result, smart grid needs to be secured considering all the potential stakeholders and actors involved, creating an even more complicated infrastructure [15].

Challenge #6. Enhancing the supply-chain security. The possibility of hostile agents infiltrating the supply chain for electronic components such as microchips, SCADA, embedded/operating systems, and altering the circuit elements of the electronic components or replacing components with altered circuitry is quite serious. The firmware of several industrial controllers or smart meters may hide backdoors, logic bombs, or other malicious software, enabling terrorists or malicious actors to gain remote access or affect the networks and equipment of the infrastructure [15].

Challenge #7. Addressing security requirements. System vendors play an important role in the process of securing the smart grid. In case that a product is developed without security in mind, the protection of smart grid environments is quite challenging. The safety of smart grid does not solely rely on the availability of products that are secure. Securing a smart grid infrastructure is a never-ending process that heavily depends on the power utilities. Grid operators should enhance the cybersecurity aspect, enforce strict security policies, increase awareness, and provide proper training to employees. Moreover, it is advisable to assess the security of the systems that are currently in use [15].

Challenge #8. Enforcing dedicated access control rules. Establishing appropriate access control guidelines and monitoring the propagation of information by computing systems during execution are essential components of information security. Thus, appropriate access control policies must be enforced on

IIoT infrastructures. This is the role of access control systems. They establish which active entities (subjects) are authorized to gain access to passive entities (objects), e.g. a resource. Each subject is characterized by a set of permissions for a specific object. Only authorized subjects are allowed to manipulate objects in authorized operations. However, there is a plethora of access control models that a smart environment can utilize [17–20]. The choice of the most appropriate scheme tailored to the specific smart environment where these rules will be applied falls within the purview of security experts.

6   Conclusion

Traditional electrical grid has a great number of limitations, such as multiple new-generation resources that must connect and disconnect from the distribution grid remotely; an operation that the electric grid cannot handle properly. Furthermore, energy demand dramatically changes daily. The utility companies cannot accurately estimate the demand, resulting either in generating more energy which goes to waste or generating less energy resulting in rolling blackouts. Finally, sustainable and renewable resources are driven in the energy market. However, these new energy sources have a few constraints, such as availability and localization, which traditional power grid is unable to accommodate.

All the above-mentioned limitations could be topped with the modernization and digitalization of the current power grid which is referred to as smart grid. With the rise of an intelligent electrical grid, actors that participate will be benefited in several ways, such as providing greater efficiency in energy consumption, gaining flexibility in the type of energy sources used, improving environmental performance, etc. The most important advantage is the independent and immediate addressing of technical issues without human intervention. Smart grid utilizes distributed generation and enhanced energy storage improving efficiency. The use of smart meters enables the participation of demand response in the market domain. Considering the above, the most notable advantage is the employment of intelligent sensing and measurement technologies, along with wide-area monitoring systems.

However, the rise of smart environments emerges new challenges regarding information security. The complexity will be even increased with a broad number of stakeholders participating in the operation of a smart grid. Moreover, because of the smart grid's heavy reliance on computer networks and other associated systems, a great attack surface is created, making smart grid vulnerable to various cyberattacks. There are numerous malicious actors with varying motives ranging from internal employees to nation backed hackers. Attackers try to exploit vulnerabilities in order to attack in the grid aiming the confidentiality, integrity, and availability of the system. Indicative examples of such attacks are: modification of the electricity usage (LAA), false data injection attacks in SCADA systems, interception of wireless communication, tampering of the data transmitted from the smart meters, and social engineering.

For this reason, security experts should implement adequate defense mechanisms in order to prevent cyber-attacks and data breaches. The defense mechanisms could be classified in good security practices such as conducting audits in SCADA systems, establishing an asset inventory, educating, and training personnel, as well as secure supply chain. Moreover, there is a number of practices that each critical infrastructure should incorporate based on standards, e.g. IEC 62443, ISO/IEC 27001, ISO/IEC 27019, etc. Indicative examples of technological applications that enhance security include antivirus, intrusion prevention systems, intrusion detection systems, industrial protocol filters, data loss prevention, security information and event management.

In this chapter we focus on information flow control in smart environments. Especially, we chose and described eight business processes of a smart grid. We depicted each business process as a directed graph. These graphs present how the information flows in the infrastructure regarding the operation that is executed. In the graph there are multiple details, such as the data category that is transmitted (e.g. sensor data, configuration data etc.) and the type of request (read or write). Our approach of simulating a smart grid environment as a graph consists of two fundamental steps. At first, eight use cases that describe different operating scenarios were investigated and analyzed. The selected scenarios cover different business processes, ranging from managing power outages to sending acceptance or opt-out responses to program event messages at customer's premise. The next step was to transform this information into a directed graph for each use case. Each graph represents the flow of information through the systems and processes of several business processes. Finally, we overview the main challenges that a smart grid environment face.

References

1. The Power Grid System, 6.5 The Power Grid System—EME 810: Solar Resource Assessment and Economics. https://www.e-education.psu.edu/eme810/node/592. Accessed 11 Oct 2023
2. Hentea, M.: Building an Effective Security Program for Distributed Energy Resources and Systems. Wiley, Hoboken (2021)
3. Flick, T., Morehouse, J.: Securing the Smart Grid: Next Generation Power grid Security. Elsevier, Amsterdam (2010)
4. Smart Grid and Enabling Technologies. Wiley IEEE. https://dokumen.pub/smart-grid-and-enabling-technologies-wiley-ieee-1nbsped-1119422310-9781119422310.html. Accessed 01 Dec 2023
5. Smart Grid security certification in Europe, ENISA. https://www.enisa.europa.eu/publications/smart-grid-security-certification-in-europe. Accessed 01 Dec 2023
6. Deebak, B., Al-Turjman, F.: Sustainable Networks in Smart Grid. Academic Press, Cambridge (2022)
7. Annor-Asante, M., Bernardi, P.: Development of smart grid testbed with low-cost hardware and software for cybersecurity research and education. Wireless Pers. Commun. **101**, 1357–1377 (2018)
8. Knapp, D., Raj S.: Applied Cyber Security and the Smart Grid: Implementing Security Controls into the Modern Power Infrastructure. Newnes (2013)

9. Feng, Y., Qian, Y., Qingyang H.R.: Smart Grid Communication Infrastructures: Big Data, Cloud Computing, and Security (2018)
10. Nakamura, S., Ogiela, L., Enokido, T., Takizawa, M.: An information flow control model in a topic-based publish/subscribe system. J. High-Speed Netw. **24**(3), 243–257 (2018)
11. Gopstein, A., et al.: NIST framework and roadmap for smart grid interoperability standards, release 4.0. Department of Commerce, NIST, Gaithersburg (2021)
12. Pillitteri, V., Brewer, T.: Guidelines for Smart Grid Cybersecurity (2014)
13. Hedin, D., and Sabelfeld, A.: A perspective on information-flow control. In: Software Safety and Security, pp. 319–347. IOS Press (2012)
14. SmartGrid Resource Center & Use Case Repository, EPRI. https://smartgrid.epri.com/repository/repository.aspx. Accessed 01 Dec 2023
15. Security Aspects of Smart Grid, ENISA. https://www.enisa.europa.eu/topics/critical-information-infrastructures-and-services/smart-grids/smart-grids-and-smart-metering/ENISA_Annex%20II%20-%20Security%20Aspects%20of%20Smart%20Grid.pdf/@@download/file. Accessed 04 Dec 2023
16. Daki, H., El Hannani, A., Aqqal, A., Haidine, A., Dahbi, A.: Big data management in smart grid: concepts, requirements, and implementation. J. Big Data **4**(1), 1–19 (2017)
17. Gouglidis, A., Mavridis, I.: domRBAC: an access control model for modern collaborative systems. Comput. Secur. **31**(4), 540–556 (2012)
18. Salonikias, S., Gouglidis, A., Mavridis, I., Gritzalis, D.: Access control in the industrial internet of things. Secur. Priv. Trends Industr. Internet Things 95–114 (2019)
19. Mattas, A., Mavridis, I., Pangalos, G.: Towards dynamically administered role-based access control. In: 14th International Workshop on Database and Expert Systems Applications. IEEE (2003)
20. Anagnostopoulou A., Mavridis I., Gritzalis D.: Risk-based illegal information flow detection in the IIoT. In: Proceedings of the 20th International Conference on Security & Cryptography, Rome. ScitePress (2023)

Social Engineering: The Human Behavior Impact in Cyber Security Within Critical Information Infrastructures

Sokratis Nifakos[1]([✉]) [iD], Krishna Chandramouli[2] [iD], and Natalia Stathakarou[1] [iD]

[1] Department of Learning, Informatics, Management and Ethics, Karolinska Institutet, 171 77 Solna, Sweden
sokratis.nifakos@ki.se
[2] Multimedia and Vision Research Group, Queen Mary University of London, Mile End Road, London 1 4NS, UK

Abstract. Following an increase in the frequency and ingenuity of attacks launched against critical information infrastructures with the intention of causing service disruption, there is a strong need to understand the level of awareness among employees working in these environments, the cybersecurity awareness programs and training activities that are in place as well as the methods used in order to measure cybersecurity and privacy awareness of these employees. The organizational resilience against cybersecurity relies on robust data governance policies spanning from data security, privacy and IT infrastructure security among others. While these strategies have offered protection against traditional cyberattacks (e.g., Distributed Denial of Service (DDoS)), the emergence of threats, such as ransomware (e.g., WannaCry) attacks have shown a tremendous increase in the frequency of data breaches in Critical Information Infrastructures (CIIs). Moreover, traditional risk assessment methods have focused on evaluating the security profile of IT systems but the emergence of social engineering dictates the need of better understanding of the cyber risks impact caused by insecure human behavior. Authors' aim is to present to the reader a clear overview of social engineering and the importance of counting human vulnerabilities as a critical cybersecurity problem that needs close attention and better understanding of the cyberattacking methods used as well as the actions could be taken in order to prevent, manage and eliminate them.

1 Introduction

In today's societies, the landscape of cybersecurity is continuously evolving, with new threats emerging that challenge traditional defence mechanisms. One of more often and increasingly sophisticated threat is social engineering. Unlike conventional cyberattacks that directly target system vulnerabilities, social engineering exploits a different kind of vulnerability – the human element. At its core, social engineering is a manipulation technique that manipulates individuals to get access to confidential or personal information that may be used for fraudulent purposes. This method relies heavily on human interaction and often involves tricking people into breaking normal security procedures. The

N. Pitropakis and S. Katsikas (Eds.): *Security and Privacy in Smart Environments*, LNCS 14800, pp. 173–184, 2025.
https://doi.org/10.1007/978-3-031-66708-4_8

concept is rooted in psychological manipulation, capitalizing on basic human tendencies such as trust, the desire to be helpful, or fear of authority. As technology advances, so do the tactics employed by cybercriminals, making social engineering a prominent tool in their "toolbox". Unlike automated cyber threats that leverage technical flaws, social engineering attacks are uniquely adaptable, targeting the constant in cybersecurity – people. With tactics ranging from phishing emails and pretexting to baiting and quid pro quo, social engineers skilfully navigate the complex landscape of human emotions and decision-making processes.

1.1   The Importance of Understanding Social Engineering in CIIs

Critical Information Infrastructures (CIIs) represent the backbone of essential services in society, including healthcare, finance, and government operations. The robustness and security of these systems are important to societal stability and trust [1]. However, the human factor within these infrastructures often presents a significant security gap as the increasing sophistication of social engineering attacks poses a substantial threat to these critical systems. The success of such attacks in CIIs can lead to severe consequences, ranging from data breaches and financial losses to critical disruptions in services and national security threats. Understanding and addressing social engineering in CIIs is not merely a technical challenge but a comprehensive strategy that encompasses psychology, organizational behavior, and cybersecurity. It demands a proactive approach to cybersecurity, one that anticipates and mitigates the risks posed by human vulnerabilities. As cybercriminals continue to refine their techniques, the need for robust and dynamic strategies to counter these threats becomes more critical. This understanding forms the basis of our exploration in this chapter, highlights the importance to evolve cybersecurity practices to address these uniquely human-centered threats. This chapter aims to provide a comprehensive understanding of social engineering, its impact on CIIs, and strategies for mitigation. We begin by exploring the concept of social engineering, highlight its various forms and the psychological principles that underpin these tactics. Following this, we present real-life examples of social engineering attacks, highlighting their impact on CIIs and the lessons learned [2]. We then categorize the risks associated with social engineering and contrast these with traditional cyber threats, providing a framework for understanding the unique challenges posed by these human-focused attacks. Subsequent sections focus on the threat landscape, discussing emerging threats such as ransomware and their connection to social engineering [3]. We also explore mitigation strategies, emphasizing the role of cybersecurity awareness programs and training methods. The chapter further explores the challenges of measuring cybersecurity and privacy awareness among employees in CIIs, aiming to present the importance of these factors in combating social engineering threats [3]. Finally, we discuss organizational resilience and data governance, exploring how robust policies and a culture of security can be used as defense against social engineering.

2   Understanding Social Engineering

2.1   Definition and Scope

Social engineering, in the context of information security, refers to the psychological manipulation of people into performing actions or getting access to confidential information. It's a term that covers a wide range of malicious activities accomplished through human interactions. It differs from traditional hacking in that it exploits human psychology rather than technical hacking techniques. The evolution of social engineering reflects a transition from scams to sophisticated cyber-attacks that target individuals and organizations. The common forms include:

1. **Phishing**: Perhaps the most well-known form, phishing involves sending fraudulent communications that appear to come from a reputable source, usually via email. Its goal is to steal sensitive data like credit card numbers and login information.
2. **Pretexting**: Here, an attacker creates a fabricated scenario (or pretext) to steal their victims' personal information. In this case, the attacker usually starts by establishing trust with the victim.
3. **Baiting**: Similar to phishing, baiting involves offering something enticing to the victim in exchange for private information. This could be as simple as a free download of a movie or software that turns out to be malicious.
4. **Tailgating**: This physical form of social engineering involves someone without proper authentication following an authorized person into a restricted area.

2.2   Historical Context

The history of social engineering is as old as human conflict, but its applications in cybersecurity have gained prominence in today's societies. Early examples of social engineering in the digital world were often simple cons, like the infamous "Nigerian Prince" email scams. However, as technology advanced, so did the methods of social engineers.

The emergence of the internet and email gave rise to phishing attacks. One notable early case was the "Love Bug" in 2000, a malicious program spread via email with the subject line "ILOVEYOU." This simple yet effective tactic caused an estimated billion in damages worldwide. As internet usage and digital data storage increased, so did the sophistication of these attacks and social engineering became a key tool for cybercriminals, with large-scale data breaches often involving some element of human manipulation.

2.3   Psychological Principles

The success of social engineering attacks relies heavily on exploiting basic human psychological principles:

1. **Trust Exploitation**: Humans have a natural tendency to trust, especially when dealing with seemingly authoritative or reputable sources. Social engineers manipulate this trust to deceive victims.

2. **Authority Influence**: People are often more willing to comply with requests from those in positions of authority. Cybercriminals impersonate police, bank officials, or company executives to exploit this tendency.
3. **Urgency Creation**: Creating a sense of urgency or fear is a common tactic. Attackers often tell victims that they need to act fast to avoid some negative consequence, pushing them to act without thinking critically.
4. **Social Proof and Conformity**: People are inclined to do things they see others doing. Social engineers exploit this by fabricating scenarios where 'others' (which could be fake profiles or entities) appear to be engaging in or endorsing certain behaviors.

In understanding these psychological points, it becomes evident why social engineering is such an effective tool for cybercriminals as it leverages human traits, turning them into vulnerabilities and the tactics continue to evolve, making it imperative for cybersecurity strategies to consider and address the human factor effectively.

2.4   Real-Life Examples of Social Engineering

In this section we present two popular cases of social engineering and we try briefly to present an impact analysis and lessons learned.

The Twitter Bitcoin Scam (2020): In July 2020, a massive security breach of Twitter led to a series of high-profile accounts being compromised and used to perpetrate a Bitcoin scam. Influential personalities like Barack Obama, Elon Musk, and Bill Gates had their accounts taken over by attackers who tweeted fraudulent messages urging followers to send Bitcoin to a specific address, promising to double any amounts sent. This scam was primarily a combination of spear-phishing and social engineering, where Twitter employees were targeted and manipulated to provide access to internal systems.

The Target Data Breach (2013): One of the most significant data breaches in history, the Target breach affected 41 million customer payment card accounts and exposed contact information for more than 60 million customers. The breach was initiated through a phishing email sent to a subcontractor of Target, which led to the installation of malware on Target's security and payments system. This incident is a prime example of how a simple phishing attack can escalate into a massive data breach.

Impact Analysis
Data Breaches: Both cases resulted in substantial data breaches, highlighting the severe security implications of social engineering. Personal and financial information was compromised, leading to widespread privacy concerns.
Financial Losses: The financial repercussions were significant. For instance, Target incurred costs of over $200 million as a result of the breach. The Twitter Bitcoin scam, though not as financially devastating, still saw victims sending thousands of dollars to the attackers.
Erosion of Public Trust: The incidents severely impacted public trust in these companies. The perception of safety and reliability, crucial for such platforms, was undermined. The long-term reputational damage can often exceed the immediate financial losses

Lessons Learned
Employee Training and Awareness: Both incidents highlight the importance of continuous employee training in recognizing and preventing social engineering attacks. Regular and updated training sessions can significantly mitigate such risks.
Robust Internal Security Protocols: Strong internal security measures are essential. This includes limiting access to sensitive systems and data, implementing two-factor authentication, and regularly updating security protocols.
Incident Response Plan: Having a well-prepared incident response plan is crucial. Quick and effective responses can limit damage and restore security and public trust more rapidly.
Supply Chain Security: The Target breach highlights the need for comprehensive security across the supply chain. Organizations must ensure that their partners and vendors adhere to stringent security standards.
Public Communication and Transparency: Effective communication following an attack is vital. Being transparent about the breach and its impacts, as well as the steps taken to resolve it, can aid in rebuilding public trust.

Social engineering in cybersecurity is facing a range of risks that impact individuals and organizations in various ways. One of the primary risks is the unauthorized disclosure of sensitive information, including personal data, trade secrets, financial details, and security credentials. This breach of confidentiality can lead to serious consequences such as identity theft or unauthorized access to restricted areas within an organization's network. Moreover, financial loss is another significant risk, often directly targeted by social engineers and these losses can be from small-scale scams that trick individuals into sending money, to large-scale frauds that drain significant corporate funds or lead to substantial regulatory fines. This financial aspect is critical, as it can have far-reaching effects on both individuals and large organizations. However, the impact of social engineering extends beyond financial loss to reputational damage. An organization suffering from a social engineering attack may experience a severe blow to its reputation, especially if the incident exposes customer data or highlights security inadequacies. This can lead to a loss of trust, affecting customer relationships and business prospects, often more damaging in the long run than immediate financial losses. For instance, an operational disruption is another outcome of social engineering attacks, aiming to interrupt the regular operations of an organization and these disruptions, while not always financially motivated, can result in indirect financial losses and hamper business continuity, showcasing the diverse objectives of social engineering strategies. Additionally, legal and compliance risks arise from breaches resulting from social engineering, especially given the stringent data protection laws in various regions. Organizations may face lawsuits, fines, and sanctions following an incident, adding a legal dimension to the cybersecurity challenges they face. Understanding the human vulnerabilities exploited by social engineering has become crucial and includes the natural human tendency to trust, especially when faced with authoritative or legitimate figures, habits that make routine behaviors predictable and exploitable, a lack of awareness about the various forms of social engineering, and emotional manipulation tactics like fear, urgency, or curiosity. In comparison to traditional cybersecurity threats such as Distributed Denial of Service (DDoS) attacks, social engineering presents a different challenge. Traditional threats target an organization's technological infrastructure, while social engineering attacks target the human element. Traditional threats often require technical expertise to exploit system vulnerabilities, but social engineering leverages psychological vulnerabilities, making it less predictable and harder to prevent with technical solutions alone. Focusing on the psychological vulnerabilities that exploit the efficacy of social engineering attacks within cybersecurity highlights a fascinating intersection of human psychology and information security. At the core of social engineering lies the exploitation of innate human traits—trust, curiosity, fear, and the proclivity for obedience to authority—tactics that have been psychologically codified and understood within the field of social psychology. These vulnerabilities are not mere artifacts of individual psychology but are embedded within every human socialization and communication. The principle of authority, for instance, is exploited in pretexting attacks where attackers pose as trusted figures to elicit sensitive information, exploiting the human tendency to comply with requests from perceived authority figures. Similarly, the urgency and fear invoked in phishing emails play on the psychological concept of 'loss aversion,' a predisposition to strongly prefer avoiding losses to acquiring equivalent gains, thus precipitating hasty decisions to avoid perceived threats. Moreover, the concept of 'social proof,' or the tendency to see an action as more desirable when others are doing it, is manipulated in baiting scenarios where the victim is lured with the promise of a reward that others are purportedly achieving. This exploitation of psychological principles presents the sophistication of social engineering strategies, which, unlike traditional cyber threats that target technological vulnerabilities, aim at the 'soft targets' of human cognition and emotion. The challenge in countering these threats lies in the inherent adaptability and complexity of human psychology. Traditional cybersecurity measures, while effective against direct technological assaults, are less equipped to guard against the foundational manipulations of social engineering. Therefore, understanding and mitigating these psychological vulnerabilities require a multidisciplinary approach that encompasses cybersecurity, psychology, and behavioural science, aiming not only to strengthen technological defences but also to enhance human awareness and resilience against these insidious attacks. For instance, the proliferation of social media, has exploited the scale and efficacy of exploiting psychological vulnerabilities through social engineering. Social media platforms, with their repositories of personal information and interconnected networks, serve as ground for attackers to collect data, craft personalized attack vectors, and execute sophisticated manipulation campaigns. The creation of fake profiles and the dissemination of engineered content exploit the human tendency towards trust and social validation, transforming social media into a potent tool for social engineering. The psychological manipulation inherent in these attacks often leverages the concept of 'online disinhibition effect,' where individuals feel more vulnerable to persuasion and manipulation due to the perceived anonymity and distance of online interactions. Attackers exploit this phenomenon by creating highly tailored and compelling narratives that resonate on a personal level with their targets, effectively bypassing rational barriers to skepticism and caution. This is further compounded by the 'echo chamber' effect prevalent on social media, where users are exposed primarily to information that aligns with their existing beliefs and values, making them more susceptible to confirmation bias—a psychological predisposition to interpret new evidence as confirmation of one's existing beliefs or theories [7]. Social engineers exploit this bias by crafting messages that align with the target's expectations and beliefs, thereby increasing the likelihood of the intended deception being accepted. Moreover, the rapid dissemination of information on social media platforms allows for the swift and wide-reaching spread of fraudulent schemes, amplifying their impact. Phishing attacks, misinformation campaigns, and the manipulation of public opinion through fake news are just a few examples of how social engineering leverages psychological vulnerabilities on a mass scale, exploiting the fundamental human need for connection and belonging to propagate deceit and fraud. To counteract these sophisticated tactics, a deep understanding of both the technological and psychological landscapes is crucial [6]. Cybersecurity measures must evolve to include strategies for identifying and mitigating the influence of manipulative content on social media, incorporating advanced algorithms and artificial intelligence to detect indicative of fake accounts or malicious content. Simultaneously, there is a pressing need for digital literacy programs that empower users with the knowledge to critically evaluate online information, recognize potential social engineering attempts, and practice safe online behaviours [8]. This dual approach, blending technological innovation with psychological insight, represents the most effective strategy for safeguarding against the threats of social engineering. Expanding further on the phenomenon of social engineering attacks via social media, it becomes important to understand the mechanisms through which these attacks are detected, a challenge that invites a confluence of cognitive psychology, computational methods, and information theory. The scientific community has been developing theories and frameworks that target the detection and mitigation of such attacks, focusing on the dual aspects of technological innovation and human cognitive resilience [6]. One of the cornerstone theories in detecting social engineering attacks is the Information Processing Theory [9], which provides insight into how individuals seek, process, and memorize information. In the context of social media, this theory can help understanding why certain malicious content or fake accounts might bypass an individual's usual scrutiny. By understanding the cognitive shortcuts, or heuristics, that individuals employ when processing vast amounts of information online, cybersecurity experts can develop more effective educational tools and warning systems that aid in the early detection of potential threats. For instance, incorporating cues within social media platforms that prompt users to critically assess the information before accepting it as true could leverage the System 1 and System 2 thinking processes described by Daniel Kahneman [8], encouraging deeper, more analytical processing of information that is less susceptible to manipulation. Additionally, the theory of Cognitive Load [7] claims that individuals have a limited capacity for processing information, which attackers exploit by overwhelming targets with information or creating a sense of urgency that clouds judgment. Recognizing this, platforms can implement design choices that reduce unnecessary cognitive load, allowing users to focus more acutely on the legitimacy of the interactions they engage in. Techniques such as flagging content that has been identified by other users as suspicious or using algorithms to detect patterns indicative of phishing attempts (such as the rapid creation of new accounts spreading similar messages) can alert users to potential threats, leveraging collective vigilance. Moreover, the application of Machine Learning (ML) and Natural Language Processing (NLP) technologies has been instrumental in automating the detection of social engineering content [3]. These technologies analyze large number of datasets to identify anomalies in account behaviour, linguistic patterns, and network structures that suggest manipulative intent, all while adapting to the evolving tactics of attackers. The integration of ML and NLP not only enhances the scalability of detection efforts but also introduces a level of precision in identifying sophisticated attacks that may bypass traditional detection methods [6].

3 Mitigation Strategies

Cybersecurity Programs
Now moving to the field of combating social engineering, cybersecurity awareness programs play a pivotal role. These programs are designed to educate employees about the variety of cyber threats, with a significant focus on social engineering tactics. Typically, these programs include a range of activities such as informational briefings, regular security updates, and incident reporting protocols. However, the effectiveness of these programs can be mixed. One critique is that while they excel in imparting theoretical knowledge, they often fall short in simulating real-life scenarios [4]. Employees might be able to recognize a phishing attempt in a controlled environment but fail to do so in the actual workplace. Additionally, these programs can sometimes be too generic, lacking customization to the specific threats faced by an organization or industry.

Training Methods
Effective training is crucial in equipping individuals with the skills and knowledge to recognize and respond to social engineering threats. Different training methodologies offer varying degrees of effectiveness:

- **Simulations**: These involve creating scenarios that mimic real-life social engineering attacks. Simulated phishing emails or pretexting calls can be used to test and train employees in a safe environment. The advantage of simulations is their practical nature, providing hands-on experience. However, their success depends on how realistically these simulations mimic actual threats.
- **Workshops**: Interactive workshops can be effective, especially when they encourage participation and discussion. These sessions can be tailored to address specific types of social engineering relevant to the organization. Workshops also provide a platform for sharing experiences and best practices. The limitation, however, lies in their reach – it can be challenging to conduct workshops for all employees, especially in larger organizations.
- **E-Learning**: This method offers flexibility and scalability, allowing employees to complete training at their own pace. E-learning modules can be updated regularly to include the latest information on social engineering tactics. However, the challenge with e-learning is ensuring engagement and retention of information, as it can sometimes become just another box-ticking exercise.

Organizational Culture

The role of organizational culture in promoting or hindering cybersecurity cannot be overstated. A culture that prioritizes security and encourages vigilance can significantly reduce the risk of social engineering attacks. This involves creating an environment where security is everyone's responsibility, not just the IT department's.

Here we list some key aspects of a security-conscious culture include:

- **Open Communication**: Encouraging employees to communicate openly about potential security threats without fear of retribution.
- **Regular Updates**: Keeping everyone informed about new threats and security practices.
- **Rewarding Vigilance**: Recognizing and rewarding employees who identify potential threats can reinforce positive behavior.
- **Leadership Example**: Leaders must exemplify good security practices, as employees often take cues from their superiors.
- **Continual Learning**: Promoting an atmosphere of continual learning and improvement in cybersecurity matters.

4 Measuring Cybersecurity and Privacy Awareness

Measuring the effectiveness of cybersecurity and privacy awareness is a critical aspect of promoting an organization's defences against cyber threats [1]. The process involves several methodologies, each with its unique advantages and challenges. Surveys and questionnaires are essential in assessing employee understanding of cybersecurity practices, these tools can explore valuable insights into knowledge gaps and prevailing misconceptions among staff. Regular distribution of such surveys aids in tracking awareness levels over time, identifying trends, and making necessary adjustments to training programs. Another method is quizzes and tests that complements surveys by calculating the retention and application of cybersecurity knowledge [4]. These assessments can range from straightforward questions about common threats like phishing to complex scenarios requiring deeper understanding. The responses to these tests offer direct insight into the practical awareness levels of employees. Simulated attacks, such as controlled phishing or pretexting attempts, offer a real-world assessment method. By tracking employee reactions to these simulated threats, organizations can gain a clear understanding of how well their staff would perform under actual attack conditions [4]. The rate of response to these simulations is a direct measure of practical awareness and preparedness. Moreover, incident tracking is another critical method, focusing on real security breaches and near-miss events. Analyzing these incidents provides a practical lens through which the vulnerabilities of the workforce can be understood, guiding future training efforts to address these specific weaknesses. The impact of awareness programs is analyzed through several factors. A key indicator is the reduction in security incidents following these initiatives. Observing changes in daily employee practices, such as improved password hygiene or increased consciousness in reporting suspicious activities, also signifies a rise in awareness. Direct employee feedback further provides an insight into the effectiveness and relevance of the training content. However, quantitatively measuring the impact of these programs is difficult with many challenges. The subjectivity in responses to surveys and tests, the variability in the complexity of cyber threats, and maintaining engagement levels over time are significant obstacles. Additionally, cultural and individual differences among employees can affect the absorption and application of cybersecurity training as well as tracking long-term behavioral changes, as these changes are gradual and influenced by a myriad of external factors [6]. In essence, while assessing cybersecurity and privacy awareness is complex, it's an indispensable part of an effective cybersecurity strategy. An approach that combines various quantitative and qualitative methods, tailored to the specific needs and context of the organization, is typically most effective in providing a comprehensive understanding of awareness and preparedness levels [6].

In this table we summarize some of the assessment methods:

Surveys and Questionnaires: These are commonly used to gauge employees' understanding of cybersecurity practices and policies. Well-structured surveys can provide insights into knowledge gaps and misconceptions. They can be distributed periodically to track changes in awareness levels over time.
Quizzes and Tests: Implementing regular quizzes or tests can assess the retention of information provided in training sessions. These assessments can range from basic questions on phishing to more complex scenarios requiring the application of cybersecurity knowledge.
Simulated Attacks: Conducting controlled phishing or pretexting attacks can be an effective way to test how employees would react in real situations. The response rate to these simulated attacks offers a direct measure of their practical awareness.
Incident Tracking: Keeping track of actual security incidents, including near misses, is crucial. Analyzing these incidents helps in understanding the areas where employees are most vulnerable and can guide future training and awareness efforts.

Data Governance Policies

Data governance and policies are fundamental in the fight against social engineering. These policies provide a framework for managing and securing an organization's data, ensuring that it is used responsibly and protected from unauthorized access. Good data governance involves defining who can access data, under what circumstances, and how it should be protected. It also includes policies on data encryption, access control, data classification, and regular audits to ensure compliance. The importance of these policies in protecting against social engineering is based on their ability to limit the exposure of sensitive information. By controlling access to data and monitoring how it is used, organizations can reduce the risk of information falling into the wrong hands and ffurthermore, to help in quickly identifying and responding to breaches, minimizing potential damage.

Enhancing IT Security

While human-focused defenses are crucial, they must be complemented by strong IT infrastructure security to build a comprehensive defense against social engineering threats [5]. Strategies for enhancing IT security include among others:

1. **Multi-Factor Authentication (MFA)**: Implementing MFA adds an additional layer of security, making it harder for attackers to gain unauthorized access even if they have compromised credentials.
2. **Regular Software Updates and Patch Management**: Keeping software and systems up to date is essential in protecting against known vulnerabilities that can be exploited by attackers.
3. **Network Segmentation**: By segmenting networks, organizations can limit the extent of access an attacker can gain from a single point of entry.
4. **Advanced Threat Detection Systems**: Employing systems that use machine learning and artificial intelligence can help in detecting unusual patterns and potential threats more effectively.
5. **Incident Response Planning**: Having a well-defined incident response plan ensures that the organization can react swiftly and effectively in the event of a security breach.

Author Contributions.   **Sokratis Nifakos:** Main author
   Krishna Chandramouli: Reviewer of the final draft provided comments in different sections. He also contributed in the structure of the chapter.
   Natalia Stathakarou: Contributed to the sections related to training and education providing information about different methods in raising awareness.

Conflict of Interest.   The authors declare that the research was conducted in the absence of any commercial or financial relationships that could be construed as a potential conflict of interest.

References

1. Nifakos, S., et al.: Influence of human factors on cyber security within healthcare organisations: a systematic review. Sensors **21**, 5119 (2021). https://doi.org/10.3390/s21155119
2. Shah, M.U., Iqbal, F., Rehman, U., Hung, P.C.K.: A comparative assessment of human factors in cybersecurity: implications for cyber governance. IEEE Access **11**, 87970–87984 (2023). https://doi.org/10.1109/ACCESS.2023.3296580
3. Graf, R., Skopik, F., Whitebloom, K.: A decision support model for situational awareness in national cyber operations centers. In: 2016 International Conference On Cyber Situational Awareness, Data Analytics And Assessment (CyberSA), pp. 1–6. London, UK (2016). https://doi.org/10.1109/CyberSA.2016.7503281
4. Warasart, M., Piriyasurawong, P.: Synthesis of gamified social collaboration via mesh community of practice to enhance cybersecurity awareness. In: 2022 Joint International Conference on Digital Arts, Media and Technology with ECTI Northern Section Conference on Electrical, Electronics, Computer and Telecommunications Engineering (ECTI DAMT & NCON), pp. 359–363.Chiang Rai, Thailand (2022). https://doi.org/10.1109/ECTIDAMTNCON53731.2022.9720416
5. Abu-Amara, F., Tamimi, H.: Cyber shield security awareness program. In: 2021 8th International Conference on Computing for Sustainable Global Development (INDIACom), pp. 422–425. New Delhi, India (2021)
6. Sharif, K.H., Ameen, S.Y.: A review of security awareness approaches with special emphasis on gamification. In: 2020 International Conference on Advanced Science and Engineering (ICOASE), pp. 151–156. Duhok, Iraq, (2020). https://doi.org/10.1109/ICOASE51841.2020.9436595

7. de Jong, T.: Cognitive load theory, educational research, and instructional design: some food for thought. Instr. Sci. **38**, 105–134 (2010). https://doi.org/10.1007/s11251-009-9110-0
8. Tversky, A., Kahneman, D.: Judgment under uncertainty: heuristics and biases. Science **185**, 1124–1131 (1974). https://doi.org/10.1126/science.185.4157.1124
9. Gurbin, T.: Enlivening the machinist perspective: humanising the information processing theory with social and cultural influences. Procedia Soc. Behav. Sci. **197**, 2331–2338 (2015). https://doi.org/10.1016/j.sbspro.2015.07.263. ISSN 1877–0428,

Distributed Ledger And Privacy Enhancement Approaches

An Overview of Security Issues for Blockchain Based Distributed Applications

Konstantinos Papageorgiou[1], Alexandros Fakis[3], Georgios Spathoulas[1,2(✉)], and Athanasios Kakarountas[1]

[1] Department of Computer Science and Biomedical Informatics, University of Thessaly, Lamia, Greece
{kopapageorgiou,kakarountas}@uth.gr
[2] Department of Information Security and Communication Technology, Norwegian University of Science and Technology (NTNU), Gjøvik, Norway
georgios.spathoulas@ntnu.no
[3] Department of Information and Communication Systems Engineering, University of the Aegean, Samos, Mytilene, Greece
alfa@agean.gr

Abstract. Blockchain technology has gained significant attention in recent years due to its potential to revolutionize various industries by providing a decentralized and secure environment for conducting transactions and implementing distributed applications. Decentralized applications (DApps) are based on arbitrary programs (called smart contracts) that can execute a transaction if some specific predefined conditions are met, in order to eliminate the need to rely on trusted third parties. However, as blockchain-based distributed applications continue to proliferate, so do the security challenges associated with this emerging technology. This paper aims to provide an overview of the security issues faced by DApps built on blockchain platforms. An analysis of the security issues for all different layers (smart contract, Dapp, user wallet) was performed. Attacks that take advantage of a single layer or a combination of those are discussed and solutions are proposed for the detection and mitigation of those attacks.

Keywords: blockchain · security · web3 · distributed applications · countermeasures

1 Introduction

In recent years, blockchain technology has garnered significant attention due to its groundbreaking nature, offering innovative solutions to previously unresolved challenges. The increased enthusiasm stems from the growing demand for

---

K. Papageorgiou, A. Fakis, G. Spathoulas and A. Kakarountas—These authors contributed equally to this work.

N. Pitropakis and S. Katsikas (Eds.): *Security and Privacy in Smart Environments*, LNCS 14800, pp. 187–203, 2025.
https://doi.org/10.1007/978-3-031-66708-4_9

decentralized systems, particularly in economic frameworks. Although the integration of blockchain into our daily lives has expanded, it has also brought forth a range of evolving threats. The intricate nature of blockchain technology can be ascribed to its six distinct layers, each contributing to its complexity. As we enter the realm of blockchain threats, it becomes imperative to explore these layers and understand the multifaceted landscape where potential risks may emerge.

Blockchain is a vast network that amalgamates elements of networking, data storage, and applications, creating a decentralized infrastructure that fundamentally transforms the way information is shared, verified, and processed across various entities. A plethora of attacks pervade each distinct layer of the system, each capable of engendering diverse adverse consequences, encompassing financial, privacy, integrity, and trust implications for the affected entities. This chapter outlines the prevailing threats associated with each layer of the blockchain architecture. Subsequently, the corresponding countermeasures are elucidated to proactively mitigate and prevent potential risks.

The rest of this chapter is outlined as follows. Section 2 provides a concise introduction to blockchain technology, elucidating its overarching architecture composed of six distinct layers. In Sect. 3, an in-depth analysis of threats specific to each layer is presented, followed by Sect. 4, which delves into the corresponding countermeasures designed to address and mitigate these risks. The chapter is concluded in Sect. 5.

2   Background

Web3 represents a fundamental change in the way we build the data-driven Internet and is achievable using current technology to achieve immediate advantages. The realization of Web3 is dependent on knowledge creators and data developers embracing Web3 libraries and concepts. At the core of the Web3 framework are essential elements, including blockchain networks characterized by decentralized but interconnected nodes, Web3 libraries, emerging specialized languages like Solidity, identity storage solutions in the form of wallets, smart contracts, and specialized service providers. The adoption of these foundational components is pivotal for the successful transition to a Web3-enabled environment [1].

2.1   Blockchain

Blockchain can be seen as a significant advance beyond distributed database technology. This technology involves a transaction database accessible to various users. In essence, distributed ledger technologies (DLTs) are designed to manage databases by distributing data, and blockchain serves as a potential form of DLT for this purpose [2]. The blockchain serves as a continuously expanding decentralized database that securely records transactions within a network. These transactions are validated and affirmed by the network nodes, forming an immutable ledger. This ledger, resistant to tampering, meticulously logs every transaction, ensuring robust data integrity. Once a transaction is etched into the blockchain, it becomes impervious to alteration or deletion. This intrinsic feature eliminates the need for a third-party authority to oversee and authenticate interactions between participants.

The transparency of the blockchain is a fundamental characteristic, as the ledger is accessible to all nodes that make up the network. This accessibility ensures that every transaction's details are disseminated across the entire network, promoting openness and accountability. Consequently, blockchain technology mitigates the risk of fraud and enhances trust among participants.

A noteworthy aspect of blockchain technology lies in its accommodation of anonymous nodes. This anonymity addresses the need for confidentiality in various scenarios, making it an appropriate solution for situations where privacy is paramount. Participants can perform transactions without revealing their identities, which adds an additional layer of security.

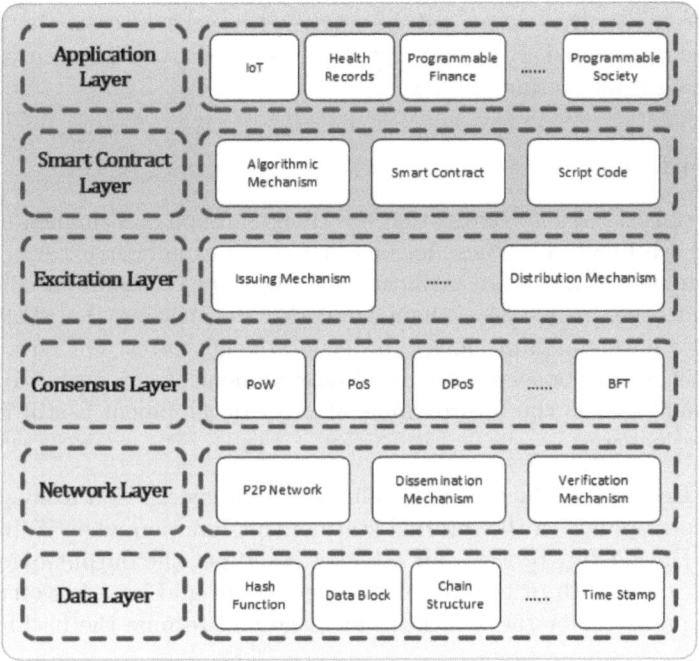

Fig. 1. Blockchain general architecture

Despite its transformative impact on facilitating network transactions and fostering trust in decentralized systems, blockchain encounters technical challenges and limitations that require resolution. Figure 1 depicts the layers of the general architecture of a blockchain network.

- **Data Layer:** In this layer, the time-stamped blocks are generated. At this level, cryptographic techniques, such as hash functions and chain structure, are used to manage the security of these data blocks.

- **Network Layer:** This layer consists of data distribution and validation mechanisms and distributed network mechanisms.
- **Consensus Layer:** This layer provides different consensus algorithms to achieve agreement among the network nodes.
- **Excitation Layer:** Nodes involved in the management of applications in the network receive incentives according to the mechanisms listed in this layer.
- **Contract Layer:** This layer is the backbone of network applications as it provides programs that execute transactions when predefined conditions are met.
- **Application Layer:** This layer concerns the applications that end-users use and interact with the network.

3   Dapp Security Issues

In this section, security issues will be analyzed in each layer of the blockchain as previously presented. Every layer will be analyzed from the security point of view, starting from the bottom up.

3.1   Data Layer

This layer consists of data blocks that are time-stamped and chained using cryptographic hash functions. These hashed blocks are usually structured in Merkle Tree, which allows the secure verification of the data blocks in an efficient way. Bitcoin and other major blockchain networks are utilizing the SHA-256 hash algorithm to hash the data blocks and generate signatures. For now, this algorithm is still secure, but with the rapid development of new technology, it may become insufficient in the future. Some attacks' development is still in progress and is listed below.

- *Length Extension Attack:* The hashing function generates a fixed size output which corresponds to the internal state of this function when it finishes executing. During this process, an attacker could get the output digest and by knowing the length of the initial message, could add an additional input to generate a MAC for the attacker's message and resume the hashing process [3].
- *Quantum Collision Attack:* On a collision attack, a malicious entity could try to find a second input that produces the same hash value. Mendel F. presented an improved algorithm that found a 31-step collision attack on SHA-256 [4]. In the future, with the power of quantum computers, collisions could potentially make SHA-256 insufficient.

3.2   Network Layer

The network layer, also known as the propagation layer, is of significant importance within the blockchain ecosystem, as it is instrumental in maintaining its distributed nature. This layer encompasses the networking methods, message dissemination mechanisms, and authentication procedures employed by the blockchain system. Additionally, crucial functions such as discovery, transaction handling, and block propagation fall under the purview of the network layer. However, due to its inherent characteristics and the pivotal role it assumes in facilitating data transmission, the network layer is vulnerable to a myriad of potential attacks.

- *Gas Exhaustion Denial of Service Attack:* This form of attack is primarily observed within Solidity smart contracts that operate on the Ethereum Virtual Machine (EVM). In these smart contracts, each operation requires a gas fee for completion. While multiple operations can occur within a single block, the cumulative gas of transactions within it is subject to a predefined limit and, when this limit is exceeded, it results in the failure of all pending transactions. Exploiting this limitation, an attacker can initiate multiple transactions simultaneously, each with the highest possible gas price. This strategy is likely to lead to reaching the gas limit of Ethereum blocks, depleting gas resources, and causing the contract to stop operating, thus denying the service to legitimate users [5].
- *Sybil Attack:* A Sybil attack in the context of blockchain [6] can have severe consequences and may lead to various other types of attacks, including DoS, DDoS, majority attacks and mining pool attacks. This form of attack poses a threat to the integrity of decentralized consensus mechanisms, as the attacker could potentially gain control over a substantial portion of network nodes. Such attacks can be particularly damaging in Proof-of-Work (PoW) based blockchains, where an attacker's control over a majority of nodes may lead to a 51% attack. When attempting to connect with the network, other nodes are likely to encounter a controlled majority fork consisting of forged blocks connected to an attacker node. Consequently, the attacker gains the ability to supervise and manage the entire network, resulting in genuine miners being unable to process blockchain transactions effectively. This situation can significantly reduce the system's throughput.
- *Eclipse Attack:* An eclipse attack is a type of attack that can be used to manipulate the consensus process in a blockchain network. In this attack, a malevolent actor strategically acquires a substantial number of IP addresses to exert influence over the consensus process within a blockchain network. The attacker achieves this by isolating a targeted node from the broader network, disrupting its communication with peers through tactics such as flooding the node with excessive traffic or employing network address spoofing. Having successfully isolated the victim node, the attacker gains the opportunity to inject false or misleading information into the node's perception of the blockchain. This manipulation can be exploited to alter the victim's voting power or to coax the victim into accepting invalid transactions. The efficacy of such an attack depends on the ability to exploit the structure of the blockchain network and compromise the integrity of neighboring nodes associated with the victim [7].

3.3   Consensus Layer

To achieve overall agreement on the state of the distributed ledger among network nodes, a consensus mechanism is required [8]. It is the core of every Layer-One. However, these consensus mechanisms have a variety of security issues. Some of the consensus algorithms are presented below, as well as potential security issues that may exist.

3.3.1   Proof-of-Work (POW)

This consensus mechanism is one of the initial consensus algorithms and is used by major blockchain networks, such as Bitcoin and Ethereum. In this protocol, a block is generated by calculating a hash. To calculate the hash, a nonce value must be found. The difficulty in finding that value depends on the criteria that the hash must meet. When the right hash is calculated, the generation of the block is broadcasted among the network nodes. Most of the time, the peer nodes choose the longest chain to expand the generation of new blocks [9]. The process of solving this puzzle requires significant computational power as the difficulty increases.

The security in this consensus mechanism is heavily dependent on the assumption that no more than 50% of the total computational power of the network is used by a single entity. If an entity possesses 51% of the computational power, it can control the network by sustaining the longest chain [10]. This vulnerability allows this entity to execute malicious attacks against the network. Some of the known attacks are presented below.

- *Double-Spending Attack:* If a malicious entity has powerful hashing power, they can use the same cryptocurrency for multiple transactions and create a longer chain of blocks than the actual incorruptible chain [11].
- *Selfish Mining Attack:* In this particular attack, if a malicious miner solves the hash puzzle and generates the new block, he can refuse to broadcast it on the network, and a chain fork will be created. He could then continue to mine the next blocks to create a longer chain. When the network chooses the longest chain that is created by a selfish miner, he can potentially earn block rewards leaving honest miners to a dead end [12].

3.3.2   Proof-of-Stake (PoS)

While the PoW mechanism is based on computational power, this leads to significant energy consumption. The proof-of-stake mechanism resolves this issue by introducing a different approach. In PoS, blockchain participants are referred to as "validators". Validators can stake an amount of their tokens to have a chance to validate new transactions. If the transaction is fraudulent and is validated, the validator not only loses the reward for proper validation, but also loses the amount of tokens that were at stake. Although this approach discourages malicious nodes from attacking the network because it will lead to financial loss, it has some security drawbacks.

- *Long-Range Attack:* This attack aims to take over a blockchain network. If an attacker owns a significant amount of tokens at stake, more than the forging balance, he can start from the Genesis block of the blockchain network and keep persuading nodes to create a new branch of the chain in the attacker's favor, which is identical with the real chain of blocks [13].
- *Nothing-at-Stake Attack:* This attack refers to a situation in which validators have no financial incentive to mine on one branch of a fork during a consensus process. Then, the validators could choose to mine on multiple branches and generate conflicting blocks without the risk of a significant financial loss. This can lead to network instability and security issues [14].
- *Coin Age Accumulation Attack:* In the initial iteration of the protocol, the longer the user's stake remained without any time constraints, the greater the influence it would have. Without limitations, an attacker could amass a significant stake over time, potentially enabling him to gain control of the network. The coin age mechanism served as an amplification method for the stakes held by validators in the system [15].

3.4   Excitation Layer

The Excitation Layer in blockchain refers to a critical component that stimulates network activity and consensus mechanisms. It plays a pivotal role in maintaining blockchain security and integrity by incentivizing participants through rewards, fostering a vibrant ecosystem of validators, and ensuring the robustness of decentralized networks. However, crucial threats, as presented below, have the potential to jeopardize its regular operations.

- *BDoS Attack:* A Blockchain Denial of Service (BDoS) represents an incentive-driven attack, wherein the malevolent actor exploits the incentive mechanism. The nefarious attacker allocates resources to produce a block and selectively reveals only the proof of mining, withholding the publication of the block. The honest miners perceive this action as an advantageous maneuver by the malicious actor, resulting in a decreased incentive for the miners to engage in mining activities. Consequently, as miners abstain from mining, the entire blockchain system may come to a standstill. Incentive-based attacks have the potential to impose a specific sequence of transactions or lead to the omission of certain transactions [16].
- *Pool hopping attack:* The practice of pool hopping involves miners strategically optimizing their profits through the frequent switching of mining pools, a tactic dependent on the prevailing profitability of the pools. Although advantageous for individual miners, the broader repercussions of pool hopping manifest in compromised stability and security within the blockchain network. This maneuver amplifies the susceptibility of blockchain networks to various attacks, particularly those directed at consensus mechanisms like Proof of Work (PoW). The consolidation of mining power within pools, facilitated by pool hopping, heightens the feasibility for malicious entities to assume control over a substantial portion of the network's hash rate. Consequently, this

empowerment exposes the network to potential threats, including, but not limited to, 51% attacks or selfish mining attacks.

3.5 Smart Contract Layer

A smart contract is a program that executes a transaction when certain predefined conditions are met. It carries out the terms of an agreement without the need for someone else to oversee it. This could involve things like transferring money, providing services, etc. Smart contracts are stored in a blockchain network and connected to various payment methods, including cryptocurrencies like bitcoin.

Their main job is to automatically carry out predefined tasks, processes, or transactions based on certain conditions. Like all computer programs, smart contracts can contain vulnerabilities in their code that, if exploited by an attacker, could potentially lead a user to financial loss or compromise their data privacy. Most of the known vulnerabilities are presented below.

- *Dangerous Delegatecall Attack:* Solidity is a programming language used to develop smart contracts. It offers a low-level interface called `delegatecall` that is used to interact with other contracts. A delegatecall execution takes place in the context of the caller contract. When a delegatecall executes from contract A to contract B, it modifies the state of contract A with the usage of the functions from contract B. In this situation, there is a potential risk that if the callee contract is malicious, the caller contract could change its state in a harmful way or the attacker could take over the ownership of the caller contract.
- *Ether Frozen:* This vulnerability concerns smart contracts that accept ether but do not provide a way for users to withdraw it. If the smart contract has a fallback function that does not allow the ether to withdraw, then the ether is locked in the contract indefinitely.
- *Integer Over/Underflow:* An overflow occurs when the value that needs to be stored exceeds the accepted range of the register. The same applies to negative numbers. If the number is smaller than the accepted range, it underflows. This vulnerability could be exploited by an attacker and potentially drain a smart contract of a significant amount of tokens. A prime example of this vulnerability is the BeautyChain contract, in which the attacker managed to bypass some security checks by utilizing the integer overflow and stole large amounts of BEC tokens [17].
- *Re-entrancy Attack:* This is a common vulnerability in smart contracts that allows the execution of untrusted external code. This occurs when a smart contract makes a call to an external contract, which then responds by calling the original contract. This sequence has the potential to cause an infinite cycle. A reentrancy attack leverages this flaw in a smart contract, allowing an attacker to iteratively activate a contract function. This continuous invocation can result in an indefinite cycle, raising the possibility of fund embezzlement.

- *Timestamp Dependency*: This vulnerability occurs when a smart contract relies on the `block.timestamp` function for crucial operations, such as initiating transactions or deriving random numbers necessary for certain processes. The value of the timestamp is decided by the node responsible for producing a block that includes a transaction that executes the smart contract code. The block timestamps are flexible, allowing nodes to accept timestamps within a given timeframe. A malicious entity, particularly miners, might exploit this vulnerability by changing the block timestamps when a smart contract uses block.timestamp for ETH transactions. They can create favorable conditions for their success by altering these timestamps.
- *Unchecked External Call*: The concept of this vulnerability is that the return value of a message call is not checked. In this case, even if the invoked contract throws an exception, the application continues to execute. Whether the call fails accidentally or is deliberately manipulated by an attacker, the resulting program flow can have unintended consequences. There are various low-level call methods in Solidity meant to act on raw addresses, including `call`, `callcode`, `transmit` and the `delegatecall` that was previously mentioned. In particular, these methods do not throw exceptions; instead, when a call meets an exception, they return false.

3.6   Application Layer

The application layer, located at the summit of blockchain technology, serves as the pivotal interface connecting end-users with the underlying software, thereby facilitating communication within the blockchain network. The constituent elements of this layer, including User Interfaces (UI), Application Programming Interfaces (API), and Decentralized Applications (DApps), render it susceptible to diverse forms of attacks.

The compromise of security in numerous trading platforms and user accounts poses a significant jeopardy to the asset security of users utilizing blockchain wallets. In the following, we list several major threats associated with these vulnerabilities.

- *Replay Attack:* A legitimate user initiates a transaction on the blockchain network, including all essential information of the transaction such as the sender's address, the recipient's address and the amount of cryptocurrency being transferred, using the transferProxy function as shown in Listing 1. At a later time, a malicious user may maliciously retransmit identical transaction data to the network for validation. In the absence of protective measures such as a unique transaction identifier or timestamp, validators lack a definitive means to determine whether the transaction has been previously replicated [5].

```
1  function transferProxy(address _from, address _to, uint256
        _value, uint256 _fee)
2  {
3      require(balances[_from] >= _fee + _value);
4      require(balances[_to] + _value >= balances[_to]);
5      require(balances[msg.sender] + _fee >= balances[msg.
          sender]);
6      balances[_to] += _value;
7      balances[msg.sender] += _fee;
8      balances[_from] -= _value + _fee;
9  }
```

Listing 1. Ethereum's transferProxy function

- *False Top-Up Attack:* The outcome of a transaction within a blockchain typically manifests in the form of a status field. This status field functions as a Boolean variable, with true and false values assigned based on the occurrence of an exception during the transaction. How transactions and the potential failure of those are managed varies between exchanges, wallets, and platforms. However, if inadequacies exist in determining a successful token transaction, a security vulnerability known as a false top-up may be exploited. This type of attack has the potential to deplete the entire amount of funds within a blockchain. As an example, in the Ethereum blockchain, when the sender's balance is less than the specified value of the transaction, the system returns a false value instead of triggering an exception [5].

- *Transaction Order Dependence Attack:* The Transaction Order Dependence Attack is essentially a form of race condition exploit. In this scenario, a smart contract offers miners a gas fee as an incentive to solve a mathematical problem before the corresponding transaction is officially recorded on the blockchain. Upon submission of the solution, the transaction enters the mempool, awaiting inclusion in the subsequent block. During this interim period, a malicious user monitoring the mempool can illicitly obtain the transaction data, including the provided solution. The malicious user can then reinitiate the transaction using the same information, but specifying a higher gas price. Typically, transactions with elevated gas fees in the mempool receive prioritization, allowing malicious users to prevail and acquire a higher gas fee than the original. This phenomenon has the potential to introduce various complications within smart contracts [5].

4 Dapp Security Countermeasures

This chapter aims to provide a comprehensive overview of the countermeasures that can be employed in each layer of the blockchain architecture, addressing specific threats and vulnerabilities identified in the previous discussion. By understanding and implementing these countermeasures, stakeholders can contribute to creating a more secure and trustworthy blockchain ecosystem.

4.1   Data Layer

As the potential security issue in this layer comes from the hash function used to produce signatures on the blockchain network, we propose to utilize alternative signature schemes.

- *Aggregate Signature:* An aggregate signature comprises four algorithms: Key-Gen, Sign, Combine, and Verify. The initial two, Key-Gen and Sign, mirror those in a standard digital signature system. The Combine algorithm accepts a vector containing n triples, each composed of a public key pk_i, a message m_i, and a signature σ_i. This algorithm then produces a single signature, denoted σ, which serves as a collective signature that includes all the messages provided. Known as an aggregate signature, σ is the same length as a signature generated for an individual message. Finally, the Verify algorithm takes as input a vector of n pairs, each consisting of a public key pk_i and a message m_i, alongside a singular aggregate signature σ. The output is valid only if σ has been generated as an aggregation of n valid signatures corresponding to the pairs of public keys and messages provided [18].
- *Ring Signature:* This signature algorithm represents a digital signature scheme with distinct characteristics. Unlike traditional signature models, a ring signature eliminates the need for designated managers and consists exclusively of ring members. In the execution of this signature scheme, the signer adopts a randomized approach by selecting public keys from multiple members of the ring. The process involves combining these public keys with the corresponding private keys, along with random numbers and other cryptographic techniques, to generate the signature. Remarkably, the verifier of the signature is only capable of confirming that the signature originates from the specified signature set. The identity of the individual signer remains undisclosed [19].

4.2   Network Layer

Incursions targeting the network layer assume critical significance, as any deviation from normal functioning within the blockchain's network may lead to a complete cessation of network operations.

- *Aggregate Signature:* The safeguarding against gas exhaustion denial-of-service attacks is contingent upon the adherence to best practices by contract developers. Given the varying gas consumption associated with each instruction, identifying which operation incurs the highest gas fee poses a challenge. A notable example is the iterative execution of the "sload" instruction, denoting "storage load," within a loop. To optimize gas utilization and reduce transaction costs for users who interact with the smart contract, it is advisable to minimize storage operations within loop structures. Additionally, the integration of rate-limiting mechanisms presents an alternative strategy. Such mechanisms serve to govern the frequency of specific operations, thus impeding malicious entities from excessively executing resource-intensive operations.

- *Sybil attack:* In the system, each participating node actively monitors the behavior of other nodes, scrutinizing whether any node exclusively forwards blocks from a specific user over an extended period. Such conduct is flagged as malicious, signaling a potential Sybil attack that could disrupt the system's throughput by hindering the verification of genuine user blocks. To counteract this, nodes engaging in such behavior are blacklisted and notifications are disseminated to other nodes, preventing propagation of transaction blocks from Sybil nodes. A secondary countermeasure involves the validation of real-world identities for each participant. By linking digital identification to tangible identifiable individuals, the creation of pseudonymous accounts becomes more challenging. This additional layer of identity verification improves the overall security of the system. Furthermore, social connections and relationships between nodes can serve as a solution to establish trust within the network and isolate potential attackers. Using the social fabric of the interactions between the nodes, the system can create a more resilient defense against Sybil attacks. As a final countermeasure, imposing a minimum economic cost for participation in the network can act as a deterrent for malicious users who attempt to control a significant number of nodes. This economic barrier adds an additional layer of security by discouraging nefarious actors from engaging in Sybil attacks due to the associated financial investment required.
- *Eclipse attack:* While a Sybil attack serves as a precursor for the successful execution of an Eclipse attack, countermeasures designed to thwart Sybil attacks do not necessarily furnish effective safeguards against the latter. This limitation comes from the potential for attackers to manipulate the overlay maintenance algorithm, thereby facilitating an Eclipse attack. One strategy to mitigate the risk of Eclipse attacks involves adjusting parameters such as increasing the maximum number of connections to a node and imposing restrictions on the number of hosts associated with a single IP address, as suggested by Marcus et al. (2018) [20]. Additionally, users are advised not to depend solely on a single node for initial synchronization; instead, they should establish connections with multiple nodes to obtain a more diverse and resilient perspective of the network.

4.3   Consensus Layer

In the previous section, we analyzed two of the most known consensus algorithms (PoW and PoS) and the drawbacks they may have. New consensus algorithms have been developed that improve security and energy consumption.

- *Delegated Proof-of-Stake (DPoS):* The DPoS concept revolves around the idea that nodes with voting power should be able to choose block verifiers, similar to block creators. In other words, stakeholders have the option of delegating the responsibility of creating blocks to selected delegates that they support, eliminating the need for them to create blocks independently. This delegation results in a significant reduction in computational power consumption, potentially down to zero. In the DPoS system, the voting power of shareholders plays a crucial role in achieving consensus through a democratic and equitable process. Unlike creating blocks themselves, stakeholders exercise their influence by supporting delegates who act as block creators. This makes DPoS a highly efficient and cost-effective consensus mechanism compared to the Proof of Stake (PoS) protocol [21].

- *Robust Proof-of-Stake (RPoS):* This is another variation of the PoS protocol that utilizes the number of coins to select miners without giving weight to the age of the coin as PoS does. This way, it eliminates the chance of a coin age accumulation attack, and also, a Nothing-at-Stake attack. The experimental test shows that the RPoS consensus protocol improves PoS security, is energy efficient and can support a significant transaction rate [22].

4.4   Excitation Layer

In this subsection, we delineate countermeasures that address the threats posed to the excitation layer, as presented in the preceding section.

- *BDoS Attack:* In a constructive approach, we posit several potential countermeasures to mitigate BDoS attacks. Initially, honest miners can opt for non-attacker blocks within a fork by employing a heuristic time-based detection system. Subsequently, the adoption of alternative reward mechanisms, as outlined in previous studies [23,24], can serve to compensate miners for lost races, thus rendering BDoS ineffective. It is important to note that, while these measures are effective, additional strategies, such as the implementation of advanced cryptographic techniques or decentralized consensus protocols, could further strengthen the resilience of blockchain networks against BDoS attacks [25].
- *Pool hopping attack:* Singh et al. [26] introduced a model based on smart contracts as a preventive measure against pool hopping attacks. In this model, miners are required to submit a deposit in the form of coins, and the deposit amount increases proportionally with each instance of pool hopping. This design incentivizes miners to remain within a specific pool, as leaving from the mining pool would entail a substantial loss of coins due to the increased deposit associated with increased hopping occurrences.

4.5   Smart Contract Layer

The vulnerabilities described in the previous section can be devastating and compromise the user's privacy or lead to financial loss. To address these issues, we provide some potential solutions to prevent them.

- *Dangerous Delegatecall:* This vulnerability can be avoided by employing a "stateless library". In other words, this refers to a library contract that exclusively provides pure or view functions and refrains from altering the state in client contracts. If the library contract lacks functions that modify the state, potential attackers are unable to misuse delegatecall to take control of the contract.

- *Ether Frozen:* In order to mitigate the risk of the Ether Frozen vulnerability, developers should carefully incorporate Withdrawal Mechanisms into their system. These mechanisms should have functions such as `transfer` or `send`. Additionally, conducting detailed code testing and audits is crucial to uncover and address any coding vulnerabilities. Developers should also be mindful of maintaining a balance between incoming and outgoing Ether traffic. The fallback function manages incoming traffic, while the functions `call`, `send` and `transfer` are responsible for handling outbound traffic.
- *Integer Over/Underflow:* To prevent overflow and underflow attacks, programmers can incorporate the SafeMath library into their code. It is essential to utilize secure math libraries or suitable data types that include overflow detection. The SafeMath library not only supports fundamental arithmetic operations, but also verifies preconditions and postconditions to ascertain if an overflow has taken place. If an error is detected, the library rejects the transaction and marks the transaction status as 'Reverted'.
- *Re-entrancy Attack:* To safeguard against re-entrancy attacks, smart contract developers employ various strategies. One approach involves the use of a mutex lock, which is activated when a function is called, preventing multiple simultaneous calls to the same function. Subsequent calls to the function are halted until the lock is released. Another protective measure is the implementation of a guard condition, acting as a flag set prior to external function calls and checked afterward. If the flag is raised, the contract refrains from executing the external call, thereby preventing re-entrancy. Moreover, developers can monitor the depth of the call stack to ensure that the contract is not invoked recursively. If the depth exceeds a predefined threshold, the contract terminates its execution. The "require" statement is also employed to scrutinize the contract's state before permitting the execution of a function. Finally, developers can update the state variables within the smart contract before initiating external functions or interacting with external contracts.
- *Timestamp Dependency:* Avoiding the vulnerability related to timestamp dependence is fairly straightforward. Employ a strong algorithm to generate random numbers. When dealing with functions that rely on time-triggered data, evaluate whether the time-dependent event's scale can fluctuate within a 15-second range while still preserving integrity. If this is the case, using a block.timestamp for that function may be considered secure.
- *Unchecked External Call:* To address this flow, developers have the option to employ the `transfer` function instead of `send`. This is advantageous because `transfer` will undo any changes if the external transaction encounters an issue. If developers opt for low-level call methods, it is essential that they manage the potential failure of the call by verifying the return value.

4.6   Application Layer

In the preceding section, an exposition on prevalent vulnerabilities within the application layer of the blockchain ecosystem was provided. The ensuing subsection delves into a comprehensive examination of strategies designed to avert diverse forms of attack within the framework of blockchain and smart contract development.

- *Replay Attacks:* The prevention of replay attacks can be achieved through various measures. First, one solution entails abstaining from utilizing the transferProxy function and opting for a more secure signature method. Another strategy involves incorporating variables such as nonce and timestamp to deter repetitive invocation of the function. Additionally, including the contract address in the keccak256 algorithm serves as a preventive measure against replay attacks, since the signature remains valid exclusively for the specified contract. Lastly, a widely adopted practice is to include the chain ID in the transaction data to thwart replay attacks across different blockchain networks. This ensures that a transaction designed for a specific blockchain cannot be legitimately replayed on an alternative blockchain.
- *False top-up attacks:* False top-up attacks often involve manipulating event logs or employing other deceptive mechanisms to mislead a system into falsely recognizing a successful top-up of a wallet or account. Routine monitoring of event logs is a prevalent practice in smart contract development to track specific actions or transactions. In the context of top-up operations, it is paramount to verify that the event logs accurately reflect the anticipated changes in balances. This verification process serves as a crucial step in detecting any disparities between the expected and actual outcomes of a transaction. However, it is imperative to acknowledge the inherent risks associated with potential manipulation of event logs, particularly within the realm of malicious smart contracts. To address this vulnerability, implementing checks and assertions within the smart contract becomes a prudent practice. These measures serve to ensure the normal behavior of the contract, and in the event that specified conditions are not met, an exception can be triggered. This mechanism effectively interrupts the execution of subsequent instructions, acting as a safeguard against false top-up attempts. This proactive approach enhances the security of the smart contract, fortifying it against potential exploitation and manipulation.
- *Transaction order dependence attacks:* Transaction order dependence attacks, commonly known as "front-running" attacks, typically stem from vulnerabilities within the blockchain ecosystem itself, particularly notable in Ethereum contracts. Addressing such attacks involves employing techniques that increase the complexity for attackers to discern the true intent of a transaction. This may include strategies such as obfuscating transactions, concealing them as internal transactions, or using other privacy enhancement mechanisms.

Another viable approach to mitigate front-running is the implementation of off-chain solutions. These solutions facilitate faster execution and diminish the visibility of transactions until they are settled on-chain. By doing so, the impact of front-running attacks is reduced. Furthermore, the adoption of advanced cryptographic signature schemes or zero-knowledge proofs represents a sophisticated method to improve transaction security and conceal transaction details until confirmation. This adds an additional layer of protection against front-running attacks, ensuring a higher level of confidentiality and integrity in transaction processing.

5 Conclusion

The protection of the various layers of a blockchain architecture is paramount to ensure a secure and resilient blockchain ecosystem. This comprehensive overview has highlighted specific security issues within each layer and proposed effective countermeasures to mitigate potential threats and vulnerabilities. Every blockchain must be carefully built, as decisions may be irreversible. Any potential security issue can compromise the integrity of the blockchain network, lead to financial loss, or leak user's private data. By embracing and implementing these countermeasures across all layers, stakeholders can actively contribute to creating a blockchain ecosystem that is not only secure but also trustworthy.

References

1. Sheridan, D., Harris, J., Wear, F., Cowell Jr, J., Wong, E., Yazdinejad, A.: Web3 challenges and opportunities for the market. arXiv preprint arXiv:2209.02446 (2022)
2. Belotti, M., Božić, N., Pujolle, G., Secci, S.: A vademecum on blockchain technologies: when, which, and how. IEEE Commun. Surv. Tutor. **21**(4), 3796–3838 (2019)
3. Cortez, D.M.A., Sison, A.M., Medina, R.P.: Cryptographic randomness test of the modified hashing function of sha256 to address length extension attack. In: Proceedings of the 2020 8th International Conference on Communications and Broadband Networking, pp. 24–28 (2020)
4. Johansson, T., Nguyen, P.Q. (eds.): EUROCRYPT 2013. LNCS, vol. 7881. Springer, Heidelberg (2013). https://doi.org/10.1007/978-3-642-38348-9
5. Duan, L., Sun, Y., Zhang, K., Ding, Y., et al.: Multiple-layer security threats on the ethereum blockchain and their countermeasures. Secur. Commun. Netw. **2022** (2022)
6. Douceur, J.R.: The Sybil attack. In: Druschel, P., Kaashoek, F., Rowstron, A. (eds.) IPTPS 2002. LNCS, vol. 2429, pp. 251–260. Springer, Heidelberg (2002). https://doi.org/10.1007/3-540-45748-8_24
7. Heilman, E., Kendler, A., Zohar, A., Goldberg, S.: Eclipse attacks on Bitcoin's peer-to-peer network. In: 24th USENIX Security Symposium (USENIX Security 15), pp. 129–144 (2015)
8. Bach, L.M., Mihaljevic, B., Zagar, M.: Comparative analysis of blockchain consensus algorithms. In: 2018 41st International Convention on Information and Communication Technology, Electronics and Microelectronics (MIPRO), pp.1545–1550. IEEE (2018)
9. Yang, X., Chen, Y., Chen, X.: Effective scheme against 51% attack on proof-ofwork blockchain with history weighted information. In: 2019 IEEE International Conference on Blockchain (Blockchain), pp. 261–265. IEEE (2019)

10. Gervais, A., Karame, G.O., Wüst, K., Glykantzis, V., Ritzdorf, H., Capkun, S.: On the security and performance of proof of work blockchains. In: Proceedings of the 2016 ACM SIGSAC Conference on Computer and Communications Security, pp. 3–16 (2016)
11. Tosh, D.K., Shetty, S., Liang, X., Kamhoua, C.A., Kwiat, K.A., Njilla, L.: Security implications of blockchain cloud with analysis of block withholding attack. In: 2017 17th IEEE/ACM International Symposium on Cluster, Cloud and Grid Computing (CCGRID), pp. 458–467. IEEE (2017)
12. Bonneau, J., Preibusch, S., Anderson, R.: Financial Cryptography and Data Security. Springer, Cham (2020). https://doi.org/10.1007/978-3-031-47754-6
13. Azouvi, S., Danezis, G., Nikolaenko, V.: Winkle: foiling long-range attacks in proof-of-stake systems. In: Proceedings of the 2nd ACM Conference on Advances in Financial Technologies, pp. 189–201 (2020)
14. Garcia-Alfaro, J., Navarro-Arribas, G., Hartenstein, H., Herrera-Joancomartí, J.: Data Privacy Management, Cryptocurrencies and Blockchain Technology: ESORICS 2017 International Workshops, DPM 2017 and CBT 2017, Oslo, Norway, 14–15 September 2017, Proceedings, vol. 10436. Springer, Cham (2017). https://doi.org/10.1007/978-3-031-25734-6
15. Deirmentzoglou, E., Papakyriakopoulos, G., Patsakis, C.: A survey on long-range attacks for proof of stake protocols. IEEE Access 7, 28712–28725 (2019)
16. Mirkin, M., Ji, Y., Pang, J., Klages-Mundt, A., Eyal, I., Juels, A.: BDoS: blockchain denial-of-service. In: Proceedings of the 2020 ACM SIGSAC Conference on Computer and Communications Security, pp. 601–619 (2020)
17. ImmuneBytes: Explained: Overflow And Underflow Vulnerability in Smart Contracts - ImmuneBytes (2023). https://www.immunebytes.com/blog/explained-overflow-and-underflow-vulnerability-in-smart-contracts/
18. Boneh, D.: In: Tilborg, H.C.A., Jajodia, S. (eds.) Aggregate Signatures, p. 27. Springer, Boston, MA (2011). https://doi.org/10.1007/978-1-4419-5906-5_139
19. Li, X., Mei, Y., Gong, J., Xiang, F., Sun, Z.: A blockchain privacy protection scheme based on ring signature. IEEE Access 8, 76765–76772 (2020)
20. Marcus, Y., Heilman, E., Goldberg, S.: Low-resource eclipse attacks on ethereum's peer-to-peer network. Cryptology ePrint Archive (2018)
21. Saad, S.M.S., Radzi, R.Z.R.M.: Comparative review of the blockchain consensus algorithm between proof of stake (pos) and delegated proof of stake (dpos). Int. J. Innov. Comput. 10(2) (2020)
22. Li, A., Wei, X., He, Z.: Robust proof of stake: a new consensus protocol for sustainable blockchain systems. Sustainability 12(7), 2824 (2020)
23. Buterin, V., et al.: A next-generation smart contract and decentralized application platform. White Pap. 3(37), 2–1 (2014)
24. Wood, E.: A secure decentralised generalised transaction ledger, ethereum proj. Yellow Pap. 151(1) (2014)
25. Habib, M.A., Manik, M.M.H.: A technique to avoid blockchain denial of service (bdos) and selfish mining attack. arXiv preprint arXiv:2310.19170 (2023)
26. Singh, S.K., Salim, M.M., Cho, M., Cha, J., Pan, Y., Park, J.H.: Smart contract-based pool hopping attack prevention for blockchain networks. Symmetry 11(7), 941 (2019)

A Critical View on Blockchain Rollups

Angeliki Katsika[1], Lydia Negka[3], Georgios Spathoulas[1,2(✉)],
and Vassilis Plagianakos[1]

[1] Department of Computer Science and Biomedical Informatics,
University of Thessaly, 2-4 Papasiopoulou Street, 35131 Lamia, Greece
{akatsika,vpp}@uth.gr
[2] Department of Information Security and Communication Technology,
Norwegian University of Science and Technology (NTNU), Mail Box 191,
2815 Gjøvik, Norway
georgios.spathoulas@ntnu.no
[3] School of Forestry and Natural Environment, Aristotle University of Thessaloniki,
54124 Thessaloniki, Greece
lnegka@for.auth.gr

Abstract. Improving blockchain scalability is a pressing issue due to
increasing demand, leading to network overload and higher user fees,
which subsequently hinder widespread adoption. To address this challenge, several strategies have been suggested as solutions to the scalability trilemma, involving enhancements to the foundational layer, such as
sharding, and the utilization of off-chain techniques such as sidechains,
state channels, and rollups. The scalability trilemma arises from the
inherent trade-offs faced by Layer 1 blockchains, necessitating a delicate balance between decentralization and security. Ethereum, recognizing the limitations in its foundational layer, has strategically adopted a
roadmap with a focus on rollups to overcome scalability issues, acknowledging the necessity for enhancements in both data and computational
aspects. Blockchain rollups are increasingly seen as a promising answer
to the scalability issues encountered by blockchain technology, as they
offer a solution that can significantly increase the capacity and speed
of existing networks, while also providing a high degree of security and
flexibility. This chapter aims to comprehensively explore the complexities of blockchain rollups, examining their characteristics, functions, and
their impact on blockchain scalability. Through a detailed analysis, we
seek to illuminate how rollups solutions operate and distinguish the features that characterize their contribution to the progression of blockchain
technology.

Keywords: rollups · Layer 2 · scalability · blockchain · zero knowledge

1 Introduction

The introduction of Zero-Knowledge (ZK) rollups marks a substantial advancement in blockchain technology, introducing an innovative method that

---

A. Katsika and L. Negka—These authors contributed equally to this work.

© The Author(s), under exclusive license to Springer Nature Switzerland AG 2025
N. Pitropakis and S. Katsikas (Eds.): *Security and Privacy in Smart
Environments*, LNCS 14800, pp. 204–239, 2025.
https://doi.org/10.1007/978-3-031-66708-4_10

consolidates multiple transactions into a unified 'package.' This consolidation not only enhances efficiency and finality but also promises to reduce transaction fees, a crucial concern in blockchain ecosystems. In essence, the rollup process, whether optimistic or zero-knowledge, necessitates executing transactions on Layer 2 (L2), where they are grouped into batches by an aggregator or sequencer. Following this consolidation, the compiled information is compressed before being transmitted to the primary blockchain. On the main blockchain, these batches are processed as unified entities, constituting a single exchange embedded within a block.

This architectural design aims to improve the transaction per second capacity of Layer 1 (L1) blockchains. Central to understanding rollup mechanisms is the role of smart contracts on the main blockchain, functioning as repositories for received batches. These contracts store and verify data, guaranteeing the unchangeable recording of valid state changes. The immutability of these changes is crucial, as alterations in state directly correlate with updates in the wallet balances connected to transactions. Moreover, the cryptographic primitives integrated into the rollup structure are of significant importance. They play a pivotal role in generating proofs, validating transactions, reconstructing the transaction history, and ensuring the accuracy of the final state.

The following Sections explore the features and mechanisms of rollups, aiming to provide a clear understanding of how rollups address scalability issues in blockchain networks. This exploration will illuminate both the advantages and obstacles associated with rollup adoption, facilitating further discussion on improving blockchain scalability. As efforts for improving public blockchains scalability are mainly focusing on rollups related approaches, the present chapter aims at providing a consolidated view on what has been achieved, what are the main issues to be resolves and pave the way for future research efforts.

This chapter begins with background information in Sect. 2 to enhance the reader's understanding of blockchain scaling solutions, where a brief overview of prominent Layer 2 scaling solutions is presented. Sections 3 and 4 introduce the main categories of Rollups: Zero-Knowledge Rollups and Optimistic Rollups, respectively. Each section presents in the corresponding subsections the high-level description of the rollups Sects. 3.1, 4.1, the main functionality Sects. 3.2, 4.2 that governs them and the main security mechanisms 3.3, 4.3 that are implemented in each mechanism. In Subsects. 3.4, 4.4 the compatibility of the EVM (Ethereum Virtual Machine) is discussed for each rollup, while in Subsects. 3.6, 4.5 the prominent protocols developed are presented. In the last Sect. 5, several issues related to the maturity of rollup mechanisms, the anticipated effects of their implementation, and the role of the rollup operator are highlighted and briefly addressed. In addition, a brief comparative analysis is also presented to complete this overview. Finally, in Sect. 6 the chapter underscores the imperative for further exploration of the efficiency and applicability of rollups.

2   Background

Addressing the issue of blockchain scalability has attracted considerable interest from various research teams, leading to the development of multiple solutions, that can be categorised mainly into two categories, first-layer ($L1$), and second-layer ($L2$) scaling solutions [1]. These categories represent distinct approaches to improve the scalability of blockchain networks, with the first focusing mainly on optimizing the core parameters of the blockchain itself, such as adjusting the underlying consensus mechanism [2] and increasing the transaction processing capacity on the chain, while the solutions of the second operate independently from the core blockchain infrastructure but collaborate with it. In this section, before we delve into the specifics of rollups, we aim to enhance our comprehension of scaling solutions by presenting a brief overview of other prominent ($L2$) scaling strategies, including state channels, sidechains, Plasma, and validiums.

2.1 State Channels

State channels provide a mechanism for a predefined group of participants to secure funds, conduct transactions by mutually authorizing new states within the channel, and ultimately submit the collectively agreed final state on the underlying blockchain. State channels facilitate numerous off-chain state transitions between users while minimizing on-chain transactions, typically requiring only two finalizations on the blockchain [3]. Specifically, payment channels, a subset of state channels, constrain state transitions to payment transactions, focusing on changes in the account balance or associated UTXOs. To record a payment, the parties involved adjust their balances, mutually sign it, and uniquely identify it as the latest agreed state [4,5].

However, state channels, while offering benefits in terms of off-chain state transitions and minimized on-chain transactions, come with their set of limitations [6]. One notable challenge is the requirement for constant online presence from users to facilitate transaction distribution through the network and promptly signal any potential misbehavior from other participants. Furthermore, state channels, particularly payment channels, face capacity issues, restricting the amount that parties can either receive or send. Additionally, the limited accessibility of state channels, confined to specific parties, restricts their application in scenarios where broader participation in smart contracts is desired. These limitations underscore the need for careful consideration of use cases and the acknowledgment that state channels may not be universally suitable for all scenarios.

2.2 Sidechains

A sidechain functions as an independent blockchain, running autonomously from Ethereum, with a two-way bridge connecting it to the Mainnet. Sidechains exhibit distinct block parameters and consensus algorithms designed for efficient transaction processing. However, trade-offs are inherent in their use, as sidechains do not inherit Ethereum's security properties, and unlike Layer 2 scaling solutions, they do not transmit state changes and transaction data back to Ethereum Mainnet. Sidechains consist of validating nodes that cover transaction verification, block production, and storage of the blockchain state. An important feature is their autonomy in selecting consensus algorithms, which can include proof of authority, Delegated Proof of Stake, or Byzantine Fault Tolerance. Interoperability between sidechains and Ethereum is facilitated through blockchain bridges [7]. These bridges utilize smart contracts deployed on Ethereum Mainnet and the sidechain to govern the transfer of funds between them. It is crucial to note that while bridges aid in moving funds, the assets are not physically transferred across the chains. Instead, mechanisms such as minting and burning are used for seamless transfer of value between Ethereum and the sidechain. Nevertheless, the adoption of sidechains involves substantial trade-offs. Each sidechain independently manages its security, lacking the inherent security properties of Ethereum, and consequently introducing the potential for malicious activities.

2.3   Plasma

Plasma originated as a solution to harness certain advantages of sidechains while emphasizing on the safety of assets stored on the sidechain. By conducting transactions off the Mainnet, Plasma optimizes for speed and cost, utilizing a single operator to oversee transaction ordering and execution. However, to benefit from Ethereum's security guarantees, off-chain transactions are settled on the Ethereum Mainnet, and this necessitates the introduction of state commitments, presented as Merkle roots. These commitments serve as periodic snapshots of the Plasma chain's state, ensuring security without overwhelming Ethereum's processing capacity. At its core, plasma operates on the principle that, in case of a security lapse on the sidechain, all user assets can reliably revert to the root chain, while essential features such as cost-effective transactions and security are preserved [8].

The procedures of entering and exiting Plasma are facilitated by a contract on Ethereum. This contract manages entries and exits, tracks state commitments, and addresses dishonest behavior through fraud proofs. Any user that wants to enter the Plasma chain deposits ETH or ERC-20 tokens into the Plasma contract, which the operator then recreates on the Plasma chain. For exiting the Plasma chain, a challenge period is implemented, during which any user can challenge a withdrawal request using a fraud-proof, ensuring the legitimacy of claims. The withdrawal process also involves a bond as a guarantee of honest behavior, and if completed within the challenge period, users can retrieve their deposits from the Plasma contract on Ethereum.

2.4   Validiums

Validiums aim to enhance Ethereum Mainnet throughput by executing transactions off-chain, similar to zero-knowledge rollups, using validity proofs, such as ZK-SNARKs or ZK-STARKs, to verify off-chain transactions on Ethereum [9]. Validiums function on Ethereum through a verifier contract, and a main contract, relying on Ethereum for settlement guarantees and security. Deposits and withdrawals involve interactions between the on-chain and off-chain components as typically users submit transactions to an operator, who batches and proves them, updating the state on Ethereum, thus ensuring security and trustlessness. The validity proof for a batch of transactions determines the state update, but unlike ZK-Rollups, validiums don't publish transaction data on Ethereum, offering a purely off-chain scaling approach. Funds on validiums are managed by a smart contract on Ethereum, offering near-instant withdrawals. However, users may face frozen funds and restricted withdrawals if data availability managers withhold off-chain data. This disadvantage of validiums, a data availability challenge, is addressed by decentralizing storage or using data availability committees, each approach presenting trade-offs in terms of trust and security.

3   Zero-Knowledge Rollups

3.1   High Level Description

Zero-Knowledge Rollups, also known as ZK-Rollups or Validity Rollups, are one of the two approaches to the rollup scaling solution. The basic premise is that of batching transactions so that a large number of off-chain transactions being served by only one transaction mined on-chain. ZK-Rollups make use of validity proofs to prevent fraud without the need for dispute periods and manual actions. Those validity proofs guarantee that the transaction data published on-chain is correct. The main benefit is that this happens in a zero-knowledge environment, therefore the proof verifier does not need to repeat the computation or be aware of the data that went into it.

The sequence of transactions executed on a rollup is immutable after being verified on-chain, and this immutable track forms the rollup chain. ZK-Rollups significantly compress transaction data to take up less storage. Said data are written on Ethereum as calldata which does not affect the global state, though it is apparent on Ethereum's logs. Both practices achieve reduced fees during the use of the rollup.

Architecture: The basic components going into the ZK-Rollup architecture are the following [10], and depicted in Fig. 2:

- Layer 1
  - Main Contract: Rollup management smart contract deployed on Layer 1, responsible for monitoring rollup state updates, storing rollup blocks and tracking user deposits to the rollup.
  - Verifier Contract: Smart contract deployed on L1, responsible for verifying the validity proofs corresponding to the transaction batches that are posted on Layer 1.
- Layer 2
  - Off-chain Virtual Machine: Execution machine for the off-chain transactions that will be going into the rollup. Its actions are the ones verified by the validity proofs.
  - Rollup users: Actors that make use of the rollup, namely they submit signed transactions for inclusion in the rollup chain.

- Middlemen
    - Operators: Operators are the key component of rollup functionality. They are L2 nodes that operate as the connection between L1 and L2 rollup components and are responsible for a number of critical actions, such as:
        * receiving transactions from users
        * making initial correctness checks
        * including transactions in batches
        * submitting the changes resulting from those batches to the on-chain contracts
        * submitting the validity proofs relating to those changes to the contracts

      Operators are also referred to as Sequencers, though that term is most usually used to indicate that the operators function in a centralised manner.

Dependencies: ZK-Rollups and rollups in general fall under the category of hybrid scalability solutions. This means that while ZK-Rollups operate independently from the main chain, they do depend on it for a number of, usually security related, features [10]. Most notable out of those are:

1. **Transaction Finality.** Finalisation of the transactions included in a batch that has been submitted to the rollup entirely depends on the Validation Contract being able to verify the validity proof provided along with the batch. This is necessary to protect rollup users and their funds from malicious operator behavior.
2. **Data Availability.** It is important that all transaction data is viewable by any individual. Firstly, users rely on that information to be informed of their balance or perform actions like withdrawing their funds from the rollup. Additionally, users that may wish to verify the state of the rollup independently must be able to do so, as an additional countermeasure against malicious operators.
3. **Censorship/Freezing resistance.** Operators perform many actions crucial to rollup functionality, which on one hand increases the efficiency of the design but as a trade-off, gives them the power to refuse transactions and effectively sensor users, or to stop submitting blocks and batches and freeze the rollup chain. For a more decentralised design, the ZK-Rollup can allow any user to submit their transactions directly on the rollup management contract to combat censorship and freezing.

3.2   Basic Functionality

Entering: For a user to enter the rollup, they must make a deposit to the rollup management contract. Those deposit transactions are not exempt from the batching process, therefore the deposit request joins an entry deposit queue. The operator can then include it in a batch and submit it to the management

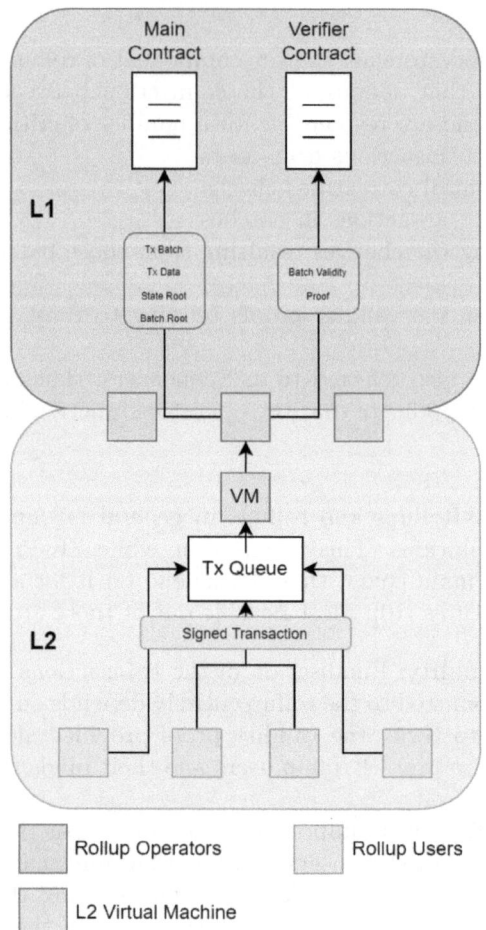

Fig. 1. Overview of ZK-Rollup main components

contract. There is a delay to this process since before the deposit transaction is submitted, the operator must wait for several deposit requests to join the queue and then submit them together.

State Updates: The basic workflow of a ZK-Rollup state update is broken down into the following steps (Fig. 1):

1. The operator responsible for the next transaction batch is selected. Several schemes can govern that selection, with a stake-based approach where the operator with the largest stake is chosen being one example.
2. User signs and submits a transaction to the operator for inclusion in a rollup batch
3. The operator processes the transaction and performs a number of checks on it:

- Ensures the sender and receiver of the transaction are valid entities
- Ensures the sender has enough funds for the submitted transaction
- Ensures the transaction is a match with the sender's rollup public key
- Other checks regarding sender information like their nonce etc.

4. Once enough valid transactions have been accumulated, the operator moves on to processing and including them in a batch.
5. The operator generates the corresponding validity proof for the transaction batch. This process is more extensively analysed in Sect. 3.3.
6. The operator computes the new state as it will ensue from the transactions in the batch. The new state is submitted on-chain and the transaction data is passed as calldata in a compressed form during the call to the contract.
7. The operator computes the batch root that will allow users to prove the existence of their transaction in a given batch.
8. The operator submits the corresponding validity proof.
9. The validity proof is verified by the smart contract. If and only if the verification is successful, the new state becomes the state of the ZK-Rollup and the transaction batch is finalised.

Balance Verification: A user who has already joined the rollup must be able to check their balance. To this end, they must submit to the contract:

- Their account data in the form of a hash
- A Merkle proof that will be successfully verified against the current state root of the ZK-Rollup

Proving Transaction Inclusion: There are built-in mechanisms in ZK-Rollups that allow users to prove the inclusion of a particular transaction in a particular batch. The elements provided to the smart contract are the following:

- The transaction details
- The batch root as calculated and submitted by the operator at the time of submission of the batch
- A Merkle proof that will successfully verify the inclusion of said transaction in the batch

The smart contract aims to check if the hash of that transaction data exists within the batch root provided, and if said batch root is valid.

Exiting: To initiate their exit from the ZK-Rollup, a user must first send all their rollup assets to a designated account for burning. Following that action, the user must wait for an operator to include that burning transaction in a transaction batch, and then the process can move on. In the next and final stage, the user must follow the process of proving the inclusion of a transaction in a batch for the burning transaction. They must also provide the contract with the Layer 1 account that will receive the assets previously deposited in the rollup.

As soon as the exit transaction is executed, the user receives their funds on L1 and has effectively left the ZK-Rollup.

3.3   Security: ZK-Proofs

A variety of different approaches exist to procure a zero-knowledge proof. In terms of ZK-Rollups, an important characteristic seems to be succinctness. A succinct proof is one whose verification resources needed (size and time) grow much slower than the resources needed for its computation. More plainly, it is necessary to have a process where the verifier does not need to replicate every step of the computation to be able to validate the proof [11]. This is a trait shared by all the alternatives for ZK-proofs used by ZK-Rollups since it is necessary to limit the computational load on L1.

Going into depth about how each of these variants operate is out of the scope of this work. A description sufficient for a high level evaluation and comparison will be provided.

SNARKs: SNARKs, or Zero-Knowledge Succinct Non-Interactive Argument of Knowledge, are the most commonly used variant of ZK-proofs used in ZK-Rollups. First introduced in 2012 [12], they have had a significant head start over other alternatives and they boast a richer community with more resources that facilitate their use.

SNARK proofs besides being succinct, are also non-interactive, requiring no back and forth communication between the verifier and the generator of the proof. Their security mechanism is based on elliptic curves, and they are not deemed to be quantum-secure [13].

SNARKs make use of a Common Reference String (CRS) to provide the parameters for proving and verifying the proofs. The information used to create those parameters cannot be compromised, or the security of the system falls apart since the generation of false proofs will be feasible. It is common in ZK-Rollups to use multi-party computation ceremonies to get around this weakness [14]. This, however, involves a trusted setup where at least one individual must behave correctly to maintain security.

Besides existing longer, SNARKs are also preferred for the small size of their proofs. This directly correlates to gas fees for their validation, and they are estimated to use as little as 24/100 of the gas that STARKs, a different ZK-proof scheme, use.

SNARKs make use of what is called a quadratic arithmetic program (QAP). QAP allows the creation of a polynomial representing the computation performed by the prover, and this in turn allows the verifier to (a) be convinced that the prover correctly executed the computation without being aware of the input or solution and (b) evaluate the polynomial at certain points to efficiently ensure the validity of the proof without knowing the details of the computation.

STARKs: A more recently coined term, STARKs, or Zero-Knowledge Scalable Transparent Argument of Knowledge [15] were an alternative created to potentially replace SNARKs. With a security mechanism based on hash functions, boasting quantum resistance, and no need for a trusted setup, they do carry several benefits compared to SNARKs. However, they come with much larger proof sizes which lead to increased time and effort to verify, and therefore considerably higher fees when those computations are performed on-chain. This drawback is large enough to hinder their adoption, and combined with the relative lack of documentation and developer support, they are currently much less frequently used than SNARKs in the context of ZK-Rollups.

STARKs also aim to represent the computation in a way that will allow a verifier to efficiently validate its integrity. The representation is done through what is called Polynomial IOP [16], an interactive oracle proof where the oracles respond to low degree polynomials. Because this setup is not usable, FRI [17] is used to convert the Polynomial IOP to an interactive concrete proof system. The final step is performing a Fiat-Shamir transform [18] to procure a STARK.

SNORKs: Succinct Non-Interactive Oecumenical Argument of Knowledge [19], or SNORKs, are an adaptation of SNARKs that allow for the required trusted setup to be universal and updatable and therefore be used for several applications.

Comparison: In Table 1 comparison can be seen between the different ZK-proof variants across many different properties.

Table 1. Comparison of properties between different Zero Knowledge Proofs

Property	STARKs	SNARKs	SNORKs
Trustlessness	✗	✓	✓
Quantum-Resistant Security	✓	✗	✗
Universality	✗	✗	✓
Interactivity	✓	✗	✗
Proof Size	Large	Small	Small

Proof Generation Process: The operator is responsible for generating the ZK-proof corresponding to the transaction batch that will be submitted to the rollup. This process is repeated for every batch.

Once there are enough accepted transactions to be aggregated, the operator then creates the following:

- A Merkle tree for all the transactions to be included in the batch
- The Merkle proofs necessary for users to prove inclusion of a transaction in the batch
- Merkle proofs that testify that each sender-receiver pairing has been checked and is valid
- State roots for all the state transitions that occurred between the previous ZK-Rollup state and the new ZK-Rollup state, after sequentially executing each transaction in the batch

The verification program, called the verifying circuit, uses those elements to loop over all transactions in the batch and generate a new Merkle root for every change in the nonce and balance of a sender or receiver. The state root keeps updating according to these Merkle roots, and its final form becomes the new ZK-Rollup state root.

Proof Verification Process: After the correctness of the state update process is successfully verified by the proving circuit, the operator can submit the validity proof to the verifier contract on the main-chain. The contract has a verifying circuit that confirms the proofs validity and therefore that a valid state update path corresponding to the batch transactions exists that leads from the ZK-Rollups previous state to the ZK-Rollup's new state.

3.4   EVM Compatibility: ZKEVM

A main drawback of ZK-Rollups is that they are not inherently EVM-compatible. Building applications that use ZK-Rollups requires building application specific program representations (circuits) for every different application. This process requires the use of specialised programming languages and a deep understanding of ZK-proofs. Therefore it is very difficult to build general ZK-dapps or migrate existing dapps to ZK-Rollups. The learning curve and limited functionality significantly hinders ZK-Rollup adoption.

An alternative to the situation is designing a universal circuit able to execute any smart contract, a ZK-compatible EVM, or ZKEVM. Native verification would solve the aforementioned problems and also allow different ZK-apps to interact with each-other at the L2 level.

The effort behind ZKEVMs is towards achieving the creation of ZK-proofs for every state transition that occurs, to verify the validity of the transition in terms of computation and use of memory/storage. The main idea is the knowledge that all programs will be run by the EVM, so the goal is to build a circuit that can verify low-level EVM functionality and use it to validate any smart contract execution. At the same time, on the user and developer side, the experience is identical to using L1.

This project of course comes with its own set of problems [20].

- First and foremost, EVM has very limited support for elliptic curve cryptography, which makes it very difficult to use various specialized protocols.
- Secondly, the sheer variety of specialised opcodes and error types that exist in the EVM raise the challenge level for the design of the universal verification circuit.
- Many inherent characteristics of the EVM, such as it being stack-based and its reliance on Keccak introduce huge overhead to the proving process.

Fortunately, recent advancements have allowed for many of the obstacles to be overcome. Polynomial commitment schemes enable more optimisation for efficiency and therefore a reduced circuit size. Lookup table arguments help deal with various previously ZK-unfriendly operations. Recursive proofs have become less reliant on pairing-friendly cyclic elliptic curves and are therefore more feasible. Already, Halo avoids the usage of pairing-friendly curves and Aztec has made proof aggregation possible for existing protocols. Hardware acceleration techniques have enabled increasingly efficient proving.

Design: The desired workflow while using a ZKEVM is for the developer to write a contract in any EVM-compatible programming language, deploy it painlessly on the ZKEVM and users can then interact with the smart contract normally. Overall, an experience equivalent to using and developing L1 apps, but with higher scalability and lower fees.

Executing a contract in L1 has the following process:

- Contract bytecode is stored in Ethereum storage
- Transactions communicating with the smart contract are broadcast over the P2P network
- Each full node of the network uses the transaction as input to execute the bytecode on the EVM. All full nodes must reach the same state from this process.

In comparison, executing a contract in L2 through a ZKEVM involves the following steps:

- Contract bytecode stored in storage
- Transactions communicating with the smart contract will be sent to a ZKEVM
- ZKEVM generates proof for every state transition after applying each transaction
- L1 contract validates the proofs and updates the state accordingly

The ZKEVM needs to generate proofs that it correctly followed the same steps the EVM would if it was natively executing the contract. If we summarize the functionality of the EVM as loading the bytecode and executing the opcodes serially and allow the opcodes to be described as reading, computing, and writing in that order, then the ZKEVM would need to provide:

1. Validity proof for the correct loading of the bytecode. This can be achieved by choosing a verifiable storage setup, like cryptographic accumulators, and designing a corresponding circuit that proves that the reading was done correctly.
2. Proof that during individual opcode execution, there was no skipping. The challenge here occurs from conditional opcodes that do not have a static outcome but instead varies per input. Providing the execution trace for a specific input and allowing the prover to verify it is a workaround.
3. Proof that each opcode was executed correctly and consistently. For an optimized approach, read/write proofs are categorized together as proofs that all needed fetch elements into a bus, and they are called state proofs. Computation proofs serve to verify that any operation performed on the bus elements was correct. These are commonly referred to as EVM proofs.

3.5   Differentiation

After analyzing the basic design and components of ZK-Rollups, an evaluation can be drawn as to the different factors that differentiate the variety of ZK-Rollup implementations.

1. As established in Sect. 3.3, there is a variety of different proof systems a ZK-Rollup could be making use of. The choice of the proof system significantly affects the size of the proof and subsequently its generation time, its verification time, and the accompanying costs. Interactivity is also a factor dependent on the choice of proof system, as is the trust requirement. As can be seen in Table 1, rollups using SNARKs benefit from short proofs and no back and forth messages, at the expense of a trusted setup. An example of rollups utilizing SNARKs is Aztec. Starkware uses STARKs that eliminate the trust requirement and bring quantum resistance at the cost of larger proofs. ZKSync makes use of PLONK, a setup that uses SNORKs for the added updatable trusted setup benefit.
2. The approach to circuit designs and execution of transactions in L2 have significant effects on the scalability benefit of the rollup, but also its user-friendliness and EVM compatibility. Alternatives include account-based circuit designs like ZKSync, UTXO-based designs like StarkWare, and ZKEVM designs, analyzed in Sect. 3.4, like ZKPorter.
3. Last but not least, there is the choice relating to data availability and how data are stored and accessed on L1. Popular options are decentralized designs like IPFS which ZKSync makes use of, availability committees as used by ZKPorter and validiums or other data sampling approaches, as used by Stark-Wave.

3.6   Examples

Polygon: Polygon ZKEVM stands as a decentralized Ethereum Layer 2 scalability solution, utilizing cryptographic zero-knowledge proofs to validate and finalize off-chain transaction computations. The architecture transparently executes smart contracts by publishing zero-knowledge validity proofs while maintaining compatibility with the Ethereum Virtual Machine (EVM). Some of the strategies used to guarantee optimal efficiency are presented in this section [21]. Firstly, the deployment of the Consensus Contract incentivizes the participation of the most efficient aggregators in the proof generation process. Secondly, the computation processes are executed off-chain, retaining only essential data and ZK-proofs on-chain. Instead of publishing the sizeable ZK-STARK proofs as validity proofs, ZK-SNARKs are used to attest to the correctness of the ZK-STARK proofs and then these ZK-SNARKs are published as the validity proofs to state changes, resulting in a significant reduction of gas cost. The ZKEVM Bridge facilitates asset transfers between L1 and L2 layers, with interoperability supporting migration between different L2 networks. The Verifier Smart Contract validates cryptographic proofs, playing a key role in ZK-Rollup architectures. The transaction life cycle involves depositing ether, finalizing transactions on L2, and achieving L1 finality through proofs and synchronized states. This comprehensive architecture ensures the effective functioning of ZKEVM in enhancing Ethereum's scalability.

The main components of Polygon ZKEVM, are the following:

- The ZKNode serves as client software for network synchronization and participant role governance. Incentivization structures for Sequencers and Aggregators, along with the ZKProver's use of advanced zero-knowledge technology, ensure efficiency and security. Polygon ZKEVM accomplishes this by engaging various entities as listed below:
    - The Users, connecting to the ZKEVM network through an RPC node (e.g., MetaMask), submit their transactions to a database known as Pool DB.
    - The Pool DB functions as the storage for transactions submitted by users, housing them in a pool until a Sequencer includes them in a batch.
    - The Sequencer, as a node, takes on the role of retrieving transactions from the Pool DB, validating their authenticity, and assembling valid transactions into batches that are submitted to L1.
    - The Synchronizer is the component tasked with updating the State DB by retrieving data from Ethereum via Etherman.
    - The Etherman is a low-level component that implements methods for all interactions with the L1 network and smart contracts
    - The State DB is a database designed for the permanent storage of state data, excluding the Merkle trees.
    - The Aggregator is the node that produces zero-knowledge proofs attesting to the integrity of the Sequencer's proposed state change, to that end, they employ a cryptographic component called the Prover.
    - The Prover is a complex cryptographic tool that generates ZK-proofs for hundreds of batches and consolidates them into a single ZK-proof, subsequently published as the validity proof. Within the ZKProver in order to speed up computations and minimize proof sizes, specialized cryptographic primitives are used such as running special languages and a cluster of state machines. Zero-Knowledge Assembly (ZKASM) language and Polynomial Identity Language (PIL) are specifically created and designed for broader adoption outside Polygon ZKEVM.
- The Consensus Contract, essential for L2 batch production, evolves from the Proof of Donation (PoD) mechanism in Polygon Hermez 1.0 to the Proof of Efficiency, enabling decentralized coordination for batch creation.
- PolygonZKEVM.sol governs the protocol, ensuring correct state transitions through validity proofs, validated by ZK-SNARK circuits. This process involves two essential procedures: transaction batching and transaction validation.
- The ZKEVM bridge, implemented as a Smart Contract operates as a decentralized bridge for the secure deposit and withdrawal of the users' assets between two layers. The Bridge L1 Contract resides on the Ethereum Mainnet, overseeing asset transfers between rollups, while the Bridge L2 Contract

is located on a designated rollup, managing asset transfers between the Mainnet and the Rollups, and ensuring Layer 2 interoperability.

StarkNet: StarkNet bundles and processes data off-chain, accompanied by STARK proofs, significantly reducing the computational resources needed for verification. The platform aims to achieve a significant cost reduction compared to Ethereum while enhancing the transaction speed. StarkNet aims to pioneer accessibility with a general-purpose smart contract platform on a fully composable network, enabling widespread development and deployment of applications. Smart contracts are scripted in Cairo, similar to Solidity, ensuring optimization and STARK scalability. Concurrently, transpilers from Solidity and other languages to Cairo are being developed for rapid deployment. Starkware, a company founded in 2017 by Eli Ben-Sasson and Alessandro Chiesa, is the driving force behind StarkNet and StarkEx [22].

Starknet's architecture involves key components:

- The Sequencer similar to validators in Ethereum or Bitcoin, receives and orders transactions, producing blocks. Sequencers in Validity rollup-based networks specialize in providing transaction capacity rather than ensuring security directly. This innovative mechanism enables Validity rollups to efficiently manage a higher transaction volume while upholding the security standards of the underlying Ethereum network. Sequencers adhere to a structured approach in transaction processing: a) Sequencing: Transactions are gathered from users and systematically ordered, b) Executing: these ordered transactions are processed, c) Batching: Transactions are efficiently grouped into batches or blocks, d) Block Production: encompassing batches of processed transactions the blocks are generated.
- The Prover: acts as the second layer of verification within the Starknet network, tasked with validating the Sequencers' work and generating proofs affirming the correctness of these processes. Key responsibilities of Provers encompass: a) Receiving Blocks containing processed transactions from Sequencers, b) Processing: Provers conduct secondary processing of these blocks, ensuring accurate handling of all transactions within. c) Proof Generation: Following processing, Provers generate proofs validating correct transaction handling, d) Sending Proof to Ethereum: Subsequently, the proof is dispatched to the Ethereum network for validation, and if the proof is accurate, the Ethereum network accepts the block of transactions.
  Due to the computational intensity of calculating and generating proofs, Provers necessitate even greater computational power than Sequencers. However, the Provers' workflow can be divided into distinct segments, enabling parallelism and streamlined proof generation. The proof generation process operates asynchronously, offering flexibility in timing. This feature permits the distribution of workload among multiple Provers, each capable of working on a different block, fostering parallelism and efficient proof generation.
- Layer 1 (L1): Hosted on Ethereum, it houses a smart contract capable of verifying STARK proofs. Valid proofs prompt an update to Starknet's state root on L1.

Starknet's state, maintained through Merkle trees akin to Ethereum, forms the architecture of the validity rollup, delineating the roles of each component.

Scroll: Scroll is addressing the scalability challenges of Ethereum by employing cutting-edge research in zero-knowledge proofs (ZK) to build a Layer 2 rollup network on Ethereum known as Scroll ZKEVM [?]. Their goal is to mirror Ethereum's behavior and ensure that all network activities are secured by smart contracts on the blockchain where the generated cryptographic proofs are published validating that the Scroll network adheres to Ethereum's rules. Ethereum smart contracts then verify the validity of every Scroll transaction through these proofs, ensuring a high level of security, decentralization, and censorship resistance. At the present time, the Scroll mainnet on Ethereum is operational, providing users with a live platform, along with a testnet for experimentation. The Scroll Protocol [23] is designed as a tripartite architectural framework, comprising the Settlement, Sequencing, and Proving Layers.

- The Settlement Layer ensures data availability and ordering within the authoritative Scroll chain. It undertakes the verification of validity proofs and facilitates the seamless exchange of messages and assets between Ethereum and Scroll and additionally, it is critical for the deployment of essential components such as the bridge and rollup contracts onto the Ethereum blockchain.
- The Sequencing Layer harbors the heart of Scroll's functionality. In this layer, an Execution Node is responsible for processing transactions submitted to the Scroll sequencer and those directed to the L1 bridge contract. Additionally, it includes a Rollup Node tasked with grouping transactions, posting transaction data and block information to Ethereum for data availability, and presenting validity proofs to Ethereum for finality.
- The Proving Layer comprises a pool of provers entrusted with generating Scroll ZKEVM validity proofs to authenticate the accuracy of L2 transactions. It also features a coordinator, responsible for assigning proving tasks to provers and transmitting the proofs to the Rollup Node for finalization on Ethereum.

When provided with an initial world state S, a transaction T, and the resulting world state S', the Scroll ZKEVM aims to generate a proof asserting that the state transition function, denoted as $f(S, T)$, indeed equals S', as specified in the Ethereum Yellow Paper. The computation initiated by an individual transaction is deconstructed into discrete machine instructions known as "opcodes." These opcodes are comprehensible and executable directly by the Ethereum Virtual Machine (EVM), which aims to validate the execution of the transition function. To this end, Scroll ZKEVM needs to prove: a) the correctness of the execution trace b) the accurate execution of each individual opcode, aligning with the Ethereum Yellow Paper specifications c) that each opcode is executed with the correct behavior, encompassing any modifications to data stores, d) the ordered list of executed opcodes is indeed the accurate sequence triggered by the transaction and last but most importantly that the execution trace starts with initial state S and results in state S'

Aztec: Aztec 2.0, an open-source Layer 2 network deployed to Ethereum mainnet, is a sophisticated multi-asset private rollup service designed to enhance financial privacy and scalability. At its core, Aztec 2.0 incorporates critical elements such as crypto primitives, elliptic curves, Ate pairing, and hash specifications, supported by innovative data structures like Notes and Commitments, forming the backbone of its private UTXO system. The protocol further incorporates a series of circuits, including the Joinsplit Circuit governing Aztec note spending, the Account Circuit dedicated to managing keys within Aztec's username system, the Rollup Circuit orchestrating the aggregation of ZK proofs in the rollup process, the Escape-Hatch Circuit serving as an outer mechanism for user fund withdrawals without a rollup proof, and the Root Rollup Circuit responsible for aggregating multiple rollup proofs [24].

In essence, the protocol's operational framework involves clients generating Joinsplit proofs to privately transfer value between accounts, with the provision for clients to craft Account proofs altering authorized key sets. The rollup provider plays a central role in batching Joinsplit proofs into a consolidated Rollup proof, validating transactions collectively. Furthermore, this rollup provider aggregates multiple rollup proofs into a Root Rollup proof, validating all transactions simultaneously, which is then placed on the blockchain, ensuring a streamlined and secure process. In instances where a client's transaction faces potential censorship from the rollup provider, a safeguard is provided through the submission of a separate Escape hatch proof directly to the blockchain [25].

The Rollup Circuits serve the primary function of consolidating numerous transactions within a rollup into a single SNARK, enhancing efficiency and facilitating verification on the Ethereum blockchain. Executed by a Sequencer, responsible for transaction ordering, the circuits employ a 'binary tree of proofs' topology to compress data, enabling parallelized proof generation.

- The Base Rollup Circuit handles state checks, updates, nullifier and commitment insertions, and kernel proof verification.
- The Merge Rollup Circuit combines two Base or Merge proofs into one.
- The Root Rollup Circuit further compresses two Base or Merge proofs into a single SNARK through hashing and verification processes.

The circuits are designed for potential deviations from symmetrical trees for efficiency, and some of them perform additional protocol checks and computations as needed for optimization.

Similar to other L2 solutions, the Sequencer stands as a central component in the development and release of new rollup blocks within the Aztec protocol. Its critical tasks involve retrieving transactions from the P2P pool, organizing them, executing public functions, guiding transactions through rollup circuits, constructing the L2 block, and submitting it to the L1 rollup contract, including any relevant contract deployment public data. Notably, the Sequencer initiates modifications to the world state database with every new block formation. However, these alterations only become permanent upon confirmation by the world state synchronizer on L1. This process involves the collaboration of various components like the block builder, prover, publisher, public processor, and simulator.

The circuits, crucial for validating execution and consolidating changes, encompass the public circuit, public kernel circuit, base rollup circuit, merge rollup circuit, and root rollup circuit, playing a crucial role in verifying and aggregating transaction outputs while ultimately establishing an L2 block verified by the L1 rollup contract.

Loopring: While the initial iterations of the Loopring protocol did not incorporate ZKSNARKs, the latest version, Loopring 3.6, has evolved into a ZKRollup Exchange and Payment Protocol [26]. This strategic shift aims to significantly enhance protocol throughput by conducting a substantial portion of the work off-chain using ZKSNARKs, reserving on-chain verification for essential processes. Emphasizing this change in design, Loopring Exchange now leverages zero-knowledge proofs to streamline operations, ensuring efficiency and scalability. Key aspects of the Loopring Exchange include [27]:

- an API to manage user off-chain requests, which includes functionalities like order submission, cancellation, and withdrawals, while on-chain requests, including account registration, password reset, and deposits, must be submitted using Ethereum transactions.
- utilization of Ethereum as both a data availability layer and a Zero-Knowledge Proof (ZKP) verification layer.
- the design choice of separating trading computations from Ethereum which contributes to lower costs and higher throughput compared to on-chain alternatives.
- a Merkle tree structure that combines several key aspects to ensure an optimal balance between complexity, proving times, and user convenience.

In the context of block organization, the protocol aggregates work into blocks, distinct from Ethereum blocks, and utilizes the Merkle tree to transition between states efficiently. Operators play a crucial role in creating, proving, and submitting blocks, with the operator contract capable of enforcing an off-chain data availability system. The operator's responsibility includes submitting blocks on-chain, with immediate verification of the correctness of the work within a block through Zero-Knowledge Proofs (ZKP). This comprehensive approach ensures robust and efficient operation of the Loopring protocol, contributing to its reliability and user-centric design.

Immutable X: Immutable X stands as a Layer Two (L2) scaling solution tailored for non-fungible tokens (NFTs) within the Ethereum ecosystem, utilizing StarkWare's potent STARK prover and rollup technology. At the core of this ecosystem is the IMX token, an ERC20 utility token native to the Immutable X protocol [28]. Users can acquire IMX through various pro-network activities such as trading, and it serves multiple functions, including fee payment, governance participation, and protocol staking. The main components of Immutable X include:

- the ZK-Rollup Scaling Engine, leveraging StarkWare's cutting-edge rollup technology to facilitate over 9,000 NFT transfers, trades, and mints per second, ensuring scalability without compromising Ethereum's security.

- the API Abstraction Layer that simplifies NFT application development by encapsulating the scaling engine in REST APIs, offering synchronous interactions and streamlining blockchain development.
- NFT-Enabled Wallets that provide a seamless experience supporting desktop Ethereum wallets
- Platform SDKs designed to further ease integration, with a Typescript SDK currently available.
- a marketplace for NFT trading, ensuring legitimacy and ease of access.
- a Transaction History Explorer for transparent verification of transactions and historical states.

The foundational element of Immutable X lies in its ZK-Rollup scaling engine, developed in collaboration with StarkWare, utilizing the StarkEx prover and verifier. The asset state in Immutable X is represented through a Vault Merkle Tree, ensuring secure state transitions through on-chain and L2 proof logic. The smart contract verifier governs deposits, withdrawals, and state updates, maintaining the root of the Merkle tree with upgradeable contract logic. Furthermore, Immutable X employs STARK proofs over SNARKs, prioritizing user security despite increased on-chain publishing costs. Addressing potential unresponsiveness, the protocol supports two data availability modes: rollup and validium. In rollup mode, state changes are published to L1, ensuring L1 security with a minimal linear cost. In validium mode, a Data Availability Committee (DAC) signs each batch, and users can withdraw assets with the assurance of data availability, even if one committee member is honest. Within Immutable X's DAC, a variety of entities contribute to the decentralized architecture, ensuring asset safety and mitigating risks associated with centralized vulnerabilities.

ZKSync Era: ZKSync Era is a user-centric and EVM-compatible ZK Rollup platform being built by Matter Labs, that supports existing Ethereum wallets, eliminating the need for separate private key registration [29]. The rollup workflow involves users performing transactions, withdrawals, and transfers, with an operator playing a crucial role in processing and finalizing these transactions through a series of well-defined stages. The ZKSync Era, at its current version, addresses the needs of most applications on Ethereum, supporting ECDSA signatures, Solidity 0.8.x, and Ethereum cryptographic primitives, but is operated centrally by the ZKSync team, which plans the transition to a decentralized system in the future. Other plans involve features like ZKPorter, offering a choice between ZKRollup accounts with high security and a 20× fee reduction compared to Ethereum or ZKPorter accounts with stable transaction fees.

Based on the code of ZKSync Era, the ZK Stack is a modular, open-source framework designed to construct custom Zero-Knowledge (ZK)-powered Layer 2 (L2) and Layer 3 (L3) solutions, known as hyperchains in [30]. This framework is freely available, offering developers the flexibility to create custom solutions within the ZK ecosystem. The concept of Hyperchains introduces a solution to enhance scalability and user experience, aiming to connect various rollups, forming a fractal tree of Hyperchains. These rollups utilize shared bridge contracts on Layer 1 (L1) and native bridges between rollups to address issues such as slow

L1 finality. Trustless validation bridges on rollups and native bridges facilitate seamless token transfers, with L1 serving as a single source of truth, preventing hard forks. To optimize user experience, the Hyperchain architecture supports cross-chain wallet management. Bridging, an integral part of the protocol, occurs during proof settlement in 1–15 min, depending on the Hyperchain.

4   Optimistic Rollups

4.1   High Level Description

Optimistic Rollups are one of the existing Layer 2 scaling approaches known as rollups. They follow the rollup premise of executing transactions off-chain, then posting the relevant data on Layer 1 in batches, increasing scalability and lowering costs. Optimistic Rollups get their name from the assumption that every transaction, batch, and proposed block is valid without requiring some kind of validity proof. Instead, a fraud-proving scheme is used to detect inconsistencies.

Architecture: A number of elements are involved in an Optimistic rollup system:

- Layer 1
  - Management Contracts: Rollup management contracts that are deployed on Layer 1 that take care of tracking user deposits, storing the blocks of the rollup chain and monitoring the state updates.
- Layer 2
  - Off-Chain Virtual Machine: the Layer 2 part of the rollup, this VM hosts all applications and computations happening off-chain. While it is completely separate from the EVM, it shares a number of its characteristics as they are fully compatible with each other.
- Middlemen
- Operators: crucial to the functionality of the rollup, operators are responsible for batching transactions and posting the batches and all the accompanying data on-chain. In a centralized setup operators are often referred to as sequencers. In the context of Optimistic Rollups they are also called aggregators.
- Verifiers: the actions of the verifiers are the basis of security for Optimistic Rollups [31]. This role is responsible for keeping an eye on the data published by operators and taking action if inconsistencies are detected. The action is initiating the dispute phase that will be analysed in Sect. 4.2.
- Bridge Contracts: Contracts used to move assets and facilitate communications between L1 and L2.

Dependencies: Optimistic Rollups have the same dependencies as ZK-Rollups upon the main chain, as they are also a hybrid solution that needs both levels to operate correctly. Therefore, transaction finality, data availability and censorship resistance are qualities that require on-chain functionality to be achieved.

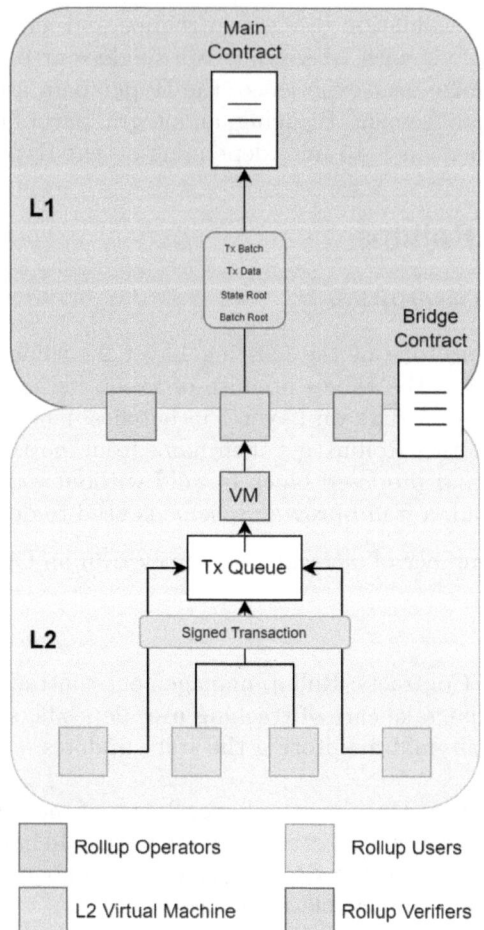

Fig. 2. Overview of Optimistic Rollup main components

There is, however, a dependency exclusive to Optimistic Rollups. As fraud proofs and dispute periods are used as a security mechanism in Optimistic Rollups, smart contracts deployed on-chain function as arbitrators that verify fraud proofs and provide resolution in the occurrence of a challenge. Optimistic Rollups therefore depend on L1 as their settlement layer.

4.2 Basic Functionality

Entering: The first step to entering an Optimistic Rollup is depositing assets to its bridge contract on-chain. The transaction to this contract will allow equivalent funds to be minted on the rollup chain and create a rollup balance for the user.

Those deposit transactions remain in queue until an operator submits them to the rollup contract, but in case of extreme delays, the users are allowed to submit the transactions themselves to prevent freezing.

State Updates: The process of using an Optimistic rollup chain is the following:

1. The operator for the next batch is determined.
2. Users sign and submit the transaction to the operator for inclusion in the batch.
3. The operator picks the transactions to include in the batch arbitrarily. The operator verifies, orders, and executes these transactions.
4. The batch and transaction data are published by the operator on the main chain as one transaction.
5. The operator computes the new state and posts the Merkle root which is the proposed rollup state root after the execution of the batch.
6. The operator additionally computes the batch root that enables users to provide proof of inclusion of a transaction in the batch.
7. The rollup contract verifies that the starting root of the batch is identical to the current rollup state root before the update. If so, then the rollup state root is updated to match the proposed state root after batch execution. This is only an initial check, and the state root can still be determined invalid later.
8. Verifiers monitor the posted data and perform validity checks on it to determine its correctness.
9. If inconsistencies are present, verifiers report them and initiate the dispute phase, analysed in Sect. ??.
10. If the dispute is successful, the transactions are re-executed and the rollup state is adjusted accordingly.

Balance Verification and Proof of Transaction Inclusion: Both the process of verifying a user's current balance in the rollup as well as proving the inclusion of a particular transaction in a particular batch are conducted in the same way in optimistic and ZK-Rollups, so the descriptions in Sect. 3 are sufficient.

Disputes: Depending on the design, it will either be possible for anyone to post new rollup blocks, or it will be an option limited to sequencers. Regardless, after the posting of a transaction batch on-chain by an operator, there is a designated time window, called a challenge period, for anyone to dispute the posted batch and the state root proposed by the operator. It is this dispute period, and the questionable validity of batches, that led the Optimistic Rollup blocks to be referred to as assertions.

Usually, a verifier is the one to spot and report irregularities, as they are expected to constantly validate posted batches. However, due to complete data availability on-chain, any user can verify posted batches. The challenger claims that executing the transactions of the posted batch does not produce the provided state root. The initiation of this challenge occurs with the verifier computing a fraud proof and giving it to the rollup contract. Fraud proofs will be further analyzed in Sect. 4.3. Overall, the communication from the challenger to the contract includes the following.

- The state transition that is being challenged.
- The rollup state root before the transition.
- The rollup state data.
- The rollup state after execution as computed by the verifier, which differs from the one submitted by the operator.

The smart contract re-executes the transition. If the contract deems the fraud proof to be valid, ergo the result aligns with the state as submitted by the verifier, then the dispute is successful, and the state of the rollup is updated according to the new valid result.

In case of a successful dispute, the operator responsible for providing the invalid state root receives a financial penalty. Likewise, if the challenge is invalid, the challenger is penalized. In general, the dispute process is quite demanding in terms of resources, and therefore there are mechanisms (Sect. 4.3) designed to limit unnecessary challenges.

If there were subsequent state updates after the successfully disputed one, all of them are reverted. One way or another, the dispute is resolved, and the progression resumes normally on the rollup chain afterwards.

Exiting: Exiting an Optimistic Rollup is a somewhat complicated process.

The sequence is initiated by an off-chain request to withdraw funds. That transaction is to be included in the next batch posted on-chain.

When the batch has been successfully posted, the user needs the proof of inclusion of their withdrawal transaction in that batch, which is necessary to begin the final step of the process. Once the proof has been submitted to the rollup contract, the waiting period of about 7 days has to expire before the user gets their funds back on the main chain.

A facilitating element in Optimistic Rollups are liquidity providers, that essentially provide instant liquidity by providing instant funds to exiting members in exchange for their pending L2 withdrawals and a fee. This is possible and risk-free on the side of the liquidity providers as they can easily verify the validity of the pending withdrawal.

4.3   Security: Fraud Proofs and Cryptoeconomic Incentives

There are several security concerns when moving processes off the main chain, and this holds true for Optimistic Rollups as well. First and foremost, how is the validity of off-chain computations ensured? How is it ensured that security mechanisms like disputes are not maliciously exploited? How are operators, and most importantly, sequencers in more centralised setups, prevented from freezing the chain and censoring users? A number of mechanisms have been built into Optimistic Rollups to combat these issues, and they are analyzed in this section.

Cryptoeconomic Mechanisms: To ensure that disputes are not used to overwhelm the system to abuse resources and delay block validation, cryptoeconomic mechanisms are used to incentivize the proper use of the challenge mechanism.

Operators, in order to gain the ability to produce blocks, provide a bond to the on-chain rollup contract. This bond is used as a stake to secure the system against any malicious operator/verifier behaviour.

In the event of a challenge, there is always one party in the wrong. Either the operator has indeed posted an invalid block, or the challenger has disputed a valid block. In each case, the "losing" party is penalised in the following way:

- For the operator posting invalid block:
  - Half the bond goes to the verifiers. This is to incentivize the constant verification of blocks.
  - The other half of the stake is burned. This is to prevent scenarios where an operator can claim back their own bond by reporting themselves via a different account.
- For the verifier challenging a valid block:
  - Half the bond of the verifier goes to the operator.
  - The other half of the bond is burned.

Fraud Proofs: Fraud proofs are the way for block verifiers to initiate disputes and prove that one or more transitions in a batch, and therefore the state transition as a whole, are invalid and the rollup chain should be rolled back. The elements of a fraud proof are as described in Sect. ??, but the process can have varying forms.

While all fraud proofs involve an interactive process that begins after the batch has been posted on chain, there are different schemes regarding the number of rounds needed to compute the fraud proof and resolve the dispute [32].

1. **Single Round Fraud Proofs:** The scheme usually described and the one most used so far is that of computing fraud proofs in a single round. It involves the straightforward process of challenging a batch and having the transaction re-executed by the contract to compare results. The drawback in this case is the high gas costs incurred by re-execution, and the vast amount of data that ends up being posted on-chain.
2. **Multi Round Fraud Proofs:** An improvement aiming to reduce the load of re-executing the complete block is multi round fraud proof computation. This entails repeated communication between the operator and the challenger, mediated by the rollup contract that will once again be the ultimate decider of the validity of the challenge. The process is called a bisection protocol and can be broken down as follows:
   - After being challenged, the operator divides the block in half, with both parts requiring equal steps to compute.
   - The challenger, knowing which half includes the disputed transaction, picks one.
   - The dividing and picking process is repeated until both remaining halves only involve one step of computation, and the challenger picks the problematic transaction.

- The contract now only has to re-execute this single step to determine the outcome of the challenge, a process much lest costly than doing so for the entire batch.

This process, of course, involves time limits for each communication, and failure to reply from either party is equal to forfeiting. Overall, the computational load on the smart contract is greatly reduced, as well as the data posted on L1.

The fraud proof mechanism is what allows Optimistic Rollups to guarantee finality for valid transactions, as long as a single, honest member of the system acts as a verifier.

4.4   EVM Compatibility

In contrast to ZK-Rollups, Optimistic Rollups are completely compatible with the EVM, since the off-chain VMs follow the constraints set by the Ethereum yellow paper, and can support the EVM at bytecode level. This comes with a significant set of benefits, mainly for developers using Optimistic Rollups:

1. Smart Contracts already deployed on L1 can easily be migrated to Optimistic Rollup chains.
2. Optimistic Rollups can utilize the same programming languages, libraries, testing environments, software as those used for the Ethereum Mainnet. Those tools have been long tested and improved on, giving Optimistic Rollups a very robust tool set to draw from.

Cross Chain Contract Calls: Thanks to Optimistic Rollups having complete EVM compatibility, cross chain contract calls are possible. Ethereum accounts can use bridging contracts to interact with L2 contracts, since it is possible for an L1 contract to invoke an L2 contract.

The difference between these calls and calls on the mainnet is that there is a delay between the call and the execution. Furthermore, it is advised that special caution is taken with gas, since a transaction going through on L1 and failing on L2 because of gas limit, the associated funds become irrecoverable [32].

4.5   Examples

Arbitrum: Arbitrum operates as an Optimistic Rollup, that records inputs, or messages, into the Ethereum chain, allowing anyone to reconstruct the state of the chain based on the public information in the inbox history [33]. This transparency enables broad participation in the Arbitrum protocol, allowing individuals to run nodes or act as validators without the need for secret information about the chain's history or state. Arbitrum advances its chain state by allowing any validator to post a rollup block on Layer 1, claiming it to be correct. A challenge period follows, during which others can contest the claim.

If unchallenged, the rollup block is confirmed. In case of a challenge, Arbitrum employs an efficient dispute resolution protocol, penalizing the dishonest party by forfeiting a deposit, part of which is awarded to the truthful challenger. The threat of deposit loss discourages cheating attempts, making it the norm that a single party posts a correct rollup block without challenges [34].

The Arbitrum rollup protocol orchestrates a chain of RBlocks, proposed by validators, which are initially unresolved and later confirmed or rejected, constituting the chain's confirmed history. Each RBlock includes essential details such as its number, predecessor RBlock number, the count of L2 blocks and consumed inbox messages in the chain's history, and a hash of the outputs generated over the chain's course. Apart from the RBlock number, all other contents are claims by the proposer, subject to validation by the protocol. Implicitly, an RBlock asserts the correctness of its predecessor and, by extension, the entire chain's historical sequence. RBlocks are subjected to deadlines, determining the duration for other validators to respond. Agreement with an RBlock's correctness requires no action, while disagreement prompts validators to submit another RBlock with differing results, initiating a potential challenge against the first RBlock's staker.

Validators in the Arbitrum ecosystem can be broadly categorized into three strategies based on their roles and engagement with the rollup protocol. First, active validators pursue a proactive stance, constantly staking to propose new RBlocks and actively contributing to advancing the state of the chain. Second, defensive validators adopt a vigilant approach, refraining from staking during normal operations but intervening when incorrect RBlocks are proposed. They either post correct RBlocks or stake on those proposed by others to defend the accurate outcome. Lastly, watchtower validators remain non-staking observers, relying on vigilant monitoring of the rollup protocol. When an incorrect RBlock is identified, they raise alarms to prompt other stakeholders to intervene and claim a portion of the dishonest proposer's stake, assuming corrective action occurs before the RBlock's deadline expires. These strategies offer a spectrum of engagement, providing flexibility for validators based on their risk tolerance and commitment to the network's integrity.

Optimism: Optimism operates as an EVM-compatible "Optimistic Rollup," a term denoting a blockchain that relies on the security infrastructure of another primary blockchain, known as its "parent". In essence, Optimistic Rollups utilize the consensus mechanism (such as PoW or PoS) of their parent chain rather than establishing an independent one. In the case of OP Mainnet, Ethereum serves as its parent blockchain [35].

The block production on the Optimism (OP) Mainnet is primarily orchestrated by a singular entity known as the "sequencer". This sequencer assumes crucial responsibilities in the network, such as offering transaction confirmations and state updates, constructing and executing Layer 2 blocks, and submitting user transactions to Layer 1. On OP Mainnet, blocks are consistently produced every two seconds, irrespective of their transaction density. Transactions reach the sequencer through two channels: those submitted on L1, labeled as deposits, are included in the corresponding L2 block identified by epoch and sequence number, ensuring L1 Ethereum-level censorship resistance. Alternatively, transactions submitted directly to the sequencer are more cost-effective but lack censorship resistance, given the sequencer's exclusive awareness of them. The Optimism Foundation aspires future plans for the eventual decentralization of the sequencer role [36].

In the Optimistic Rollup (OP Mainnet), state commitments are initially considered pending when published to L1 (Ethereum) without immediate proof of their validity. for a period known as the "challenge window" that lasts 7 days. If unchallenged within this period, they become final whereas challenged commitments can be invalidated through a process called "fault proof". Importantly, a successful challenge does not alter OP Mainnet itself; it only impacts the published commitments about the chain's state. The ordering of transactions and the state of OP Mainnet remain unaffected by a fault-proof challenge. Following the November 11th EVM Equivalence update and as part of the OVM 2.0 upgrade [37], the Optimism fault-proof process had to be temporarily disabled, meaning that the users of the OP Mainnet network need to trust the Sequencer node during the undergoing redevelopment.

Boba Network: Boba Network, launched in September 2021, functions as a Layer-2 scaling solution on Ethereum, leveraging Optimistic rollup technology [38] Its core objectives include reducing transaction and computation fees, enhancing throughput, and expanding smart contract capabilities within the Ethereum ecosystem nevertheless Boba distinguishes itself through various features, including a) the incorporation of additional cross-chain messaging, b) the adoption of unique gas pricing logic, c) acceleration of L2 to L1 exits using a swap-based system that eliminates the 7-day delay, d) introduction of a community fraud-detector for independent transaction verification, e) functioning as a DAO, and f) supporting native NFT bridging.

Additionally, Boba Network addresses the challenge of executing complex smart contract algorithms efficiently by introducing the Hybrid Compute architecture. This design allows smart contracts to trigger algorithms off-chain, akin to running an application on infrastructure like AWS Lambda. The results are then seamlessly integrated back into the on-chain smart contract, creating a hybrid model that operates both on-chain and off-chain. Rolled out in March 2022, the Hybrid Compute deployment significantly expanded the capabilities of the network, acting as a bridge between Boba Network's sequencer and external APIs, in order to facilitate calls to complex algorithms, interactions with machine learning models, engagement with real-world data, and synchronization with the states of external servers.

Metis: Metis L2, as a rollup solution, has recently introduced a significant stride toward achieving comprehensive decentralization in Layer 2 networks by unveiling a Decentralized Sequencer Pool. Designed to eliminate single-point failures stemming from centralized sequencers, this approach fortifies the robustness and resilience of the Layer 2 ecosystem, aligning with the broader industry trend towards increased decentralization in blockchain solutions [39]. The proposed solution leverages cutting-edge technologies such as the Tendermint consensus developed by Cosmos, Threshold Signature Scheme (TSS), Multi-Party Computation (MPC), Libp2p, L2 Geth, and others. When a user initiates a transaction, it is transmitted to Sequencer nodes within the network. These Sequencers, upon validating the transaction, generate a block. Multi-party computation (MPC) nodes collaboratively merge these blocks, forming a cohesive unit that is subsequently forwarded to the Ethereum main chain [40].

5   Discussion

5.1   Maturity

The exploration of rollup solutions highlights the dynamic and evolving nature of the blockchain ecosystem, driven by continuous innovation from research teams. The difficulty in categorizing or organizing emerging breakthroughs is compounded by the field's exploratory phase, where academic resources might not remain up-to-date, making it crucial to stay informed through field-related forums and media networks that host informative insights into this ever-changing and dynamic environment.

Following Vitalik's post "Proposed milestones for rollups taking off training wheels" on the Ethereum Magicians' forum [41], which categorizes rollups into three distinct stages, L2Beat proposed a comprehensive framework that not only categorizes the maturity level of a rollup clearly but also offers detailed technical insights for evaluation.

This classification system categorizes rollups into three distinct stages using the concept of the "training wheels":

- Stage 0 (Full Training Wheels) where operators effectively run the rollup, but there exists source-available software enabling the reconstruction of the state from the data posted on L1, for comparing state roots with the proposed ones.
- Stage 1 (Limited Training Wheels) where the rollup transitions are governed by smart contracts and the implementation of a fully functional proof system is achieved. Additionally, this stage is characterized by the decentralization of fraud proof submission, and provision for user exits without operator coordination, although a Security Council remains in place to address potential bugs. The Security Council, consisting of diverse participants, acts as a safety net, but its authority introduces potential risks.
- Stage 2 (No Training Wheels) is the final stage, where the rollup is entirely managed by smart contracts and the fraud proof system becomes permissionless. The Security Council's role is confined to addressing on-chain soundness errors and protecting users from governance attacks.

In alignment with this framework, the subsequent Table 2 offers a snapshot of the current stages and market shares of prominent rollup projects, offering a practical overview of the status of the ecosystem at the time of writing this text.

Table 2. Stage evaluation based on the L2Beat framework

RollUp	Stage_0	Stage_1	Stage_2	Market Share[a]
Arbitrum One	✓	✓	✗	52.90%
OP Mainnet	✓	✗	✗	28.01%
BobaNetwork	✓	✗	✗	0.07%
MetisAndromeda[b]	n/a	n/a	n/a	0.62%
Polygon ZKEVM	✓	✗	✗	0.75%
Starknet	✓	✗	✗	0.98%
Scroll	✓	✗	✗	0.29%
Aztec[b]	–	–	–	–
Loopring	✓	✗	✗	0.73%
immutable X[b]	n/a	n/a	n/a	1.33%
ZKSync Era	✓	✗	✗	3.56%

[a] Share of the sum of the total value locked in all the projects - https://l2beat.com/scaling/summary
[b] Not available due to changes or ongoing review

5.2 Effects

Rollups are meant to be an improvement upon the existing blockchain design. This improvement can and is claimed to concern a number of factors. The scale and manner in which rollups enhance those areas are presented in this section.

Scaling Effects: First and foremost, rollups are viewed as a scaling mechanism. Usually such mechanisms try to offload some of the burdens of transaction execution off of the main chain. However, for rollups, the greatest scalability effect does not seem to come from executing transactions off-chain. Instead, the most impactful factor is the excessive compression of transaction data. This allows for better use of the capacity that blocks have, allowing them to contain bigger amounts of information in that same limited space [10].

The proposal known as Proto-Danksharding or EIP-4844 [?] is expected to significantly improve rollup scalability benefits by moving on from posting transaction data as calldata, which maintains the data indefinitely without it being necessary for rollup functionality. Instead, data is posted in blobs that are deleted after a set time period, usually 1–3 months.

ZK-Rollups have another inherent advantage. The zero-knowledge proofs they make use of are recursive. While currently verification is done for each block independently, by making better use of this characteristic, multiple blocks could potentially be verified with the submission of only a single validity proof. This dramatically increases the throughput of the rollup and by extension the main chain. An estimation of ZK-Rollups transaction throughput in 2021 came to about 3800 transactions per second [42]. This number is much higher than Ethereum's limit of 25 and even higher than the Visa average throughput of 2000, but it remains a theoretical estimation at the moment.

Effects on Fees: Various mechanisms built into rollup designs affect the gas fees associated with using them.

First and foremost, batching transactions has an immediate effect on gas fees. Since there is a fixed minimum cost of about 21.000 gas units per submitted transaction [43] regardless of any included computations, submitting transactions in batches instead of individually ensures this cost will be paid once and divided amongst all the senders of the different transactions in the batch.

Additionally, publishing transaction data as calldata which prevents the alteration of Ethereum's state also has a positive effect on gas fees. Currently, the specified cost for calldata bytes is 4 gas units per zero byte and 26 gas units per non-zero byte [44].

Rollup operators take on transaction processing which comes with its own computational costs. ZK-rollup operators, in particular, are also responsible for the generation of validity proofs, which is an intensive and costly process. A fee is reserved for operators as compensation, comparable to mining fees. However, the fees charged by the operators are significantly lower, since rollups have higher processing capabilities and no network congestion issues.

There is a major drawback that ZK-Rollups bring in terms of costs, that is the significant fees associated with the verification of validity proofs on Layer 1. While the size of the proof and the computational difficulty of its verification differ depending on the kind of ZK proof used, current estimates range from 400.000 gas units per verification [45] to 600.000 [42].

5.3   Centralised vs Decentralized Operators

As already established, operators that function in a centralized manner are commonly referred to as sequencers. In a system with centralized sequencing, the sequencing occurs from a single node/entity, regardless of the fact that this node may belong or not to the rollup operator. Proposals for decentralizing sequencing aim at reducing the power of the node that commits the sequencing either by enabling multiple nodes to take up that role or by forcing that node to interact with the L1 upon which the rollup operates.

The option of a centralized sequencer offers the following advantages.

- **Efficiency**: Having a single sequencer to which all rollup participants send their transactions greatly improves performance. Communication requirements are minimized, and there are no coordination needs. The single sequencer responds to the participants with a soft confirmation that their transaction has been included in the next block. In the optimistic case, that is the only communication required.
- **Simplicity**: Having a system as simple as possible reduces the probability that things may go wrong. On the other hand, increasing the complexity of the system increases that probability as the attack surface of the system is enlarged while you have more modules/operations that may accidentally fail.
- **Control in early-stage systems**: Monitoring and maintaining a distributed system is hard, and the option to have a single node to do sequencing enables

rollup operators to better handle the process. For example, if a bug is discovered, it is much easier and faster to update a single node than a set of nodes run by independent users.

On the other hand, coming up with a decentralized sequencer has the following advantages:

- **Censorship resistance**: The main concern regarding centralized sequencers is that it is feasible for the sequencer operator to censor transactions, which is in complete disagreement with the foundational concepts of blockchain systems. Decentralized sequencing protects users of the rollup from such malicious operators, as no single node decides which transactions are included in blocks.
- **Fairer distribution of MEV**: The sequencing of transactions is strongly coupled with the extraction of MEV. Upon the assumption that a single operator is responsible for the sequencing in a rollup, then he is in place to extract all MEV and get a large economic gain from that. Although this may be in favor of rollup operators and may support a sustainable business model for them, it is not fair. Having multiple competing sequencing nodes can provide a better and more transparent model for the MEV allocation.
- **Better availability**: A single sequencing operator will be a single point of failure for a rollup ecosystem. If that fails, then the rollup operation will cease or in the best case (transactions are somehow cached), its liveness will suffer. Having a decentralized solution for the sequencing guarantees much better availability and no interruptions in the operation of the rollup even if one or some of the sequencing nodes fail.

One of the approaches proposed for rollups is to offload the sequencing process to an external service that operates in the following directions:

1. The sequencing service is decentralized, and it operates upon a network of distributed nodes that offer similar security/liveness guarantees to the L1 upon which the rollup operates.
2. The sequencing service is shared among more than one rollups and that enables more efficient cross-rollup operations
3. The sequencing service is compatible with the PBS concept of using private mempools and forcing blind commitments to the next block to control MEV distribution.

The vision behind early attempts in the space is to enable rollups to separate the data availability layer, the sequencing layer, and the execution layer. The data availability will be offered by the corresponding L1 on which rollups are based, the sequencing will be offered by the shared sequencer service, and the execution will be conducted within each rollup. The main components of the approach are depicted in Fig. 3.

A shared sequencer is a service that concurrently offers sequencing services to multiple rollups based on the same L1. Users of each rollup submit their

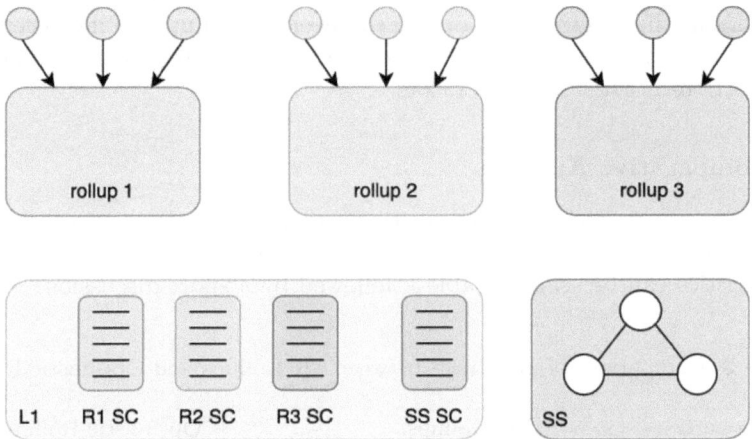

Fig. 3. Operation of a shared sequencer

transactions to the shared sequencer network either directly or through their rollup. The nodes of the shared sequencer operate a consensus algorithm between them and use an L1 smart contract to coordinate their operation. The network produces common batches of mixed transactions for all rollups and each rollup uses the ordered subset of transactions that relate to it. Then the rollup follows the rest of the pre-established process. It executes the transactions and posts the result to the L1 contract of the rollup (along with the validity proof in the case of a ZK rollup).

The main advantages of the shared sequencer scheme are as follows:

- It brings all the advantages analyzed in the previous section: censorship resistance, fairer MEV allocation, and enhanced availability.
- On top of that, the shared decentralized sequencer offers rollup interoperability. While the introduction of rollups has provided the basis for scaling public blockchains, a by-product of it is the fragmentation of the blockchain applications in different siloed ecosystems (different rollups). Achieving cross-rollup operations such as moving tokens between those are hindered by all issues that exist for inter-blockchain interactions. Sharing the same sequencing service facilitates such processes as the relevant transactions in the different rollups (e.g. burning tokens and minting tokens transactions in two different rollups) can be bundled together by the sequencing service providing atomicity properties to the whole operation.

With regard to the disadvantages, the main one is that there is no ready-to-use option in the space. Currently there exist two competing companies working towards such schemes, Espresso and Astria, both of which are in the early stages of development of their solution.

Existing rollups shall probably work towards achieving decentralization of their sequencing processes. While the technical maturity of the offered solutions is not at the desired level, opting to develop a rollup-specific decentralized sequencer will add more complexity and will require more research and development efforts from the rollup operators.

5.4   Comparative Analysis

This section touches on various positive and negative properties of both ZK and Optimistic Rollups. A comparison between the two with respect to some characteristics can be seen in Table 3, followed by a short discussion.

Table 3. Comparison of properties between ZK Rollups and Optimistic Rollups

Property	ZK-Rollups	Optimistic Rollups
Proofs	Validity Proofs	Fraud Proofs
Finality	Instant	Delayed
Trustlessness	✓	Reliance on honesty
Liveness Assumption	✗	✗
Data Compression	Great	Good
EVM Compatibility	✗ (ZK-EVMs in development)	✓
Specialized Hardware	✓	✗
Incentives	✗	✓

- **Proofs.** The mechanisms behind the proofs used by each rollup scheme have been analysed in Sects. 3.3 and 4.3 respectively. Comparatively, validity proofs are easy to verify on L1, while fraud proofs usually require the re-executing of an entire block. However, fraud proof verification does not need to be performed in every update, contrary to validity proofs that have to be verified every time a batch is posted.
- **Finality.** Thanks to validity proofs, ZK-Rollups can claim instant finality for every batch that is posted and verified by the rollup verification contract. That includes the ability of ZK-Rollup users to instantly withdraw funds from L2 to L1. Optimistic Rollups cannot make similar allowances due to timeout periods associated with the dispute mechanism that accompanies the use of fraud proofs.
- **Trustlessness.** ZK-Rollups can boast complete trustlessness, as the only entity involved in the validation of posted batches is the verification smart contract. Optimistic Rollups have a dependency on the existence of at least one honest verifier checking the posted batches and utilizing fraud proofs to correct irregularities. For larger systems, it is a challenge to motivate verifiers to properly validate batches [46].

- **Liveness Assumption.** In the general design, both ZK and Optimistic Rollups make no liveness assumptions, since any user is able to compute and post the next batch of the rollup chain thanks to complete data availability. However, in practice, most designs utilize sequencers and therefore are susceptible to the chain freezing if the sequencer acts maliciously or is forced offline.
- **Data Compression.** While both rollups post compressed transaction data as calldata and therefore clearly lower the size of data posted on-chain, ZK-Rollups achieve greater compression compared to Optimistic Rollups.
- **EVM Compatibility.** Optimistic Rollups benefit greatly from native EVM compatibility. In contrast, the ZK-Rollup ecosystem pools significant funds into developing EVM-compatible virtual machines and is currently restricted by the necessity of using specialised programming languages, libraries, and wallets to interact with them.
- **Specialized Hardware.** ZK-Rollups have significant demands in terms of hardware used by the nodes to be able to compute validity proofs. Optimistic Rollups on the other hand, have no such restrictions and therefore enable any node to take on the role of an operator or verifier.

6   Conclusion

Rollup technology, while highly promising, is confronted with various unresolved challenges that demand careful consideration and effective solutions. An ongoing exploration is underway to assess the efficiency and comprehensive applicability of ZKEVMs, with uncertainties about their suitability for diverse applications. Additionally, there is a need to delve into various "Data Availability solutions" within rollups, recognizing the crucial role of data distribution in ensuring security and trustlessness. Lastly, comprehending and mitigating potential failure scenarios, including those involving centralized sequencers and vulnerabilities to denial-of-service (DOS) attacks, is crucial for fortifying the resilience and reliability of rollup technologies. While current L2 implementations are in their early stages and may be considered immature, the industry and ecosystem show significant activity and momentum in advancing these technologies. The ongoing research, development, and adoption efforts signify a collective commitment towards refining L2 Rollups, underlining their potential to play a substantial role in the scalability evolution of the Ethereum network.

References

1. Hafid, A., Hafid, A.S., Samih, M.: Scaling blockchains: a comprehensive survey. IEEE Access **8**, 125244–125262 (2020)
2. Snowflake to avalanche: a novel metastable consensus protocol family for cryptocurrencies team rocket (2018). https://api.semanticscholar.org/CorpusID:198184325
3. Negka, L.D., Spathoulas, G.P.: Blockchain state channels: a state of the art. IEEE Access **9**, 160277–160298 (2021)

4. Poon, J., Dryja, T.: The bitcoin lightning network: scalable off-chain instant payments (2016)
5. Lee, S., Kim, H.: On the robustness of lightning network in bitcoin. Pervasive Mob. Comput. **61**, 101108 (2019). https://doi.org/10.1016/j.pmcj.2019.101108
6. STATE CHANNELS (2023). https://ethereum.org/en/developers/docs/scaling/state-channels/
7. Singh, A., Click, K., Parizi, R.M., Zhang, Q., Dehghantanha, A., Choo, K.-K.R.: Sidechain technologies in blockchain networks: an examination and state-of-the art review. J. Netw. Comput. Appl. **149**, 102471 (2020). https://doi.org/10.1016/j.jnca.2019.102471
8. Poon, J.: Plasma: scalable autonomous smart contracts (2017). https://api.semanticscholar.org/CorpusID:13266881
9. Ethereum Validium (2023). https://ethereum.org/en/developers/docs/scaling/validium/
10. ZERO-KNOWLEDGE ROLLUPS (2023). https://ethereum.org/en/developers/docs/scaling/zk-rollups/
11. Buterin, V.: An approximate introduction to how zk-SNARKs are possible (2021). https://vitalik.ca/general/2021/01/26/snarks.html
12. Bitansky, N., Canetti, R., Chiesa, A., Tromer, E.: From extractable collision resistance to succinct non-interactive arguments of knowledge, and back again. In: Proceedings of the 3rd Innovations in Theoretical Computer Science Conference, pp. 326–349 (2012)
13. Pinto, A.M.: An introduction to the use of zk-SNARKs in blockchains. In: Pardalos, P., Kotsireas, I., Guo, Y., Knottenbelt, W. (eds.) Mathematical Research for Blockchain Economy. SPBE, pp. 233–249. Springer, Cham (2020). https://doi.org/10.1007/978-3-030-37110-4_16
14. Bellés-Muñoz, M., Baylina, J., Daza, V., Muñoz-Tapia, J.L.: New privacy practices for blockchain software. IEEE Softw. **39**(3), 43–49 (2021)
15. Ben-Sasson, E., Bentov, I., Horesh, Y., Riabzev, M.: Scalable, transparent, and post-quantum secure computational integrity. Cryptology ePrint Archive (2018)
16. Kattis, A., Panarin, K., Vlasov, A.: RedShift: transparent SNARKs from list polynomial commitment IOPs. IACR Cryptology ePrint Archive 2019/1400 (2019)
17. Ben-Sasson, E., Bentov, I., Horesh, Y., Riabzev, M.: Fast Reed-Solomon interactive oracle proofs of proximity. In: 45th International Colloquium on Automata, Languages, and Programming (ICALP 2018). Schloss Dagstuhl-Leibniz-Zentrum fuer Informatik (2018)
18. Canetti, R., et al.: Fiat-Shamir: from practice to theory. In: Proceedings of the 51st Annual ACM SIGACT Symposium on Theory of Computing, pp. 1082–1090 (2019)
19. Herskind, L., Katsikouli, P., Dragoni, N.: Privacy and cryptocurrencies-a systematic literature review. IEEE Access **8**, 54044–54059 (2020)
20. Zhang, Y.: zkEVM (2021). https://hackmd.io/@yezhang/S1_KMMbGt
21. Polygon zkEVM Architecture (2023). https://wiki.polygon.technology/docs/zkevm/architecture/
22. StarkNet Documentation. https://docs.starknet.io/documentation/
23. Scroll Docs EVM Overview. https://wiki.polygon.technology/docs/zkevm/
24. Aztec Yellow Paper. https://hackmd.io/@aztec-network/ByzgNxBfd#Background
25. Aztec Protocol. https://docs.aztec.network/
26. Wang, D., Wang, A., Zhou, J., Finestone, M.: Loopring: a decentralized token exchange protocol (2018). https://loopring.org/resources/en_whitepaper.pdf

27. Loopring v3 Design (2021). https://github.com/Loopring/protocols/blob/master/packages/loopring_v3/DESIGN.md#introduction
28. Immutable X Whitepaper. https://uploads-ssl.webflow.com/646557ee455c3e16e4a9bcb3/6499367de527dd82ab7475a3_Immutable%20Whitepaper%20Update%202023%20(3).pdf
29. zkSync Documentation. https://docs.zksync.io/userdocs/tech/#zk-rollup-architecture
30. zkSync Era Hyperscaling. https://era.zksync.io/docs/reference/concepts/hyperscaling.html
31. Thibault, L.T., Sarry, T., Hafid, A.S.: Blockchain scaling using rollups: a comprehensive survey. IEEE Access **10**, 93039–93054 (2022)
32. OPTIMISTIC ROLLUPS (2023). https://ethereum.org/en/developers/docs/scaling/optimistic-rollups/#how-optimistic-rollups-work
33. Kalodner, H., Goldfeder, S., Chen, X., Weinberg, S., Felten, E.: Arbitrum: scalable, private smart contracts. In: Proceedings of the 27th USENIX Security Symposium, 15–17 August 2018, pp. 1353–1370. USENIX Association (2018)
34. Arbitrum Docs Inside Arbitrum Nitro (2023). https://docs.arbitrum.io/inside-arbitrum-nitro
35. Optimistic Rollup Overview (2021). https://github.com/ethereum-optimism/optimistic-specs/blob/0e9673af0f2cafd89ac7d6c0e5d8bed7c67b74ca/overview.md
36. Optimism Docs Rollup Protocol (2023). https://community.optimism.io/docs/protocol/2-rollup-protocol/#block-production
37. OP Mainnet's Security Model (2023). https://community.optimism.io/docs/security-model/
38. Boba Network GitHub repository year = 2022, note = https://github.com/bobanetwork/boba/tree/develop
39. Metis Whitepaper. https://drive.google.com/file/d/1PHsyvCJOhnUR37l0X18DuutQHXOvrNFx/view
40. Metis Documentation. https://docs.metis.io/dev/
41. Proposed milestones for rollups taking off training wheels (2022). https://ethereum-magicians.org/t/proposed-milestones-for-rollups-taking-off-training-wheels/11571
42. Schaffner, T.: Scaling public blockchains. A comprehensive analysis of optimistic and zero-knowledge rollups. University of Basel (2021)
43. Signer, C.: Gas cost analysis for Ethereum smart contracts. Master's thesis, ETH Zurich, Department of Computer Science (2018)
44. Buterin, V., Dudley, R., Slipper, M., Norden, I., Bakhta, A., Conner, E.: EIP-1559: fee market change for ETH 1.0 chain (2019). https://eips.ethereum.org/EIPS/eip-1559#specification
45. Fekete, D.L., Kiss, A.: Toward building smart contract-based higher education systems using zero-knowledge Ethereum virtual machine. Electronics **12**(3), 664 (2023)
46. Roșca, I., Butnaru, A.-I., Simion, E.: Security of Ethereum layer 2s. Cryptology ePrint Archive (2023)

Decentralized Identity Management and the European Identity Reform

Panagiotis Rizomiliotis[1]([⊠])[iD] and Maristel Hairetaki[2]

[1] Harokopio University, 17676 Athens, Greece
prizomil@hua.gr
[2] University of Piraeus, 18534 Piraeus, Greece

Abstract. In this work, we present the advances in identity management and the paradigm shift from centralized to decentralized solutions. In this context, we introduce and analyze EUDI wallet, the European Commission's proposal as it has been recently reported and agreed in the final eIDAS 2.0 document.

Keywords: decentralized identity · self-sovereign identity · eIDAS 2.0

1 Introduction

As an outcome of the Internet revolution, a fundamental shift in the perspective of digital interactions between entities has emerged. This shift has given rise to the idea of decentralized identities an upcoming method that offers control, security and privacy to individuals and businesses over their personal information. Amidst the ongoing digital transformation, there is a constant reshaping of the way businesses and individuals interact, the effective management of digital identities becomes requisite.

A digital identity is a collection of data that can establish a virtual representation of an entity in the digital realms. Any entity that is considered to have a digital presence can also establish a digital identity; including individuals, organizations, applications and even devices in some cases. When a digital identity is established, it enables the seamless authentication and assessment of an entity on the web without requiring human intervention.

A digital identity not only comprises data that can uniquely characterize an entity but also information gathered from interactions and relationships from the entity's digital environment. Every entity is not limited to only one identity, it can maintain numerous sub-identities each containing a set of attributes tailored around a specific context or established to be utilized in a unique situation. Identity verification is one of the main purposes digital identities are used, the entity can easily be verified during processes like online transactions, accessing government services and interacting with organizations that require proper identification. Besides verifying an entity, digital identities can also serve as Access management mechanisms, granting or restricting entry to authorized individuals based on the appropriate attributes.

N. Pitropakis and S. Katsikas (Eds.): *Security and Privacy in Smart Environments*, LNCS 14800, pp. 240–255, 2025.
https://doi.org/10.1007/978-3-031-66708-4_11

With the introduction of digital identities in the majority of web applications, it is important for an efficient Identity Management System (IMS) to be established that would satisfy multiple services, standards and technologies simultaneously. Therefore, through the central implementation of an IMS that can concurrently address multiple factors, it becomes possible to enable cross-network identity management.

A crucial security concern of IMSs is the risk of data breaches that could compromise the private information belonging to each entity, particularly in the context of verifiable credentials. Verifiable credentials serve as the electronic counterparts to physical identification documents like a driver's license or a passport. The solution to this challenge lies in the concept of decentralized digital Identities (DIDs), a novel type of IMS where the uses is the sole owner of their identity without the need for any external entities to oversee the authentication process.

Europe faces unprecedented changes, in which digital transformation is fundamental elements for the future of Europe. In this context, the European Commission has decided to support legally and technologically the adaption of a common identity management system for the EU citizens. As a starting point, the eIDAS regulation was adopted in 2014 [1]. While the importance of the regulation was widely recognized, the proposed solution has not been accepted, in practice, by the EU citizens.

Thus, the European Commission, five years after the adoption of eIDAS, proceeded with an evaluation of the regulation. Based on an extensive impact assessment which was carried out by a group of experts, analyzing in detail eIDAS, an amendment was proposed, known as eIDAS 2.0 [2]. eIDAS 2.0 proposes a radically different electronic identity management model by introducing the concept of the EU digital identity wallet into the EU framework for digital identity. The new approach is user-centric and privacy friendly. The EU citizens using a mobile application will be able to gain control over their personal data while using online services across Europe. The new architecture is an implementation of the so-called Self-Sovereign identity (SSI) model [4].

In our work, we highlight the importance of the new identity management system. Moreover, we attempt an evaluation of the challenges that new model faces and we identify possible threats that may undermine the whole project. In Sect. 2, we provide a short background on centralized and decentralized identity management systems. In Sect. 3, the eIDAS regulation is presented and in Sect. 4, the new wallet based architecture is introduced. In Sect. 5, the opportunities that the common identity system can offer to the European digital economy are presented and in Sect. 6, the challenges and threats that this common European effort can face are analyzed. Finally, in Sect. 7 we present some thoughts on the eIDAS 2.0 final document and the role of browser providers.

2   Advances of Identity Management Systems

Significant progress has been made during the past years, regarding the improvement of identity management systems. The numerous improvements derive from the need of greater security and privacy aspects, considering the identity management of individual personnel. One way for this to be achieved is to incorporate multiple layers of verification as well as establish a more user-centric approach of identification, reducing the risk of identity theft and exposure.

Among the most notable developments is the adoption of biometric authentication, such as fingerprints, facial and iris recognition, adding a new layer of security protection and at the same time reducing the need of password memorization. Another way to incorporate multiple layer of verification during the authentication process is by using Multi-Factor Authentication (MFA). This technique is slowly becoming a norm, as it is incorporated into most web services.

Blockchain technology, on the other hand, has been a primary tool for decentralized identity solutions, offering more control over personal information stored in the chain while reducing the risk of data breaches. The Self-Sovereign Identity technology gives even more authority to individuals concerning their digital identities. Last but not least, the emerging concept of zero trust security models challenges the old concept of trust in networks and adds a new specter of trust in out increasingly interconnected world.

Password less logins are taking over the traditional password authentication approach thus ensuring more safety and comfort for users. These changes have derived from data privacy laws which have been proposed so that user experience is enhanced in a secure manner that lowers the possibility of identity theft and exposure. Therefore, we can say that all these innovations form the foundation of digital identity management that seeks to ensure security, privacy, and user friendliness.

2.1 Centralized Solutions

Traditionally, identity management systems have been following a centralized model [3], revolving around a singular entity that controls and manages all aspects of user identities. Such solutions are common across many fields, including but not limited to corporate, government and online services.

The user directory, also referred to as an Identity and Access Management (IAM) system, is a key implementation of a centralized identity management system. Users' profiles, authentication credentials, and access rights can be stored in this directory. Similarly, such systems allow for changes to be made to user profiles, by making it possible for modifications in permissions and characteristics throughout the lifecycle of the user identity. In many centralized identity management systems, users are granted some roles, which determine the scope of items they can access. Overall, such integrations enable seamless user access to multiple resources using a single set of credentials, enhancing convenience and productivity.

Single Sign on (SSO) is designed as part of centralized identity management that allows for a continuous and transparent user experience, where multiple resources can be accessed using the same set of credentials. In many cases, such functionalities are accompanied by multi-factor authentication and password policies in order to ensure higher security levels. When choosing an identity management, organizations need to consider what they really need and the level of security required.

With the primary characteristic of centralized solutions being that they consist of a single point of control for both authentication and authorization, there are many drawbacks regarding privacy and trust issues. The main factor for these issues being that the concentrated power withing the identity provider can give rise to concerns about extensive data collection and potential surveillance, raising many privacy issues for users. There is high accumulation of user data within just a single entity which makes prone to potential attempts for data breaches and identity spoofing, thereby jeopardizing the confidentiality and integrity of users' personal data.

Therefore, there has grown the need for individuals to manage their personal data through effective methods that ensure privacy protection. Propositions have been made stating that decentralized technology offers a new way of moving away from the centralized model, towards systems that are privacy driven and promote user's confidentiality. Individuals can have more control over their personal data by using decentralized identity systems thus minimizing privacy risks that the centralized identity systems are prone to. The decentralized approach seeks to foster convenient and private information management while maintaining trust on users' behalf.

2.2 Decentralized Solutions

Decentralized identity was introduced at the Internet Identity Workshop in 2015 [5], as a concept that revolutionizes the traditional centralized identity system by removing central registration authorities' control. The primary objective of decentralized identity management systems is the creation of a trustworthy digital identity ecosystem that is centered around users, thus making it a decentralized infrastructure. In the latest years there has been a remarkable surge in adoption and development of decentralized identity infrastructures globally, with government and companies actively participating in its advancement. This collective effort signifies the growing recognition and commitment towards fostering a more secure, user-centric, and globally accepted approach to digital identity.

Decentralized systems aim to empower users to independently manage their identities and selectively disclose personal information to various service providers or parties, without the need of any centralized entity. This type of architecture provides users with enhanced control over their digital identities, giving them full control to choose who may access their data. There is also a privacy improvement, that derives from the exposure limitation of personal information, reducing the risk of attacks like data breaches. Self-sovereign identity (SSI) [4] is a decentralized solution that grants individuals complete authority over their digital identities. Users have the capability to generate and manage their digital identity credentials, selectively sharing them with service providers or other users. Another widely used solution is the Decentralized Identifiers

(DIDs). These types of identifiers are distinctive, persistent, and globally resolvable, offering association with individuals, organizations, data models or entities. DIDs are registered on distributed ledgers or decentralized networks, offering a means to establish ownership and control of an identity without dependence on a central authority.

While in a decentralized architecture system, non-disclosure of personal information is a necessity. Verifiable credentials can provide a solution to this, by enabling the verification of identity claims. Individuals can collect verifiable credentials from various issuers and later share them with service providers when needed. All in all, decentralized identity management solutions offer the promise of enhanced user control, heightened privacy protection, and bolstered security. The selection between centralized and decentralized identity solutions hinges on the specific use case, regulatory obligations, and the preferences of both users and service providers.

Advantages of Decentralized Solutions. Decentralized identity can pose key advantages for both organizations and individuals, thus making them a preferred identity management solution. Organizations are able to issue and verify credentials with ease, while reducing the risk of data breaches as they store less user personal information. At the same time individuals have full control and ownership over their digital identities, decreasing the possibility of identity fraud to occur.

Organizations use decentralized identities, they are able to verify information with great ease, without the need of contacting the issuing party. For example, during the recruitment process of a new employee, the verification of various documents such as IDs, certificates, driving licenses to check whether they are valid could be a time-consuming process. While using decentralized solutions, this process could be done instantly, and be as easy as scanning a QR code for example. This process could also prevent the submission of fraud paperwork from individuals.

Individuals on the other hand, using decentralized identities, are able to have full ownership over their personal data. They can choose who would gain access to their information, while they are also able to specify what they will share each time. This allows for control sharing and prevents from the spread of their data without their knowledge, as authorization has to be given by the owner in order for information to be shared upon request.

3   The eIDAS Regulation

3.1   Introduction to eIDAS

In 2014, the European Union adopted the Regulation 910/14 on electronic IDentification, Authentication and trust Services (eIDAS) [1]. The eIDAS regulation's objective was to create a common regulatory framework for secure and trustworthy cross-border electronic transactions in the internal EU market. eIDAS provides the legal validity of trust (qualified) services, like electronic signature and electronic seal, timestamping, and certificates for website authentication. However, the most challenging and interesting contribution was the common European framework for EU citizens electronic cross-border identification. The EU citizens or residents of one Member State would be able to access public services of another Member State electronically.

Within the eIDAS framework, Member States were invited to submit for approval one or more national electronic identification (eID) schemes. The approved schemes are called *notified* and since 2018, all Member States were obliged to recognize the notified eID schemes. The schemes are categorized based on the level of assurance of the authentication mechanism into three levels: low, substantial, high. Almost all Member States have notified at least one eID scheme, and 17 have notified at least one eID scheme of assurance level high.

The eIDAS supports a federated approach. Each country operates one or several eIDAS nodes as a gateway to public services and the national notified eID schemes. When a citizen from country A wants a public service from country B, the eIDAS nodes are responsible for the citizen authentication and the communication between the citizens and the service providers (see Fig. 1). The communication between the eIDAS nodes is standardized.

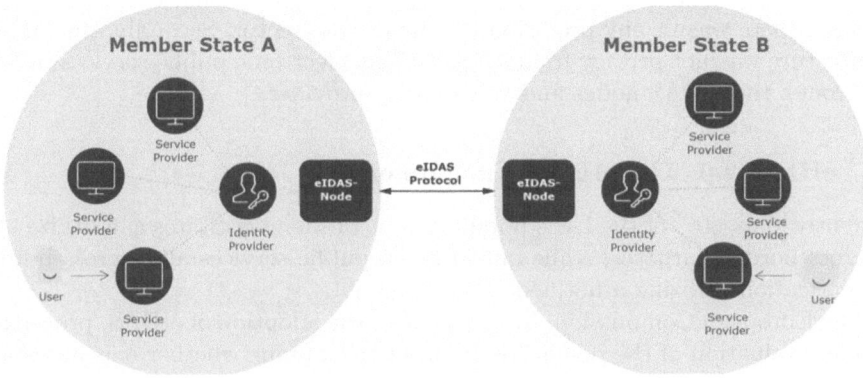

Fig. 1. eIDAS interoperability Architecture [11].

eID Schemes and Authentication Mechanisms. The Member states have been left alone to decide on the eID scheme that they prefer and on the interaction with their citizens.

A variety of authentication mechanisms are used to provide substantial level of assurance. All of them are two factor based. The vast majority of these schemes use a long-time sustained password, stored and provided by the EU citizen, combined with a one-time password (OTP). This password is valid for a short period of time and it is usually send via SMS message or via a messaging application to a registered SIM card. There also some implementations that are using either software OTP generators installed in the citizen's phone or physical electronic device for the OTP generation. In any case, this class of authentication solutions doesn't depend on advanced cryptographic algorithms. They rely mainly on secure random OTP generators and on secure transmission of user's password (mainly protected by TLS protocol).

High assurance eID schemes are based on public key cryptography. The user is authenticated by proving the possession of a private key. The private key is store by the user in a secure device, like a physical smart card or a crypt SIM, while corresponding public key is stored in a certificate. The smart card used is either dedicated for the eID scheme or the national identity card is used. A challenge-response type of protocol is used for authentication. The user (prover) proves the ownership of a private key, and the verifier obtains reliably the correct public key leveraging a Public Key Infrastructure (PKI). The vast majority of the eID schemes are using one of the two most widespread authentication standards, i.e. either SAML 2.0 [12] or OpenID Connect [13]. Very few (only two) support FIDO [14].

Unfortunately, the proposed common European identification approach and electronic proof of identity had limited acceptance. Two were the main reason: firstly, the eIDAS regulation focused on the public sector, i.e. while the Member States were obliged to offer cross-border public services using the notified eID schemes, there wasn't any provision for the private sector. Secondly, the eIDAS architecture was not privacy friendly. Accessing electronic public services had to go through the eIDAS nodes and the identity provider.

3.2   eIDAS 2.0: The eIDAS Amendment

Currently, only 60% of the EU's population in 14 Member States can utilize eID for cross-border purposes, while only 14% of public services allow cross-border authentication utilizing a notified eID scheme [9].

The European Commission, five years after the adoption of eIDAS, proceeded with an evaluation of the regulation in order to examine whether it is necessary to adjust the scope of the regulation or to modify its specific provisions. That is, taking into account the experience gained from the implementation of the regulation, but also the technological, market and legal developments, it was investigated to what extent the eIDAS framework remains suitable for the purpose for which it was created.

As part of the evaluation, an open consultation was conducted during the pandemic, from 24 July to 2 October 2020. The aim of the consultation was to collect comments and observations on the barriers presented to the development and adoption of trust services and eID schemes in Europe, taking into account the technological developments, market evolution, and citizens' expectations.

Thus, an extensive impact assessment was carried out by a group of experts, analyzing every detail of eIDAS, including the expected burdens and benefits for relevant stakeholder groups, such as public authorities, trust service providers, conformity assessment bodies, electronic identity providers, supervisory bodies. That is that, for the entire eIDAS ecosystem.

Based on the data collected during this evaluation, an expert committee reached the conclusions which were reflected in a proposal to amend eIDAS, the so-called eIDAS 2.0 (6/2021) [2]. After almost two years of public debate and consultation, the European Council and the European Parliament issued two amendments of the eIDAS 2.0 proposal, in December 2022 and in February 2023, respectively. On the 29th of June 2023, trilogue negotiations between the European Parliament and the Council of the EU have led to a political agreement on the main points of the eIDAS amendment. On the 8th of the November 2023, there was the agreement on the final version of the proposal [7].

The eIDAS 2.0 proposal recognizes that the eIDAS Regulation was an important step towards an common electronic identity for citizens of EU Member States. However, the variety of national identification schemes led to lack of uniformity and limited success of the initiative. The central innovative element of this agreement is the adoption of a personal digital wallet in the form of a secure and easy-to-use mobile application. This new wallet is called *EU Digital Identity (EUDI) Wallet* and it is expected to enable all EU citizens, residents and businesses to have reliable access to public and private online services across Europe.

4 EUDI Wallet and a New European Identity Architecture

The European Commission highlights the importance of secure privacy friendly cross-border transactions and, for that, the eIDAS 2.0 amendment proposes the introduction of the concept of the EUDI Wallet (EUDIW) into the EU framework for digital identity. The wallet is expected to facilitate the adoption of a new user-centric electronic identity management model that will offer privacy by design. It will also expand the digital identity-based ecosystem to include other data items associated with an individual, such as diplomas, driver's licenses and more. The new user-centric approach and the proposed ecosystem appear in Fig. 2.

EU Citizens, using EUIDW, will be able to upload and manage their Personal Identification Data (PID) as well as other (Qualified) Electronic Attribute Attestations ((Q)EAA). The identity and attribute providers will be certified, while the service providers will be chosen from a trusted list. The service providers are called *relying parties*. The wallet owner will be able to prove its identity and attributes to a relying party without interacting with an identity/attribute provider.

At the same time, using her EUDIW, the citizen will be able to manage the details of her personal identities (private and public), and to share any parts of her personal information she wishes, and only with her expressed consent. Also, since the service providers must be registered, the citizen knows exactly to whom she has shared personal information and for what purpose. That is, citizens will have full control over their identity with the full convenience of their mobile

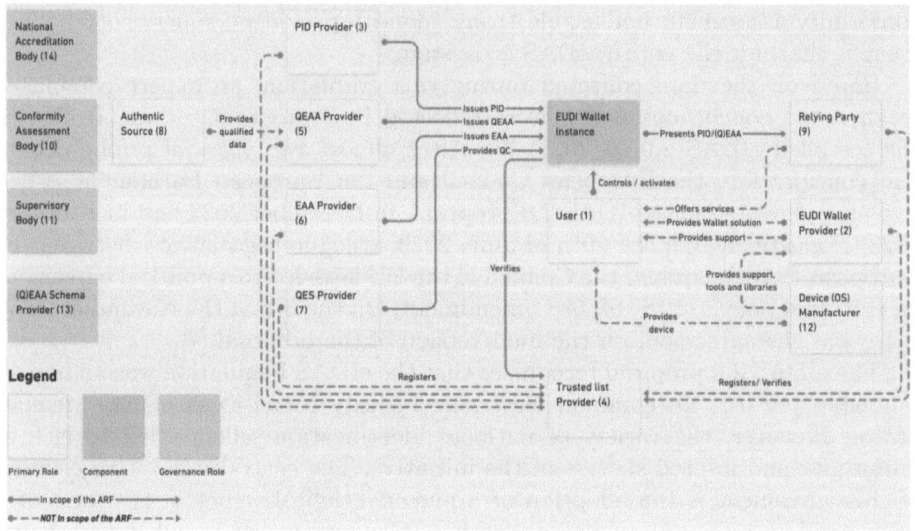

Fig. 2. EUDI Wallet Roles [15].

device. The EU, with eIDAS 2.0, expects to revolutionize digital identification by promoting the most modern and privacy-friendly identity management model, the self-sovereign identity model.

Member States will have two years in their disposal to offer a certified EUDIW solution to their citizens for free. The EUDIW issuance procedure must meet the highest assurance level possible, and its use will be voluntary for the citizens. To ensure that member states will be ready to provide the EU digital identity wallet within the deadline set in the Regulation upon its adoption, the Commission is working with the member states in three parallel ongoing actions. More specifically:

- The EU Digital Wallet Architecture Reference Framework (ARF) which aims to make the implementation of the technology more binding, facilitating its adoption and interoperability (the first version of the Toolbox was published on GitHub in February 2023, [15]).
- the open source digital wallet model implementation project that member states can use as a basis for their national wallets
- four Large Scale Pilots (LSP) are implemented alongside these other efforts in order to provide valuable insights into the practical implementation of digital wallets.

The requirements and specifications of the ARF, once completed, will be mandatory. The whole project is particularly demanding, as the technical solution will have to be dynamically adapted to the legal framework being developed. While member state will be, once again, left alone to design their national eID scheme, it is clear, that the technological and security restrictions that the ARF enforces on the national EUDI wallets will also restrict the eID schemes design choices.

5   Digital Europe and Trustworthy Digitization

The EUDI Wallet is an ambitious initiative that aims to create a unified electronic identity system for citizens, public sector and businesses in the European Union that will enable them to access online public and private sector services across borders.

Due to world's rapid digitization, there is an increasing demand for digital credentials for citizens to identify and authenticate themselves online. The EUDIW's ambition is to offer the technical infrastructure that will make identification, authentication and data sharing, possible with a high level of security and trustworthiness on a massive scale.

For many online interactions with sensitive public services and in certain sectors, like health and finance, government certified identity information it is exactly what it is needed. According to McKinsey [8], EUDIW can unlock value equivalent to 3% upto 13% of GDP by 2030.

The EUDIW will allow people to use their digital identities to access online services across borders. Since already the initial adoption of the regulation, back in 2014, public sector has been identified as the main beneficiary of the unified identity management. Requesting birth certificates, medical certificates, reporting a change of address, or filing tax returns, are only a few indicative services.

The amended version of the regulation has extended the application area. The EUDIW is currently tested in real-life situations encompassing different sectors, like healthcare, financial services, education, and transportation. The personal digital wallet promises significant benefits for citizens, as service providers can rely on a single trusted and secure digital relationship with their customers and they will be able to provide better personalized services.

It goes without saying that new technical measures, and procedures are needed for establishing trust frameworks in various different sectors and the European Commission has identified a set of high priority use cases. For that, more than 250 private companies and public authorities across the 27 member states and several associated countries, like Norway, Iceland, and Ukraine, are participating in the four EUDIW pilots.

5.1   Bank Sector

The EUDIW will provide new opportunities for banks and financial institutions to develop innovative services offering personalized products. Services that range from on-line opening of a bank account, to the provision of new payment means and the acceptance of payment in a retail context using EUDIW will be provided.

Customers will also be able to choose where their financial data is stored, which organization has access to it, and be able to keep track of the sharing of their data.

Moreover, a bank could offer additional value-added services such as improved fraud detection (Know Your Customer policies) or optimized cross-selling opportunities based on purchasing behavior analysis of all transactions performed by its customers over time and based on purchase history. They will provide better services to enable customers make more informed financial decisions. It is expected that EUDIW will reduce costs, and increase revenue.

5.2 eHealth Sector

EUDIW has the declared ambition to become a fundamental element for cross-border use cases in the field of social security. The citizens will have a secure and paperless way to manage medical data and prescriptions through mobile phones.

As a first step, the wallet will be engaging in the execution of the portable document A1 (PDA1) and the European Health Insurance Card (EHIC) as part of the European Social Security Pass (ESSPASS) project [16].

Furthermore, EUDIW will support ePrescriptions, i.e. storing a medical prescription that can be used in any Member State.

5.3 Digital Credentials

The revision of the eIDAS regulation extends its scope of competence beyond identity, encompassing attribute management via the EUDIW. Each citizen has several identities which are treated as a set of attributes. Using the wallet, a citizen can select a subset of all the available attributes and compose a new identity, a different one per service provider.

For every use case, each everyday electronic service requires different attributes values from the user. The possession of a digital driving license needs to be proven for renting a car on-line, while the electronic attestation of age is needed to buy alcoholic drinks. The EUDIW will be used for checking into a hotel, buy an air-ticket and issuing the boarding pass, or even for applying for a university in another EU Member State.

In this context, issuance and management of educational credentials, professional qualifications, mobile driving licence and digital travel credentials have been identified as fundamental elements for EUDIW wide adaption and acceptance.

6 Challenges and Threats

One of the main objectives of eIDAS 2.0 is to offer a user-centric identity solution that supports security and privacy by design. In the final version of the amendment there are several provisions in this direction. However, it is still unclear whether, in practice, the vision is realistic or just an illusion.

We make a first assessment of the new architecture taking into consideration the first version of the ARF [15], as well as our experience gained from the existing eID schemes.

6.1 Privacy and Design

Pseudonymity: Many people need anonymity online in order to exercise their human rights, particularly freedom of speech. The EUDI Wallet will give the option to use a pseudonym generated by the Wallet. The pseudonym is stored locally in the Wallet and re-used when necessary.

However, the right to use pseudonyms can be restricted by national laws. It is not clear how the citizens will be protected either from a relying service provider who rejects pseudonyms without a reasonable legal ground or from a Member state that does not strongly support the right to anonymity. It is probable EU citizens will be entitled of varying levels of privacy for the same service across Europe.

Unobservability: According to the last agreed version of the eIDAS 2.0, and more precisely in Recital (11c), the EUDIW providers must of *Unobservability*. Unobservability ensures that a user may use a resource or service without others, especially third parties, being able to observe that the resource or service is being used [10]. The EDIW providers should ensure that by not collecting data and not having insight into the transactions of the users of the Wallet.

However, so far, safeguards that are defined in the legislation are missing from the ARF. At the same time, the eIDAS 2.0 agreed document states that in some cases the providers could be granted access to the information necessary for the provision of a particular Wallet service.

EUDIW Non-discrimination: The use of the Wallet will be voluntary. In The final text of the eIDAS amendment ensures that no one is obliged to use the Wallet. It remains challenging to guarantee that everyone who decides not to use the Wallet will not have either to pay a higher price or to be excluded from access to any service.

Citizen's Privacy Awareness: The eIDAS amendment provides an arsenal of weapons that the citizen can use to protect her privacy, Besides, the afore-mentioned protection mechanisms, the Wallet offers access to a full transaction history of every request for information the user has ever received and to the information the user has shared. Moreover, the user will be able to learn more details about the companies requesting the user's information. More precisely, all relying parties must be registered in the country they are established in. They have to identify themselves and provide information about the use case for which they ask user's personal data.

It is clear that the user of the Wallet will be powerful. However, it is notorious known that users are the weakest links of the security and privacy chain. Users tend to underestimate the impact of personal information sharing either due to ignorance or due to lack of technological skills.

Mobile Device Security: The EUDIW will be a user friendly application. Mobile phones have been selected as the mean for identity management mainly due to their pervasiveness. At the same time, cell phone is becoming single point of failure. Citizens privacy violation, that can even lead to identity stealing, depends on the security of the device and on the users choices. Recently, the use of mobile spyware, like Pegasus and Predator, to illegally monitor EU journalists and politicians has demonstrated the vulnerability of mobile devices [17].

It is promising that the code of the Wallet application will be open source licensed and thus, available to public scrutiny. However, such public assessment will not be possible for the software of the back end. The eIDAS 2.0 regulation gives Member States the choice to keep close the source code in the back end for (national) security reasons.

Finally, each Member State will be responsible for the certification of their national EUDIWs. However, it is still unclear how the security assessment will be performed and how the same maximum possible level of security will be guaranteed for all the different EU Wallets. For the first few years, there will not be any EU-wide security standard. There will only be national security certification schemes and in practice there will be different security levels between Member States. The back end software might not receive any security certification at all.

The New Ecosystem and the Role of Member States: In the new ecosystem, there are several main and critical roles: PID providers, QEAA and EAA providers, EUDIW providers and device manufacturers, QES providers, relying parties, conformity assessment bodies, national accreditation bodies, and supervisory bodies. All of them must interact with each other and across Europe, following national and European laws, directions, guidelines, standards and certifications.

The definitions provided in the eIDAS 2.0 are very generic to offer flexibility. At the same time they leave sufficient space for various interpretations. Given that the ecosystem is new, it needs organizational and technical orchestration in order to work. So far, it is not clear how this will be achieved. Governance tools for the decentralized digital trust infrastructure and standard protocols must be mandated.

6.2   Privacy Requirements and Cryptographic Solutions

In the final version of eIDAS 2.0 and in the first draft version of the ARF working document, two privacy requirements are defined: selective disclosure and unlinkability. Two informal definitions are provided in the ARF:

- selective disclosure: capability of the EUDI Wallet that enables the user to present a subset of attributes provided by the PID and/or (Q)EAAs
- unlinkability: lack of information required to connect the user's selectively disclosed attributes beyond what is disclosed.

In order to support selective disclosure, the first version of the ARF mandates the use of two standards. Namely, the ISO/IEC 18013-5 [18] mobile driving license (mDL) and IETF SD-JWT [19] are selected. The ISO mDL is a mature standard, while IETF SD-JWT is still a draft. Both of them are based on salted hashes for selectively disclosing attributes.

Salted hashes is an easy to implement solution leveraging cryptographic hash functions and digital signature algorithms as a proof mechanism. However, salted hashes technique doesn't support by design unlinkability. To cope with this design limitation, it is proposed to use redundancy. The attributes provider must generate several signed versions of the same set of attributes using different random strings (salt) each time. The wallet owner will a different version of the same set every time the set must be presented to a service provider.

This approach, besides the obvious waste of resources, it is based on a strong security assumption. It is assumed that the attributes providers will not collude with the service providers.

On the other hand, there are very powerful cryptographic solutions, like zero-knowledge proof (ZKP) systems [6]. ZKPs have emerged as a powerful tool in addressing privacy issues within digital identity systems. ZKPs support by definition both unlikanbility and selected disclosure, while they can minimize the disclosed information. This capability has garnered significant attention and has found successful applications in various domains to enhance privacy and security.

Digital identity systems often require users to verify their identity for various purposes, such as accessing online services, participating in financial transactions, or interacting with government agencies. However, traditional identity verification methods often involve sharing sensitive information, raising concerns about data privacy and the potential for identity theft or misuse. By leveraging ZKPs, digital identity systems can overcome these privacy challenges. ZKPs allow individuals to provide proof of their identity while keeping their personal information confidential. The concept behind ZKPs is that a prover can convince a verifier of the truthfulness of a statement without revealing any additional information beyond the validity of the statement itself.

In the context of digital identity systems, ZKPs enable individuals to authenticate themselves without disclosing their actual identity attributes. Instead, they provide a cryptographic proof that verifies the validity of their claims. For example, an individual can prove they are of legal age without revealing their exact date of birth, or demonstrate that they hold a credential without exposing the details of that credential. This privacy-preserving aspect of ZKPs is crucial for protecting user privacy and preventing unauthorized access to sensitive personal information. It allows individuals to maintain control over their identity data, reducing the risk of identity theft or surveillance.

ZKPs also offer advantages in terms of efficiency and scalability. Traditional identity verification methods often involve complex and time-consuming processes, such as manual document verification or data sharing among multiple parties. ZKPs streamline this process by enabling efficient and secure identity proofs, reducing the need for extensive exchange of data and simplifying the verification process. The adoption of ZKPs in digital identity systems is gaining momentum, with various projects and initiatives exploring their potential. By leveraging ZKPs, these systems can enhance privacy, improve user control over personal data, and mitigate the risks associated with centralized identity management.

Unfortunately, ZKP systems don't have mature standards and currently is not promoted by the ARF working group. It only appears in Recital (6b) as an example of the privacy-preserving technologies that can be applied.

7   Discussion

One of the main controversies was Article 45, on the "Requirements for qualified certificates for website authentication". Aiming to provide more information about the websites, the EC enforced the idea of "extended validation" under the name "Qualified Website Authentication Certificates (QWACs). In a nutshell, the owner of a domain name must be visible in the web browser.

There were several reactions from the browser providers, mainly arguments against QWACs, and an open discussion on the importance of this type of website of certificates. However, there was a new paragraph in the amendment that created a huge wave of reaction from the security community. According to a version of eIDAS 2.0, every web browser in the world will be forced to trust, unconditionally, root certificates from all European Trust Service Providers.

That is that, European Trust Service Providers are not obliged to fulfill the Certification Authority Browser Forum security requirements, and they only need to be approved by their own government, breaking the complete trust architecture of the world wide web. According to the law, when a certificate is found to be used for surveillance, the browser will still have to trust it. Fortunately, in the last version of eIDAS 2.0, there was a distinction between QWACs and TLS certificates. Thus, browser providers can choose to follow more strict rules to protect web traffic.

There is no doubt that the new eID architecture that the EC is envisioning for the EU citizens can be very beneficiary of the EU economy. At the same time, it needs a lot of effort to become a success story. The citizens need to believe in the new model, embrace it and support it.

Of course, in order to trust a new path, you need to first to trust the leader that you follow and certainly the problems and suspicion that Article 45 caused did not help.

References

1. Regulation (EU) No 910/2014 of the European Parliament and of the Council of 23 July 2014 on electronic identification and trust services for electronic transactions in the internal market and repealing Directive 1999/93/EC
2. Proposal for a Regulation of the European Parliament and of the Council amending Regulation (EU) No 910/2014 as regards establishing a framework for a European Digital Identity. COM/2021/281
3. Fang, J., Yan, C., Yan, C.: Centralized identity authentication research based on management application platform. In: First International Conference on Information Science and Engineering. IEEE (2010). https://doi.org/10.1109/ICISE.2009.382

4. Preukschat, A., Reed, D.: Self-sovereign identity: decentralized digital identity and verifiable credentials. Manning (2021)
5. Jing, Y., Li, J., Wang, Y., Li, H.: The introduction of digital identity evolution and the industry of decentralized identity. In: 3rd International Academic Exchange Conference on Science and Technology Innovation (IAECST). IEEE (2021). https://doi.org/10.1109/IAECST54258.2021.9695553
6. ZKProof Standards. https://zkproof.org/
7. European Digital Identity - Provisional Agreement. https://www.europarl.europa.eu/cmsdata/278103/eIDAS-4th-column-extract.pdf
8. McKinsey and Company. Digital ID: The opportunities and the risks. https://www.mckinsey.com/industries/financial-services/our-insights/banking-matters/digital-id-the-opportunities-and-the-risks
9. Attitudes towards the Impact of Digitalisation on Daily Lives. https://europa.eu/eurobarometer/surveys/detail/2228
10. Common Criteria Project, Common Criteria for Information Technology Security Evaluation, Version 3.1, Part 2: Security Functional Requirements (2006). www.commoncriteriaportal.org
11. "How does it work? The eIDAS solution. https://ec.europa.eu/digital-building-blocks/wikis/pages/viewpage.action?pageId=467109866
12. SAML 2.0. http://docs.oasis-open.org/security/saml/Post2.0/sstc-saml-tech-overview-2.0.html
13. OpenID. https://openid.net/
14. FIDO Alliance. https://fidoalliance.org/
15. European Digital Identity Architecture and Reference Framework. https://ec.europa.eu/newsroom/dae/redirection/document/83643
16. ESSPASS Pilot Project. https://ec.europa.eu/social/main.jsp?langId=en&catId=89&newsId=10341&furtherNews=yes
17. Pegasus and similar spyware and secret state surveillance, Parliamentary Assembly, Doc. 15825, 20 September 2023
18. ISO/IEC 18013-5 mobile driving license. https://www.iso.org/standard/69084.html
19. IETF Selective Disclosure for JWTs (SD-JWT). https://datatracker.ietf.org/doc/draft-ietf-oauth-selective-disclosure-jwt/

Designing Secure and Privacy-Aware IoT Services in the Health Sector

Costas Lambrinoudakis[1] (ID) and Christos Kalloniatis[2]([envelope]) (ID)

[1] Department of Digital Systems, University of Piraeus, 150 Androutsou St, Piraeus, Greece
clam@unipi.gr
[2] Department of Cultural Technology and Communication, University of the Aegean, University Hill, Lesvos Island, Greece
chkallon@aegean.gr

Abstract. The chapter discusses the intricate challenges of privacy and data protection in eHealth/M-Health systems. These systems must adhere to specific demands from organizations and users, along with the diverse legal mandates set by the GDPR, which governs the rights of data subjects and the duties of data controllers. To tackle these challenges, the chapter introduces a Privacy and Data Protection Framework. This framework outlines necessary steps and measures—technical, organizational, and procedural—to be implemented. Unlike prior studies, it integrates privacy by design principles with the recent GDPR stipulations, facilitating a robust process for identifying essential technical security and privacy needs. Additionally, the framework suggests a method for verifying that these identified requirements meet the objectives outlined in the Data Protection Impact Assessment (DPIA), conducted in compliance with the GDPR.

Keywords: Privacy · security · DPIA · GDPR · health applications · Framework

1 Introduction

Europe's medical advancements and demographic shifts are rapidly aging its population, presenting numerous challenges for the EU's future healthcare policies. To cater to an increasing number of individuals, there's a growing reliance on Information and Communication Technology (ICT) within healthcare, leading to the rise of eHealth systems [1]. These systems, utilizing ICT, enhance various aspects of healthcare including prevention, diagnosis, treatment, monitoring, and data management [2, 3]. Research in eHealth particularly focuses on remote health-monitoring systems and equipment, propelling the widespread adoption of mobile medicine/mHealth [1, 4–6]. Supported by mobile devices and wireless networks, mHealth offers significant benefits like ease of use, infection risk reduction, and cost-effective care, while collecting vital data through wearable or implantable biosensors [7, 8].

However, these advancements also bring security and privacy concerns related to data storage, processing, and exchange with third-party systems. These concerns are major barriers to fully implementing e-healthcare solutions [9]. The "privacy paradox"

N. Pitropakis and S. Katsikas (Eds.): *Security and Privacy in Smart Environments*, LNCS 14800, pp. 256–285, 2025.
https://doi.org/10.1007/978-3-031-66708-4_12

exemplifies this, where users are worried about their medical data's protection yet rely on these devices for health reasons [10]. The introduction of the GDPR has further emphasized privacy protection, influencing user trust and adoption of wearable technologies. In general, privacy and personal data protection constitutes a major concern for e-health consumers [11], affecting wearables adoption. Indicatively, 82% of the respondents to a PricewaterhouseCoopers (PwC) survey reported that they are worried that wearable technology will invade their privacy [12]. Therefore, users' acceptance is strongly dependent on trust. Additionally, the principles enforced by the new General Data Protection Regulation (GDPR) put special emphasis on the protection of citizens' privacy by elevating the obligations of the parties collecting, distributing and processing users' data [13].

In response, privacy engineering has become crucial in Europe and beyond, forming an integral part of system development. It involves defining technical requirements to ensure minimum security levels and build trust among citizens [14]. However, existing research [15–18] often fails to integrate privacy engineering benefits with GDPR's technical and legal aspects, especially in medical wearables. Some studies, like Mustafa, Pflugel & Philip's [9], focus on specific patient groups, neglecting broader or healthier target audiences for m-health applications.

Addressing these gaps, the EU-funded BIONIC system introduces medical wearables in workplaces, setting a standard for respecting privacy rights and maintaining high data protection standards. BIONIC aims to develop a holistic, unobtrusive, autonomous, and privacy-preserving platform for real-time risk alerting and coaching, tailored to aging workforces. Its Privacy and Data Protection Framework outlines steps for meeting security, privacy, and legal requirements. By combining privacy by design principles with GDPR requirements, BIONIC establishes a robust process for identifying technical needs and validating them through the Data Protection Impact Assessment (DPIA), thus enhancing end-user trust in the software.

The rest of the chapter is organized as follows. Section 2 presents the concept of privacy within e-Health/m-Health Systems. Section 3 briefly describes the BIONIC system and its main features while Sect. 4 presents the proposed privacy Framework. Finally, Sect. 5 concludes the work presented.

2 Privacy in e-health Systems and GDPR

The extensive use and popularity of mobile medical devices in eHealth, which gather individual health data and manage health-related information, present significant challenges for eHealth systems. These systems need to effectively support the protection of individuals' personal data privacy and ensure proper access control [1]. Given that health information is an exceptionally sensitive type of personal information, coupled with ethical considerations, privacy is recognized as a primary requirement in the eHealth sector [19]. Due to the involvement of various parties from different authorities and management structures in eHealth, privacy concerns arise from the sensitive nature of the data accessed, handled, stored, and used by the applications, as well as the manner and entities with whom it is shared. This situation highlights numerous security vulnerabilities and privacy risks [20]. In this regard, Omoogun et al. in [21] support that various sensors deployed in mHealth monitoring are often designed without adequately addressing security and privacy aspects, making them susceptible to multiple types of attacks and potential data leaks. Challenges and threats like eavesdropping, imperson-ation, data integrity issues, data breaches, and collusion add further complications to managing patients' personal data in a privacy-conscious manner, particularly in con-trolling pervasive tracking and profiling [22]. Additionally, users' privacy concerns may be a great barrier to the acceptance of the eHealth technology [1, 23] and in order to adopt socially acceptable health services, the security and privacy issues need to be analyzed and addressed [3]. Privacy represents a multifaceted challenge in eHealth sys-tems, which need to satisfy specific needs from the organizations and providers utilizing them. Additionally, these systems must comply with various legal mandates stemming from the General Data Protection Regulation (GDPR) and other privacy enforcement regulations. These laws establish the rights of data subjects and outline the duties of data controllers [3]. Consequently, there is a need for a comprehensive framework that effectively balances various levels of data privacy. This framework should consider not only the specific requirements and needs of individuals and the objectives of healthcare service providers and data controllers but also the diverse regulatory standards governing privacy. This approach is essential for devising appropriate technical solutions that align with all these considerations.

3 Privacy in eHealth Under GDPR

The General Data Protection Regulation (GDPR) introduces specific definitions con-cerning individuals' data, especially regarding health/medical information. Under GDPR (Article 4–15), health data is defined as "personal data related to the physical or mental health of a natural person, including the provision of health care services, which reveal information about his or her health status." Furthermore, this type of data is classified as sensitive according to Article 8 paragraph 1 of the Directive 95/46/EC and is recognized as a special category of personal data under GDPR (Article 9). As a result, processing such data necessitates special attention and the explicit consent of the data subjects. As [24] argues, the applying of the principles of privacy by design, in order to build security and privacy into the systems, could be a solution to the privacy concerns regarding e-health data. Hence, despite the importance of privacy protection of health/medical data within e-Health area, which has been recognized in a number of works [20, 23], to our best knowledge not many research works approach it from the outset of the design or take into consideration the necessity of compliance with new GDPR requirements. Privacy needs to be considered during systems design and implementations [25]; otherwise as [1] support, plenty of limitations are posed on the system's deployment, leading to a not adequate data protection solution for the users. Additionally, literature [26] highlights that previous research mainly focuses on the security issues and especially on crypto-graphic techniques (e.g. [7]) that perform client-side encryption of data to protect against untrusted service providers, solving fragmentary some aspects within mHealth, while privacy engineering in this area is lacking. Milutinovic & De Decker in [1] p. 53 have presented a list with the privacy requirements that an eHealth system should fulfill in order to be compliant with the legal and ethical requisites and gain a wider acceptance as following: a) personal and recognizable information should be protected by strict access control policies, b) non-medical professionals should not access to patients' information, excluding authorized guardians, c) data access should be logged securely, so that later auditing is enabled and d) flexible control access policies should be revoked or expanded. From a technical aspect, Sawand & Khan in [27] p. 534 propose that eHealth monitoring systems should fulfill the following security requirements: a) Trusted Authority, which generates public and secret key parameters, is responsible for the issuing keys, granting as well differential access rights to the patients' based on their attributes and roles, b) cloud service provider, in order to provide secure communication mechanisms, as well secure data storage, processing and retrieval according to the access rights of the requesters, c) registered user, referring to the patients' registration to the trusted authority in order to define their data access and the attributes based on specific access policy and d) data access requester, referring to a doctor, a pharmacist, a researcher and a health care service, whose access rights and are defined by the patient who use the eHealth monitoring system.

Pirbhulal et al. in [28] pp. 385–386 suggest a more analytic context regarding remote healthcare systems, including not only security, but also privacy requirements as follows: a) data confidentiality, in order to prevent any disclosure during any data transmission, b) data integrity, in order to protect original data from external attacks, c) data availability, in order for the data to be available to legitimate and authenticated nodes/users, d) data freshness, ensuring that data is updated and no one authorized or not can replay old data, e) scalability, in order to reduce latency and control computational and storing overheads and g) secure key distribution, in order to allow encryption and decryption operations for accomplishing the estimated security and confidentiality. As the previous work, the study of [29], as far as the safety of Wireless Body Area Network systems concern, supports the data confidentiality, the data integrity, data freshness and the secure management requirements, but it also maintains a) the availability of the network, in order to provide at all the times access to patients' information both to healthcare professionals and patients, e.g. in case of an emergency health issue, b) data authentication, by which the applications should be capable to verify that the information is sent from a known trust center, c) dependability, referring to the systems' reliability, since data errors may lead to health-threating issues, d) secure localization, in order to prevent attackers to transmit improper details, such as fake signals about the patient's location, e) accountability, in order for the individuals' personal information to be secured, f) flexibility, referring to the individuals' flexibility of designating the control of their medical data and g) privacy rules and compliance requirement, referring to the need to secure private health information by setting privacy measures, such as rules/polices regarding the right to access to patients' sensitive data, since several regulations are enlisted for health care services. In this regard, ambitious current approaches, such as the study of [30], which developed a framework called Privacy-Protector to preserve the privacy of patients' personal data in IoT-based healthcare applications by presenting a secret sharing scheme, which devises the patients' data and stores it in several cloud servers for optimizing the secret share size and supporting exact-share repairs, while still keeping the advantages of the previous scheme, or previous studies [15, 16], focusing on the security and privacy frameworks for m-health applications that provide users with more control regarding the use of their sensitive health data within them, are not taking into consideration legal aspects of privacy that are mandatory for eHealth environments.

Especially, as far as compliance with GDPR concerns, until now few research works consider the new requirements. Authors in [25] focused their study on openEHR, a standard that embodies many principles of secure software for electronic health record and provided a list of requirements for a Hospital Information Systems compliant with GDPR, in order to examine to what extent, the openEHR may be a solution for the compliance to GDPR. Although, matches were found, the results showed that the related to the organizational processes GDPR requirements, hardly could be met by any EHR specification standard. Iwaya's et al. study in [26] p. 46 on mobile health data collection systems that have been used by community health workers, provides a list with specific privacy recommendations associated with the privacy principles and challenges emerged from the systems under the GDPR. This includes: a) Transparency-enhancing tools, guidelines for purpose specification, fine-grained access control, anonymization and pseudonymization, data validation and integrity and automated data deletion measures for the principle of Data Quality within mHealth, which refers to the process of transparent data, the purpose specification, the data minimization, the data accuracy and integrity and the data retention, b) Obtaining informed consent and check validity of consent measures for the principle of Legitimate Process, which refers to a legitimate data processing of sensitive data that takes into consideration other relevant legal basis for using personal data, c) Accurate, up-to-date, easily found and understandable information about data controller, purpose, recipients measures for the principle of Information Right of Data Subject, d) TETs for individualized information (e.g., privacy dashboards) and timely response to data subject's information requests and rectifications for the principle of Access Right of Data Subject, e) Provision of interfaces for objections and timely responses measures for the principle of Data Subject's Right to Object, f) Authentication and authorization, secure communication and storage and logging measures for the principle of Security of Data, which refers to personal information confidentiality, integrity and availability and the detection and communication of personal data breaches and g) Compliance with notification requirements and logging measures for the principal of Accountability, which refers to the implementation of safeguard data protection and compliance with data protection provisions to subjects, general public and supervisory authorities. Finally, [9], under the EU project WELCOME, studied the privacy of the mHealth applications for patients suffering from Chronic Obstructive Pulmonary Disease, and proposed the following privacy requirements within the context of GDPR: a) data patients' right to access and modification or erase by the applications in any case of inaccurate measurement, b) patients' right to information for the collected data by the applications and the time period of processing, c) limitation of collected data in accordance to the functionality of the applications in combination with the respective permissions, d) patients' fully awareness of the security measures regarding data storage or transmit to other third parties, e) patients' right to information regarding the risk and benefits of an m-health application, g) the prohibition of using collected data for marketing or profiling purposes, stated on an informed medical consent, h) appropriate security mechanisms for mobile devices, providing access only to authorized users with the proper authentication, i) access controls in order for authorized users and mobile devices to be authenticate, k) integrity of the medical data provided by the applications, l) proper security mechanisms for data storage. However, this interesting approach focuses only on a specific target group of patients, excluding larger or healthier target groups in which m-health applications are also addressed.

4 Security and Privacy by Design Methods and GDPR

Privacy by Design, as it was supported by [31], has been incorporated into the GDPR and it is considered to be the most appropriate approach for meeting the privacy and data protection expectations to a large scale, since it offers realistic solutions in order for legal requirements to be combined with the technical ones [24]. In this regard, the data controllers and processors are obligated to enforce the appropriate technical and organizational measures and procedures to ensure the protection of the data subjects' rights and to be compliant with data protection principles [32]. As [33] support, data protection by design should be managed after the specification of the processing purposes and during the processing itself. The controller is obligated to ensure the security of the system, as well as to enhance Protection Impact Assessment into the architecture of the system in order to safeguard adequately by default individuals' rights regarding their data. Hence, the proposed privacy principles by [31] have been under criticism regarding its difficulties to be implemented into the system requirements [34, 35]. Therefore, the stated necessity to embed the technological aspects of privacy within the regulatory field [14] and to bridge GDPR privacy regulations with technical solutions is even more emerged, in order for privacy engineering practice to confront more easily the privacy concerns and the compliance to the Regulation [36]. To address that need, [37], aiming at translating existing GDPR requirements into technical solution templates for compliant services, defined a catalogue of three types of privacy control patterns, namely: a) general privacy control patterns, b) patterns that affect the data subjects' rights and c) patterns regarding data controllers and processors' obligations. Although authors support that the proposed patterns provide generally applicable privacy guidelines, it is important to note that their work focused only on the following specific GDPR principles, Transparency and Traceability, Purpose Limitation, Data Minimization, Accuracy and Storage Limitation, since the principles of Lawfulness, Fairness, Integrity and Confidentiality and Accountability are considered not to be fulfilled by technical measures in a manageable time limit. Authors in [36] presented an approach based on model transformations, aiming to enable a more constructive approach to privacy by design under the principles of GDPR. Although, their work consists an interesting approach to bridge privacy legal and technical field, it focuses only on limited requirements, such as purpose limitation, or accountability of the data controller and consequently it presents specific technical privacy properties. In this sense, [34] supports that the need of holistic privacy patterns is emerged in order for the systems to achieve compliance with the new GDPR regulation, while even the notable privacy approaches, such as LINDDUN, a risk-based method for modelling privacy threats in order to support software developers in identifying and addressing privacy threats early during software development, should be combined with other goal-oriented approaches so as to be effective. [38] introduced an interesting privacy ontology that models the GDPR main conceptual cores as following: data types and documents, agents and roles, processing purposes, legal bases, processing operations, and deontic operations for modelling rights and duties, focusing on the analysis of deontic operators in order to manage the checking of compliance with the GDPR obligations. However, the study has not yet achieved to integrate the different levels of semantic representation for multiple goals and the analysis was restricted only to the Right to Data Portability. Finally, the current EU project PDP4E [14] aims to integrate data protection approaches such as LINDDUN, PRIPARE and PROPAN into systems engineering methodologies and process models, specializing them to operationalize GDPR compliance. Although, authors recognize the impact of goal-driven approaches, they focus mainly on risk-based approaches as LINDDUN and PROPAN, a threat identification method, as well as on PRIPARE methodology, derived from a previous EU project [39], which has combined articulations of risk-based methods and privacy by design principles for implementing privacy in practice. Therefore, no one pure goal- oriented methodology is considered, despite the fact that GDPR is considered as a purpose-oriented approach [34]. Thus, it is arguable to maintain that a Privacy by Design, goal-oriented privacy methodology could effectively support the implementation of the privacy technical prerequisites that the GDPR poses itself.

With respect to this and taking into account that Privacy Safeguard-PriS [40], a goal-driven Privacy by Design engineering methodology, was considered as an effective one for GDPR-compliant socio-technical systems [41], lacking thus in incorporating legal aspects, we provide the ground for PriS to be implemented in the proposed framework, by making provision for the legal concepts that GDPR has imposed. To address that, we emphasize on the interrelation between GDPR approach and PriS methodology in a high level. To address that, we present in the following figure the connections between GDPR approach and PriS methodology in a high level, while subsequently a table matching the concepts of the seven GDPR seven key data protection principles and the PriS privacy requirements concepts is demonstrated (Fig. 1).

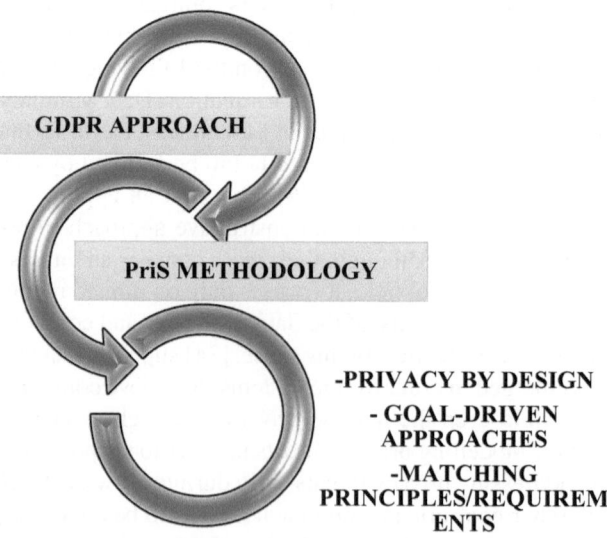

Fig. 1. Associating GDPR approach with PriS Methodology

PriS considers privacy requirements as organisational goals (privacy goals), which constraint the causal transformation of organisational goals into processes, and by privacy-process patterns describe the impact of privacy goals to the affected organisational processes. In particular, eight types of privacy goals are recognised, namely: Authentication, Authorisation, Identification, Anonymity, Pseudonymity, Unlinkability, Data Protection and Unobservability. At first, PriS was designed to support traditional privacy-aware information systems. Thus, cloud computing environments introduced a number of new privacy related concepts, leading to an extended version of PriS [42, 43] that provides a new set of privacy requirements along with the ones already stated, namely, undetectability, isolation, provenanceability, traceability, interveanability and accountability.

The first three requirements, Authentication, Authorization and Identification, are mainly security requirements, but they are included in the method due to their key role in the privacy protection. The Anonymity concerns the ability of subjects to be unidentifiable by other subjects, while Pseudonimity is referring to subjects' ability to use pseudonymous in order to ensure their anonymity. The Unlinkability is achieved when a third party cannot link the relationship between subjects, actions, messages, while Undetectability concerns the existence of a component that cannot be detected by a third party. The Unobservability is referring to the ability to hide the actions between subjects, and in particular it is satisfied if a system sufficiently realises undetectability among the respective assets and anonymity of the users accessing them, indicating an indirectly realisation through of the two previous concepts. The requirements of isolation, provenanceability, traceability, interveanability and accountability are related to data protection of the users or the systems over the cloud and therefore, they are grouped under the Data Protection requirement.

The next step is the modelling of the privacy-related organisational processes. These processes aim to support the selection of the system architecture that best satisfies them. Therefore, PriS provides an integrated way-of-working from high-level organisational needs to the IT systems that realize those [40]. On the other hand, GDPR aims a) to promote data collecting and processing organisations' and companies' work, by introducing specific rules and requirements as a primary goal of the organizations, as well as by providing direct instructions for the implementation of data protection, dealing thus with several complex aspects, such as company-level awareness raising, nature – scope - context and purposes of processing, adoption of both organizational and technological data protection processes and measures at the start of a project, cost of implementing the protection measures and documentation of processing operations [32, 33] and b) to provide EU citizens with further control on their personal data, while minimizing the threats against their data rights and freedoms [32]. Consequently, the conceptual association with PriS methodology is more than clear, since PriS promotes a set of expressions based on which the whole processes of an organization are considered, starting from the goal level and leading to the selection of the appropriate implementation techniques.

Based on the fact that the two approaches share a common conceptual ground, it is interesting to identify the associations between more specific concepts, such as GDPR principles, as the key concepts for its enforcement, and PriS Privacy Requirements, as the key concepts for the methodology to be implemented. Therefore, Table 1 represents the analysis following by matching the GDPR privacy principles with the PriS privacy requirements.

The principal of Lawfulness, Fairness, and Transparency concerns all the legitimate procedures, by which individuals' data should be collected and processing, as well as the individuals' right to be properly informed for any procedure. Consequently, this principle is matching with all PriS requirements, enabling data collecting and processing organizations to set primary privacy purposes and providing individuals their fundamental privacy rights from a technical aspect.

The principal of Purpose limitation, concerning the use of collected data only under a specific purpose designated by the data controller is associating with the requirements of data protection, anonymity, undetectability, unlinkability and unobservability, supporting data collecting and processing organizations to adopt appropriate technical and organizational measures, as well as individuals' data freedoms.

The principal of Data minimization, referring to the collected data only under a specific purpose for a specified time period, matches with the requirements of anonymity, data protection, anonymity, undetectability, unlinkability and unobservability, posing suitable measures for the organizations, while supporting individuals' data rights.

The principal of accuracy, concerning the quality of data and including individuals' rights on data access, rectification and erasure, is also associated with all with all PriS requirements, providing ground for both organizations and individuals to protect their rights and data properly.

The principle of storage limitation, referring to data limited storage under the stated purposes, is matching with the following requirements, authorization data protection, anonymity, undetectability, unlinkability and unobservability, providing data controlling and processing organizations the technical measures to ensure the storage of personal data in appropriate ways, while promoting individuals in appropriate ways.

The principles of Integrity & confidentiality and Accountability, concerning respectively the appropriate data protection technical and organizational measures and data controllers' obligation to implement measures that safeguard data protection and demonstrate compliance with data protection rules, are also matching with all PriS requirements directly and indirectly, facilitating the organizations to develop and implement the appropriate processes in order for individuals data to be protected.

Acknowledging that previous analysis is subjected under plenty limitations, the next step concerns the development of a conceptual model in order to examine the relationship between GDPR privacy principles with PriS privacy requirements more thoroughly, aiming, beyond the positive associations, to identify whether GDPR privacy principles conflict with privacy requirements and if so, which ones. Through this examination, software developers will be able to consider in a more understandable way the appropriate technical solutions that satisfy GDPR.

5   The BIONIC Project

BIONIC, funded under the European Union's Horizon 2020 program (Grant Agreement No 826304), is an innovative system designed to assist the aging workforce in the industrial sector. It aligns with the World Health Organization's principles for e-Health and m-Health, introducing wearable medical technology in a new context.

Table 1. Matching GDPR privacy principles with PriS privacy requirements

GDPR Privacy Principles	Authentication	Authorisation	Identification	Anonymity	Pseudonymity	Unlinkability	Undetectability	Data Protection	Unobservability
Lawfulness, fairness, and transparency	✓	✓	✓	✓	✓	✓	✓	✓	✓
Purpose limitation				✓		✓	✓	✓	✓
Data minimization				✓		✓	✓	✓	✓
Accuracy	✓	✓	✓	✓	✓	✓	✓	✓	✓
Storage limitation		✓		✓		✓	✓	✓	✓
Integrity and confidentiality	✓	✓	✓	✓	✓	✓	✓	✓	✓
Accountability	✓	✓	✓	✓	✓	✓	✓	✓	✓

The system's core mission is multifaceted, focusing on creating a holistic, non-intrusive, autonomous, and privacy-protecting platform. This platform facilitates real-time risk alerting and ongoing motivational coaching, tailored to the specific needs and fitness levels of the aging workforce. Given the physically demanding nature of many industrial jobs, BIONIC provides tools to help aging workers remain employed. It assesses the health impacts of their work activities and suggests exercises to mitigate these effects and promote a healthy, active lifestyle.

BIONIC incorporates a configurable Body Sensor Network (BSN), a groundbreaking concept that combines a range of connected sensors. This network builds a comprehensive, real-time data model of the human body, capturing both static and dynamic information like posture, gait, body temperature, and heart rate. The integration of these sensors into everyday or work clothing eliminates discomfort and movement restrictions associated with traditional wearables. The system also integrates commercial wearables with open APIs, like smartwatches, enhancing data continuity and self-quantification.

Depending on specific chronic musculoskeletal disorders (MSD), BIONIC develops targeted BSN modules, like belts or bandages, for focused monitoring. These modules are based on innovative biomechanical models adapted for age-related constraints and chronic impairments. The BSN's flexibility allows for sensor placement customization based on various application needs (fitness, medical, ergonomic), supporting the development of customizable or multifunctional wearables.

A significant innovation in BIONIC is the incorporation of AI on a chip, embedding predictive AI algorithms directly into the BSN. This approach minimizes data transmission to remote gateways by processing raw data at the source and extracting informative features for input into artificial neural networks. This combination of machine learning with biomechanical models enables efficient, real-time, personalized risk and physical strain assessments.

BIONIC's continuous, personalized on-site assessments provide valuable insights at both individual and age-dependent levels. It associates risks with design criteria and recommendations for position adaptations while maintaining workers' data privacy. The system offers real-time feedback through haptic, acoustic, or visual systems, with communication to external networks being optional and under the user's control.

The integrated prototype includes three distinct applications targeting workers, managers, and doctors. These applications provide access to movement data, ergonomic risk assessments, and health information, respectively, structured to support timely interventions and prevention strategies.

By focusing on usability and privacy (e.g., HCI, gamified coaching, GDPR compliance), BIONIC builds confidence and trust among older users, demonstrating the value and relevance of such technology for their health and well-being. It shows that wearable technology is not just for tech-savvy youth but offers tangible benefits for the health and well-being of older individuals. Figure 2 presents the concept diagram of BIONIC.

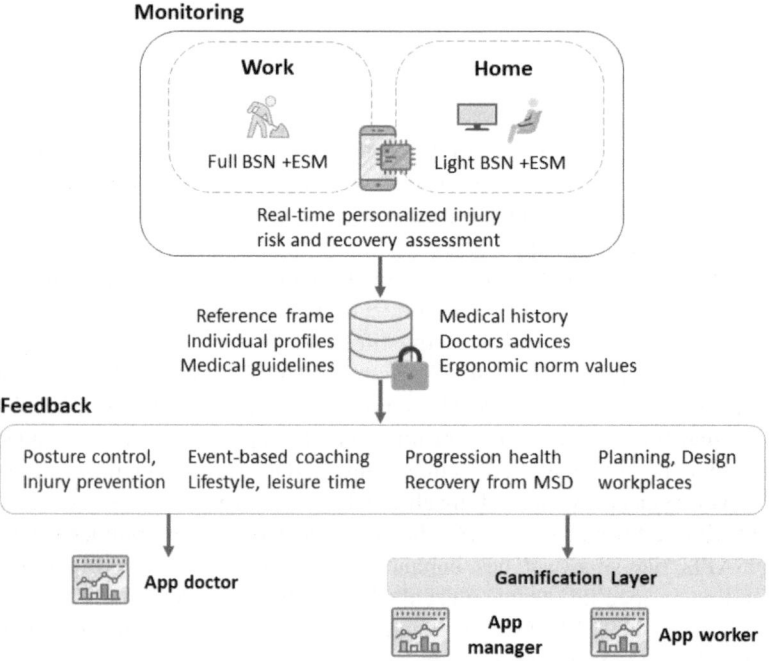

Fig. 2. BIONIC Concept Diagram

The BIONIC project offers two primary service types, each tailored to different phases of a worker's day.

1. Services During Working Hours (Full BSN Network):

- Composition: The full Body Sensor Network (BSN) is integrated into the worker's workwear, including a t-shirt, helmet or cap, and trousers.
- Functionality: Sensors on this workwear gather raw data, which is then processed by an AI chip located in the trousers. This processing yields insights into the worker's current health status.
- Integration with Smartwatch: The AI chip communicates with the worker's smartwatch, which provides additional health data (like heart rate) and delivers notifications or alarm messages to the worker.

– Mobile Device Interface: Workers also have a mobile device that enables them to manage their data, interact with BIONIC apps, and carry out coaching exercises at home.

2. Services During Leisure Time (Light BSN Network):

– Composition: The "light BSN" is a simplified version of the full network, consisting of the worker's smartwatch, a few Inertial Measurement Units (IMU) sensors, and a mobile device.
– Usage: This network is used outside working hours, primarily for self-guided exercises based on the BIONIC project's coaching app.
– Monitoring Function: Besides supporting the coaching app, the light BSN continues to monitor the worker's ergonomic and health data during leisure time. This monitoring helps capture habits and movements outside the workplace, aiding in the prevention of potential health issues and providing more tailored advice throughout the day.

These two service modes ensure continuous, context-sensitive support for workers, focusing on health monitoring and guidance both during work and in their personal time, thus promoting a comprehensive approach to occupational health and wellness.

The proposed BIONIC data flow is presented in Fig. 3. The full BSN and the light BSN parts are also visible in the same figure.

In the BIONIC project, all processed data derived from the Body Sensor Network (BSN) is accumulated in a secure storage system. This setup allows the developed applications to access this data and provide necessary feedback to workers. For the lighter BSN network used during leisure time, data storage in the secure repository is optional, based on the worker's preference.

For research and development (R&D) purposes, and limited to the project's duration, the raw data produced by the BSN network are stored in an anonymized format in a distinct database referred to as "Research Data."

The types of data collected from both the BSN and light BSN networks include:

• Kinematic Data: This might include IMU sample rate data such as joint angles.
• Kinetic Data: Typically collected at a lower sample rate, this data can include vertical ground reaction force, ground contact information, and external load indicators.
• Ergonomic Data: Depending on the chosen ergonomic tools, this could involve timestamped detected events (like picking up objects, body positions), repetitions, and possibly ergonomic scores.
• Physiological Data: This encompasses data such as heart rate, blood pressure, and body temperature.

The subsequent section of this chapter aims to outline an effective Data Protection Framework for BIONIC. This framework is designed to meet technical, legal, and organizational requirements to ensure the safety and security of the workers. It will delve into the specifics of the proposed Privacy and Data Protection Framework, highlighting how it addresses these crucial aspects.

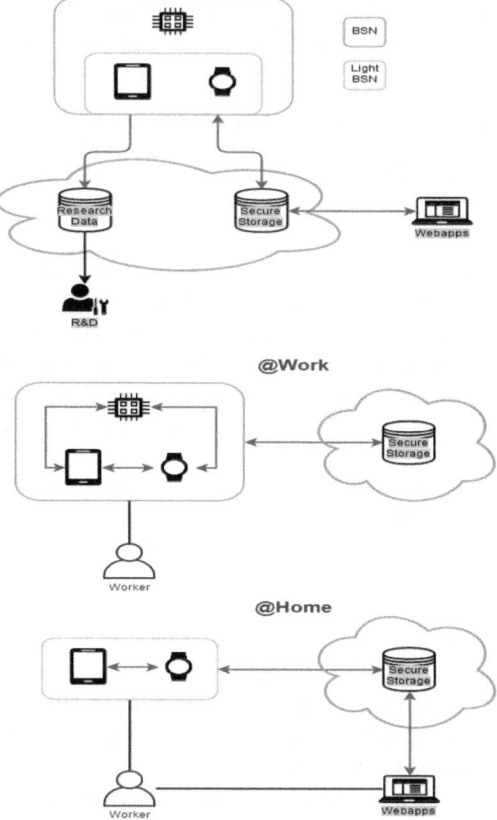

Fig. 3. BIONIC Data Flow

6 Privacy and Data Protection Framework

This section introduces a versatile and effective methodology for GDPR-compliant Personal Data Management, specifically designed to address the security, safety, and privacy concerns associated with using comprehensive and unobtrusive Body Sensor Networks by the aging workforce. This methodology is integral to their daily tasks within the system.

Key aspects of the Framework include:

- GDPR Compliance: The Framework adheres to crucial GDPR principles, including purpose limitation, data minimization, accuracy, accountability, the lawfulness of processing, and user consent. This ensures comprehensive compliance with the regulation.
- Respecting Data Owner Rights: The medical wearable applications respect all rights of the data owners (aging workers). This includes their right to object to the storage/processing of their information, their right to be forgotten, and their right to restrict data processing.

- Obligations of Intermediate Users: The Framework clearly defines the responsibilities of intermediate users, such as occupational health professionals and production managers. These users manage the system and may use the data generated within BIONIC for professional purposes.
- Technical, Organizational, and Procedural Measures: The framework takes into account all necessary technical, organizational, and procedural measures to meet the elicited security and privacy requirements. This comprehensive approach ensures a high level of confidence and trust among all stakeholders involved.

By incorporating these elements, the proposed Data Protection Framework provides a robust solution for managing personal data in a way that is both secure and respectful of individual privacy rights, particularly for the aging workforce utilizing wearable technology in their professional environment. It comprises of the following four stages as shown in Fig. 4.

6.1 Stage 1: Personal Data Handling Processes Elicitation

The purpose of this step is to define the perimeter of the Personal Data handling processes, by capturing, reviewing and formalizing the following issues:

- The categories of the personal data processed.
- The categories of the data subjects
- The purposes of each processing activity
- The identification of high risk data processing activities
- The legal basis of each processing activity (e.g. contract and/or consent and/or legitimate interest and/or statutory obligation)
- The categories of recipients to whom the personal data are disclosed
- The envisaged time limits for erasure of the different categories of data – if they exist
- The existing technical and organizational measures for the protection of personal data.

To this end, a Questionnaire for Accessing GDPR Compliance is designed and implemented. The objective of this questionnaire, addressed to BIONIC respective stakeholders in order to identify, through a systematic procedure, the aforementioned information for each purpose of data processing separately.

The questionnaire includes seven items, requiring open responses, regarding the following issues:

- Short Description for the purpose of processing
- Legal Basis for the purpose of processing, including the subcategories a) Law, b) User / patient consent, c) Contract
- The data involved in serving the specific purpose of processing, including the subcategories a) General Data, b) Personal Data, c) Special categories of Personal Data
- The necessity of the involved data for serving the specific purpose of processing
- The sources of collected data
- The transmission of personal data to third parties
- The processing of automated decision-making, including profiling

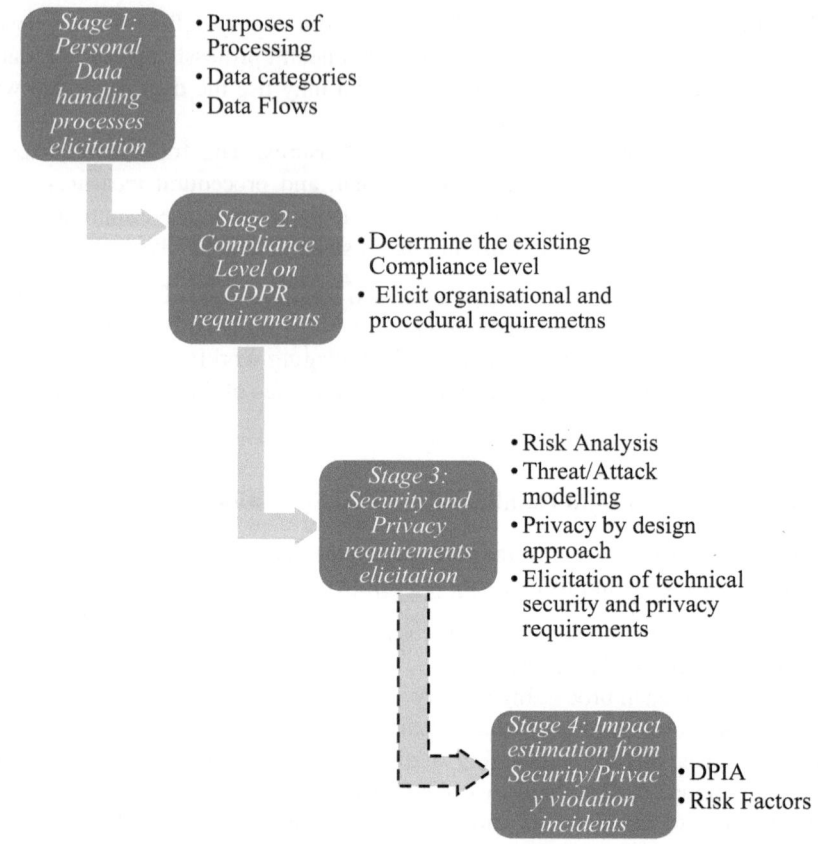

Fig. 4. GDPR Compliant Personal Data Management Methodology

Therefore its main output concerns the listing of the Purposes of Processing and the Personal Data Categories (Fig. 5).

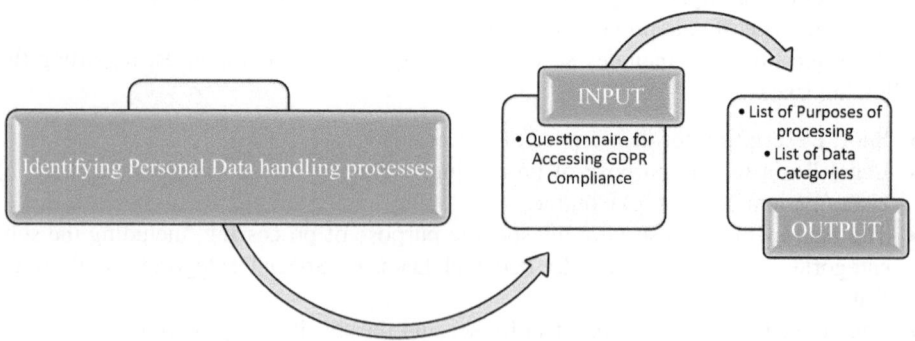

Fig. 5. Output of Stage 1

6.2 Stage 2: Compliance Level on GDPR Requirements

In the second stage, it is essential to map the organizational context following the results of stage 1. Therefore, when the above list of information has been compiled, the processing of each personal data category is being reviewed against the GDPR requirements to deduce the existing compliance level, through a GDPR gap analysis. Topics that are examined are presented indicatively as following:

- Lawfulness, fairness and transparency of the personal data processing
- The processing purpose limitation, the data minimization,
- The consent of the data subjects,
- The personal data storage limitation
- The measures for personal data protection
- Integrity and confidentiality
- The readiness of the involved stakeholders to respond to the data subjects' rights' is examined, such as the 'right of access', the 'right to rectification', the 'right to be forgotten', the right to restriction of processing', the 'right to data portability', the 'right to object'.
- Information to be provided where personal data have not been obtained from the data subject
- Automated individual decision-making, including profiling
- Data protection by design and by default
- Joint controllers
- Security of processing
- Processing under the authority of the controller or processor
- Tasks of the data protection officer
- Transfers on the basis of an adequacy decision

Moreover, the readiness of the organization to respond to the data subjects' rights' will be examined. Indicatively the 'right of access', the 'right to rectification', the 'right to be forgotten', the 'right to restriction of processing', the 'right to data portability', 'right to object'.

- Indicative activities that will be performed during the gap analysis include:
- Review of legal basis on which the organization processes Personal Information.
- Review the necessary retention periods per category of Personal Information, for various reasons such as for compliance with a legal obligation, for inquiries of auditing authorities, for legal claims, for public interest etc.
- Review of Privacy Notices.
- Review of the legal basis for marketing services.
- Legal review of all defined internal Personal Information Protection Policies and Procedures.
- Legal Review of sample employment contracts and updating with necessary legal language to allow the processing of employees Personal Information for legitimate business purposes.
- Review of standard consent forms used to collect and record data subject consent for the processing of Personal Information.

- Legal review of standard Intra- and Third-Party contracts, Procurement contracts and Supply Contracts to identify any contractual gaps in relation to Data Protection relevant clauses. If no standard contracts are used, the review will cover key activities, which should at least include all contracts related to identified high risk processing activities.
- Understand the operational policies and procedures for the IT systems
- Access the efficiency of the organization to protect the data (data protection measures)
- IT and Security Governance review
- Network architecture review

The main output of this stage concerns a Set of GDPR Compliance Requirements for each identified purpose of Processing (Fig. 6).

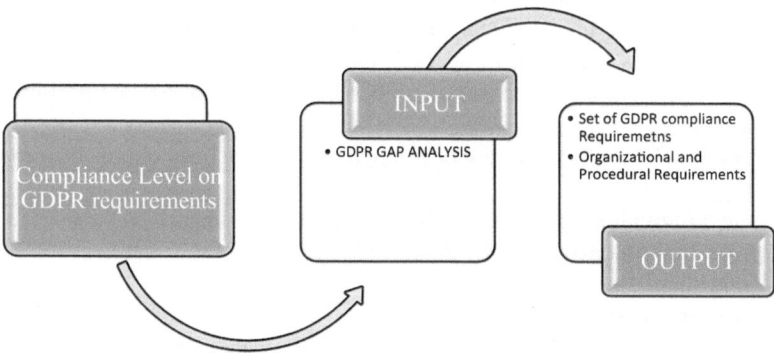

Fig. 6. Output of Stage 2

6.3   Stage 3: Security and Privacy Requirements Elicitation

Security is essentially the safeguard against intentional incidents, incidents that arise from deliberate or planned actions. It focuses on protecting assets from threats, which are defined as "the potential for misuse of protected assets." In contrast, privacy is about safeguarding the identity of the assets' owners from unauthorized access or processing by users who lack the owner's consent. Risk analysis, or Risk Assessment, is a systematic approach to examining an IT infrastructure or the interconnections between computational devices. This process involves identifying potential security and privacy threats, as well as specific vulnerabilities and potential failures that could trigger these threats. The primary objective of a security assessment is to ensure that essential security and privacy goals are incorporated into the architecture's design and implementation. A vulnerability is characterized as a security or privacy weakness that can exist in a resource, an actor, or a goal. These vulnerabilities are targets for exploitation by threats, manifesting as attacks or incidents in a particular context. A threat, then, is a situation with the potential to cause loss or pose a risk to the system's security features. Understanding these concepts is critical for developing robust security and privacy protocols in any IT system or infrastructure.

In Stage 3 of the process, a Risk Analysis is undertaken, focusing on identifying threats, vulnerabilities, and data to develop attack modeling and understand threat propagation. This analysis is essential, stemming directly from the GDPR principle of accountability. It aids in pinpointing and evaluating the risks related to security and privacy. Consequently, this analysis plays a crucial role in choosing suitable measures to mitigate these risks. By doing so, it helps to minimize the potential impact on the data subjects, reduce the risk of non-compliance, lessen the chances of legal actions, and decrease operational risks. This stage is pivotal in ensuring that the system adheres to GDPR requirements and maintains a high level of data protection and privacy. Finally, the Privacy by Design approach, specifically the Privacy Safeguard (PriS) method, is utilized to gather all identified security and privacy requirements, which include Legal, Organizational, and Technical aspects. This method ensures that these requirements are articulated in a technically robust manner and verifies their feasibility for implementation within the specific system context. The requirements elicitation methodology employed in PriS plays a key role in supporting the analysis of threats, vulnerabilities, and attacks. It aids in the logical assessment and reasoning of security and privacy requirements, as well as in the modeling of the system. This comprehensive approach ensures that the system not only meets the necessary legal standards but also upholds a high level of security and privacy integrity, making it well-equipped to handle the specific challenges and risks it may encounter.

The decision to select the Privacy Safeguard (PriS) for the security and privacy modeling of BIONIC is grounded in several key reasons:

- Alignment with GDPR Principles: PriS embodies the Privacy by Design approach, which is a fundamental principle of the GDPR. This approach mandates that data protection and privacy considerations are integrated into the development process from the outset.
- Established Methodology: PriS is one of the oldest and most thoroughly evaluated Privacy by Design methodologies. Its long-standing application and assessment ensure a reliable and tested framework for implementing privacy measures.
- Conceptual Compatibility with GDPR: At a conceptual level, the principles outlined in the GDPR find a strong correlation with the privacy requirements concepts in PriS. This congruence ensures that the application of PriS aligns well with GDPR mandates.
- Stakeholder Needs and Goal-Driven Modeling: PriS emphasizes the importance of considering stakeholders' needs and employs a goal-driven modeling approach. This aspect is particularly significant given the purpose-oriented philosophy of the GDPR, which focuses on the objectives behind data processing activities.
- Alignment with Risk-Based Analysis: PriS is adaptable to a risk-based analysis framework, which is a prerequisite for GDPR compliance. This compatibility ensures that the privacy and security modeling with PriS will be comprehensive, addressing potential risks in accordance with GDPR requirements.

The proposed steps for implementing stage 3 can be described as follows:

Substep 1: Identify System Assets and Stakeholders This step involves defining the scope or perimeter of the study, which is crucial for establishing a comprehensive understanding of the system under development. The objective is to gain a global perspective on the system's components and the interactions between them. During this phase, the following data are systematically collected and formalized:

- Essential Assets of the System: Identification of the key assets within the system that are critical to its functioning and require protection.
- Functional Description of Components and Relationships: Detailed descriptions of each component within the system and how they interact or relate to one another.
- Security Issues to Address: A list of specific security challenges or problems that the study aims to tackle.
- Assumptions Made: Documenting any assumptions that are being made during the study. This could include presumptions about the system environment, user behavior, or external factors.
- Existing Security Rules: Compilation of applicable laws, regulations, and any existing security rules derived from other studies that are relevant to the system.
- Constraints from the System: Identifying any internal or external constraints imposed by the system itself, which could affect its design, functionality, or security.

At the conclusion of this step, there will be a clear and formalized understanding of the system's components and their interconnections. This detailed mapping will serve as a foundational input for the subsequent risk analysis phase, ensuring that the analysis is grounded in a thorough and accurate representation of the system.

Substep 2: Identify Potential Security and Privacy Threats and related System Vulnerabilities. In this phase, the focus is on conducting a dedicated risk analysis to examine the security and privacy threats and vulnerabilities of the proposed system. This detailed analysis will involve a series of structured activities:

- Listing Relevant Attack Methods: In collaboration with system's stakeholders, a comprehensive list of potential attack methods targeting security and privacy will be compiled. This list reflects the diverse range of threats that the system could encounter.
- Characterizing Threat Agents: For each identified attack method, the corresponding threat agents will be characterized based on their type. This involves understanding the nature and capabilities of potential attackers.
- Identifying Vulnerabilities: The security and privacy vulnerabilities of the proposed system's entities will be pinpointed with respect to the various attack methods. This step is critical in understanding where the system might be most susceptible to security breaches or privacy invasions.
- Estimating Vulnerability Level: An evaluation of the level of vulnerability for each identified weak point in the system will be conducted. This estimation helps in quantifying the severity of each vulnerability.
- Formulating Security and Privacy Threats: This activity involves defining and articulating the specific security and privacy threats that emerge from the identified vulnerabilities and attack methods.

- Assigning Priority to Threats: The security and privacy threats are prioritized based on the likelihood of their occurrence. This prioritization is crucial for focusing resources and efforts on the most probable and impactful threats.

The culmination of this step is a detailed and prioritized list of pertinent security and privacy threats, along with the types of attacks associated with them. This output is important in guiding subsequent steps in the risk management and mitigation process for the proposed system.

Substep 3: Security and Privacy Requirements Analysis. In this stage, following the identification of threats and potential attack methods to the proposed system, the focus shifts to uncovering the system's vulnerabilities. Detecting Security and Privacy vulnerabilities is crucial as it leads to the establishment of corresponding security and privacy objectives. These objectives are essentially strategies aimed at mitigating vulnerabilities, thereby reducing potential risks to the system's entities. A key activity involves employing the Privacy Safeguard (PriS) methodology, which serves as a Privacy by Design approach. This method is vital for analyzing the security and privacy goals that must be fulfilled, based on the threats and vulnerabilities that have been identified. The stage progresses with the definition of security and privacy objectives. These objectives provide a clear strategy on how the system plans to address and reduce vulnerabilities. They are a critical component in formulating a strategic response to the identified risks. The final aspect of this stage is specifying the security and privacy requirements. These requirements offer a detailed description of the implementation of the identified security and privacy objectives. They act as specific, actionable directives that the system must follow to mitigate the identified risks effectively. By following this structured approach, the system ensures that all identified vulnerabilities are comprehensively addressed with corresponding objectives and requirements, strengthening its security and privacy posture. This method aligns with the principles of Privacy by Design, integrating these critical aspects into the system's foundation.

When identifying security and privacy requirements in this step, several actions are undertaken to ensure comprehensive coverage and effectiveness. These actions include:

- Listing the functional requirements related to security and privacy. This involves outlining all the necessary conditions and capabilities that the system must meet to ensure robust security and privacy protection.
- Justifying the adequacy of coverage of the security and privacy objectives. Each requirement is evaluated to ensure that it aligns with and effectively addresses the established objectives.
- Highlighting any coverage flaws, also known as residual risks, and providing justifications for them. This step is crucial for understanding any limitations or gaps in the security and privacy framework and planning for potential risk mitigation strategies.
- Classifying the security and privacy requirements for each specific use case. This classification helps in tailoring the security and privacy measures to the unique scenarios and functions the system will encounter.

- Where applicable, justifying the coverage of dependencies of security and privacy requirements. This involves ensuring that the interdependencies and interactions between various requirements are adequately addressed, enhancing the overall coherence and effectiveness of the security and privacy measures.

The main output of this stage concerns a) a List of Threats and Attacks, b) the provision of Legal and Organizational Measures and c) the elicitation of the appropriate security and privacy requirements (Fig. 7).

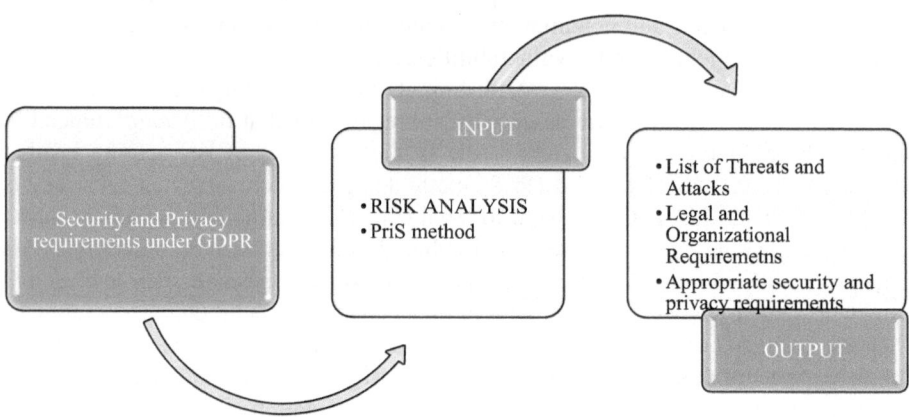

Fig. 7. Output of Stage 3

6.4   Stage 4: Impact Estimation from Security/Privacy Violation Incidents

Under Regulation (EC) 2016/679, enacted by the European Parliament and Council on April 27th, 2016, there are specific mandates for the protection of natural persons in relation to personal data processing and the free movement of such data. The regulation stipulates that if a processing activity, particularly one utilizing new technologies and considering the nature, scope, context, and purposes of the processing, is likely to pose a high risk to the rights and freedoms of individuals, the data controller is required to conduct a Data Protection Impact Assessment (DPIA) before commencing such processing. This assessment can be applicable to a group of similar processing activities that present comparable high risks.

Given the constant evolution of systems and threats, effective risk management involves identifying suitable controls for the processing of personal data. The management and prioritization of risks need to be analyzed in a manner that optimizes cost-effectiveness and supports informed decision-making, with the primary goal of safeguarding personal data. Performing an impact assessment is instrumental in implementing privacy principles effectively, ensuring that individuals retain control over their personal data. This approach is integral to maintaining compliance with GDPR and upholding high standards of data protection and privacy.

A data protection impact assessment, and hence, the criticality of data shall (in accordance with Regulation (EC) 2016/679 of the European Parliament and of the Council) particularly be required in the case of:

- a systematic and extensive evaluation of personal aspects relating to natural persons which is based on automated processing, including profiling, and on which decisions are based that produce legal effects concerning the natural person or similarly significantly affect the natural person (e.g., user profiling by web search activity monitoring for targeted advertising and promotion of products and services (hotels, restaurants, etc.),
- processing on a large scale of special categories of data referred to in Article 9(1), or of personal data relating to criminal convictions and offences referred to in Article 10 (e.g., processing of patients' medical records (special category of personal data) from healthcare organisations, including medical history, illnesses, and patient care, etc.) or
- a systematic monitoring of a publicly accessible area on a large scale (e.g., traffic monitoring for informing drivers of the fastest route, residence entries' monitoring, public transport entrance, etc.).

Moreover, the assessment shall contain, in accordance with Regulation (EC) 2016/679 of the European Parliament and of the Council, at least:

- a systematic description of the envisaged processing operations and the purposes of the processing, including, where applicable, the legitimate interest pursued by the controller;
- an assessment of the necessity and proportionality of the processing operations in relation to the purposes;
- an assessment of the risks to the rights and freedoms of data subjects;
- the measures envisaged to address the risks, including safeguards, security measures and mechanisms to ensure the protection of personal data and to demonstrate compliance with this Regulation taking into account the rights and legitimate interests of data subjects and other persons concerned.

The privacy impact assessment being conducted is founded on a comprehensive conceptual framework that emphasizes the protection of personal data and the rights of data subjects. This framework covers both the processing of data through Information Systems and non-automated means, as well as the analysis of the potential impact that incidents involving personal data may have on the individuals to whom the data pertains.

The core objective of the impact assessment is to safeguard personal data, in line with the definitions and guidelines set forth by the GDPR. This protection extends not only to the data itself during processing but also to the various elements that facilitate this processing, which are identified as Assets. The significance of these assets is determined by the potential impact that a breach of individuals' privacy could have.

A Feared Event in this context refers to any unauthorized access to personal data, undesired alterations to the data, or the loss of data. For an Information System to be compromised, there must exist a vulnerability within the system, and a relevant Threat must materialize from a source of risk.

In summary, it's important to recognize that a Threat exploits a vulnerability in an Information System, which can lead to a data protection breach. Such an incident can have varying degrees of Impact on the data subjects, affecting their rights and freedoms. Therefore, understanding and addressing these aspects are crucial in ensuring robust data protection and privacy compliance (Fig. 8).

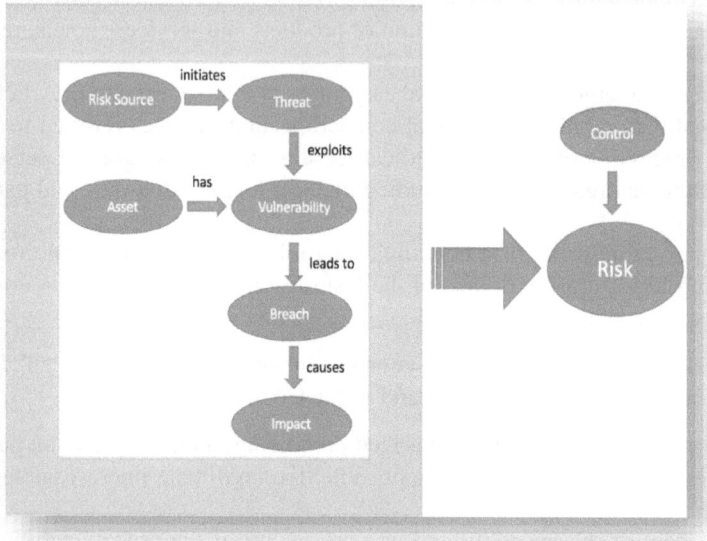

Fig. 8. Conceptual Framework of Impact Assessment

A Risk in the context of data protection and privacy can be described as a hypothetical scenario that outlines:

- The Risk Sources: These are the origins of potential risk, such as an employee who might be bribed by a competitor.
- Exploitation of Vulnerabilities: This refers to how these risk sources could take advantage of weaknesses in systems handling personal data, like exploiting flaws in a file management system that permits unauthorized data manipulation.
- Context of Threats: The specific circumstances under which the vulnerabilities could be exploited, for instance, through the misuse of email systems.
- Occurrence of Feared Events: This outlines the possible outcomes of the vulnerabilities being exploited, like illegitimate access to personal data.
- The Personal Data Involved: Identifying which specific data could be affected, such as a customer file.
- Potential Impacts on Privacy: The consequences of the feared events on the privacy of data subjects, which could range from unwanted solicitations to feelings of privacy invasion.

Understanding each of these elements is crucial in assessing and managing the risks to personal data and the privacy of individuals involved. This comprehensive approach ensures a thorough understanding of potential privacy risks and guides effective risk mitigation strategies.

The following figure summarizes all the concepts above (Fig. 9):

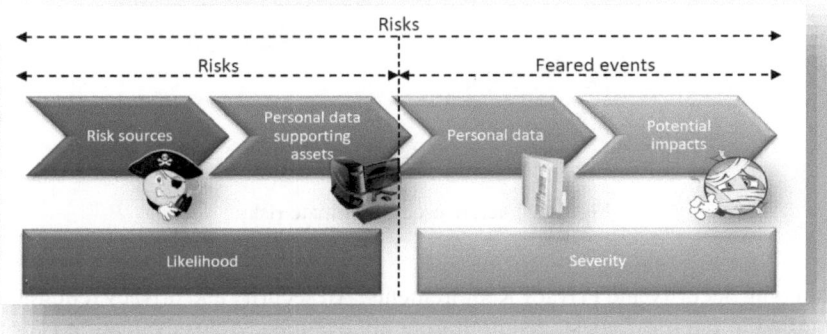

Fig. 9. Risks and Feared events

The estimation of risk level involves assessing two critical dimensions: severity and likelihood.

- Severity: This represents the magnitude of a risk and primarily hinges on the prejudicial effects of potential impacts. It essentially measures how significant or damaging the impact of a risk would be if it were to materialize. This could range from minor inconveniences to major breaches impacting the privacy and rights of data subjects.
- Likelihood: This dimension assesses the probability of a risk occurring. It is largely dependent on two factors: the level of vulnerability in the supporting assets that are susceptible to threats, and the capabilities of the risk sources to exploit these vulnerabilities. The likelihood takes into account how exposed the assets are to potential risks and the effectiveness or strength of the threats posed by risk sources.

Combining these two aspects – the severity of the potential impacts and the likelihood of their occurrence – provides a comprehensive understanding of the overall risk level. This approach enables organizations to prioritize risks effectively, focusing on those with the highest likelihood of occurrence and the most severe potential impacts. By doing so, they can allocate resources and implement controls more efficiently to mitigate these risks (Fig. 10).

At this stage, the focus shifts to applying a Data Protection Impact Assessment (DPIA) method. The primary goal is to identify the severity and likelihood of potential privacy violation incidents. This step is crucial in understanding the impact and probability of privacy risks, which guides the selection of effective countermeasures. Key activities during this stage include:

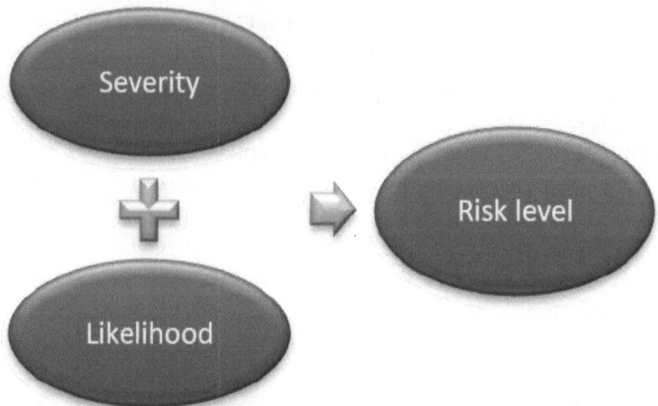

Fig. 10. Factors used to estimate risks

- Evaluating Security and Privacy Requirements: All security and privacy requirements identified in the previous stage are thoroughly examined. This evaluation is critical to ensure that there are no conflicts among the requirements and that they align well with the overall objectives.
- Assessing Privacy Risks: The DPIA plays a vital role in identifying and evaluating privacy risks. By understanding these risks, appropriate measures can be chosen to mitigate them, thereby reducing the potential impact on data subjects, the likelihood of non-compliance, the risk of legal actions, and operational risks
- Identifying Technical Countermeasures: Based on the DPIA findings, suitable technical countermeasures are determined for each identified requirement. These countermeasures are tailored to address specific vulnerabilities and risks.
- Facilitating Implementation Decisions: The insights gained from the DPIA enable developers to choose the most appropriate implementation techniques. These techniques are aimed at ensuring the security of the data (including its confidentiality, integrity, and availability) and protecting user privacy. Important aspects include user consent for data processing and transmission, as well as the fulfillment of user rights.

By undertaking these steps, the DPIA method ensures that the system not only meets the necessary legal and regulatory requirements but also upholds a high standard of data protection and privacy. This approach is integral to building a secure and trust-worthy system that respects and protects the privacy of its users.

For conducting the DPIA it is necessary to consider both the output of stage 2 regarding the organizational and legal requirements as they are derived from the Gap analysis as well as the output of stage 3 regarding the technical security and privacy requirements derived from the risk/threat analysis and the privacy by design approach.

The main output of this stage concerns a) the severity of the privacy violations incidents and b) the likelihood of the privacy violations incidents. The following figure presents the input and output of stage 4 (Fig. 11).

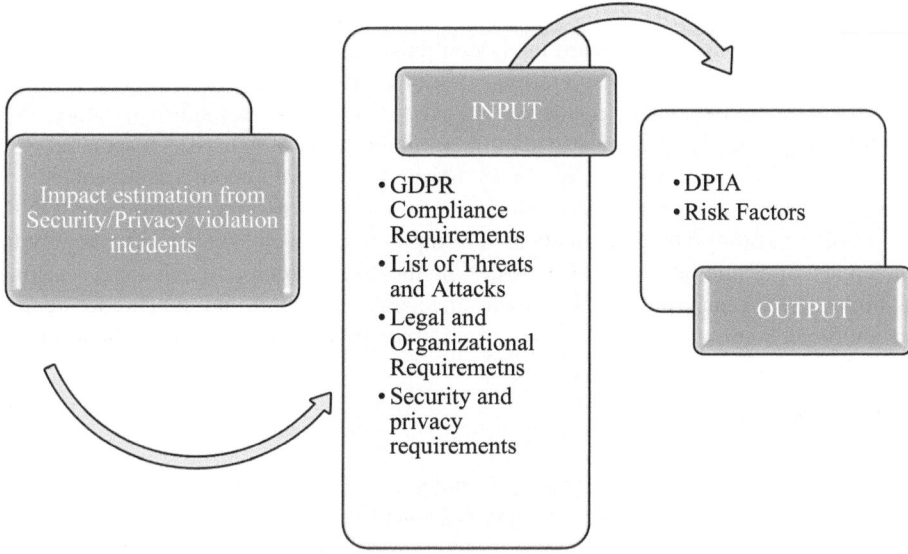

Fig. 11. Output of Stage 4

7 Discussion

The Privacy and Data Protection Framework designed specifically for the IoT environment, emphasizes GDPR compliance and integrates privacy by design principles to address the unique challenges of deploying IoT services in healthcare.

The framework addresses several challenges inherent in implementing IoT solutions in healthcare. In relation to security and privacy issues it is important to stress that IoT devices in healthcare generate sensitive data that require robust security and privacy measures. The framework's emphasis on GDPR compliance and privacy by design principles helps in mitigating risks associated with data breaches, unauthorized access, and privacy violations. Furthermore, IoT devices can generate vast amounts of data, making it challenging to ensure privacy and data protection. The framework's adherence to GDPR principles like data minimization and purpose limitation aids in managing this data responsibly.

Another important issue is that of interoperability, since with a myriad of devices and platforms, ensuring interoperability while maintaining privacy and security standards is a challenge. The framework's focus on technical, organizational, and procedural measures facilitates smoother integration. Finally, by prioritizing privacy and data protection and transparently adhering to GDPR requirements, the framework aims to enhance user trust in IoT healthcare applications.

Compared to existing frameworks, the BIONIC project's Privacy and Data Protection Framework offers a more holistic approach since it has been designed with GDPR compliance at its core, addressing legal, technical, and organizational requirements comprehensively. Furthermore, many existing frameworks are not prioritizing privacy from the initial design stages. The BIONIC framework incorporates privacy by design, ensuring that privacy considerations are integrated into every phase of system development, thus ensuring that privacy considerations are not an afterthought but are integral to the system's architecture, enhancing the system's trustworthiness. In addition it emphasizes a systematic risk analysis process to identify, evaluate, and mitigate risks associated with security and privacy, which may not be as thoroughly covered in other approaches.

Overall, the Privacy and Data Protection Framework presented within the BIONIC project offers a comprehensive and GDPR-compliant approach to addressing the unique challenges of implementing IoT services in the health sector. It advances existing methodologies by ensuring a privacy-focused, legally compliant, and stakeholder-centric design, thereby enhancing trust, security, and privacy in IoT healthcare applications.

8   Conclusion

BIONIC acknowledges that the ageing population in Europe impacts multifaceted on the EU productivity growth [45, 46] and rises healthcare costs that are leading to the necessity of developing new digital forms of health self-management systems outside the health-care services [47]. Therefore, it provides the development of medical wearables applications for personalized information and treatment to ageing workers, as a form of an m-health system [5], aiming to assist them in managing their health issues and maintaining behaviours that promote health, so as to improve the quality of their work and life. In this regard, it proposes a Privacy and Data Protection Framework that complies with GDPR legislation.

The PriS Privacy by Design method is essential in gathering necessary information throughout the entire software development lifecycle. This method places significant emphasis on the various types of personal data processed, including sensitive, biometric, and health-related data. The Framework ensures that applications adhere to data owners' rights, including the right to object to data storage/processing, the right to be forgotten, and the right to restrict processing. By adopting a privacy-by-design approach, the Framework guarantees compliance with both functional and non-functional (security and privacy) requirements specified by users. During the design phase, it takes into consideration all legal and technical demands of the General Data Protection Regulation (GDPR), addressing GDPR stipulations related to data protection, accountability, and handling of data breaches through necessary technical, organizational, and procedural measures. It ensures adherence to GDPR principles like purpose limitation, data minimization, accuracy, and lawfulness of processing, including user consent. The Framework's implementation of concepts such as privacy and data protection by design and default, accountability, and data minimization, is crucial in e-health and m-health systems. It acknowledges the varying interpretations of these terms among stakeholders from different fields and backgrounds. It also facilitates the allocation of liability among interconnected platform actors and outlines procedures for managing potential data breaches. In conclusion, the Framework fills existing methodology gaps, enhancing user trust in software by providing a robust process for identifying technical requirements and validating that these requirements meet GDPR objectives.

Acknowledgements. This research has received funding from the European Union's Horizon 2020 Framework Programme for Research and Innovation under grant agreement #826304–BIONIC.

References

1. Milutinovic, M., De Decker, B.: Ethical aspects in eHealth–design of a privacy friendly system. J. Inf. Commun. Ethics Soc. **14**(1), 49–69 (2016)
2. WHO: "E-Health", (2015). http://www.who.int/trade/glossary/story021/en/
3. Esposito, C., Castiglione, A., Tudorica, C.A., Pop, F.: Security and privacy for cloud based data management in the health network service chain: a microservice approach. IEEE Commun. Mag. **55**(9), 102–108 (2017)
4. Chib, A., Lin, S.H.: Theoretical advancements in mHealth: a systematic review of mobile apps. J. Health Commun. **23**(10–11), 909–955 (2018)
5. Drosatos, G., Efraimidis, P.S., Williams, G., Kaldoudi, E.: Towards privacy by design in personal e-health systems. In HEALTHINF, pp. 472–477 (2016)
6. Marcolino, M.S., Oliveira, J.A.Q., D'Agostino, M., Ribeiro, A.L., Alkmim, M.B.M., Novillo Ortiz, D: The impact of mHealth interventions: systematic review of systematic reviews. JMIR mHealth and uHealth, **6**(1), e23 (2018)
7. Solomon, M., Elias, E.P.: Privacy protection for wireless medical sensor data. Int. J. Sci. Res. Sci. Technol. **4**(2), 1439–1440 (2018)
8. Almarashdeh, I., et al.: Real-time elderly healthcare monitoring expert system using wireless sensor network. Int. J. Appl. Eng. Res. (2018). ISSN, 0973–4562
9. Mustafa, U., Pflugel, E., Philip, N.: A novel privacy framework for secure m-health applications: the case of the GDPR. In 2019 IEEE 12th International Conference on Global Security, Safety and Sustainability (ICGS3), pp. 1–9. IEEE (2019)
10. Lee, N., Kwon, O.: A privacy-aware feature selection method for solving the personalization–privacy paradox in mobile wellness healthcare services. Expert Syst. Appl. **42**(5), 2764–2771 (2015)
11. Zhang, X., Liu, S., Chen, X., Wang, L., Gao, B., Zhu, Q.: Health information privacy concerns, antecedents, and information disclosure intention in online health communities. Inf. Manag. **55**(4), 482–493 (2018)
12. PwC: The Wearable Future, Consumer Intelligence Series (2014). http://www.pwc.com/es_MX/mx/industrias/archivo/2014-11-pwc-the-wearable-future.pdf
13. Kurtz, C., Semmann, M., Böhmann, T.: Privacy by design to comply with GDPR: A Review on Third-Party Data Processors presented at the Americas Conference on Information Systems (AMCIS), New Orleans (2018)
14. Martin, Y.S., Kung, A.: Methods and tools for GDPR compliance through privacy and data protection engineering. In: 2018 IEEE European Symposium on Security and Privacy Workshops (EuroS&PW), pp. 108–111. IEEE. (2018)
15. Alagar, V., Periyasamy K., Wan, K.: Privacy and security for patient-centric elderly health care. In: 2017 IEEE 19th International Conference on e-Health Networking, Applications and Services (Healthcom), Dalian, 2017, pp. 1–6 (2017)
16. Alibasa, M.J., Santos, M.R., Glozier, N., Harvey, S.B. Calvo, R.A.: Designing a secure architecture for m-health applications. In: 2017 IEEE Life Sciences Conference (LSC), Sydney, NSW, 2017, pp. 91–94 (2017)
17. Zhou, J., Lin, X., Dong, X., Cao, Z.: PSMPA: patient self-controllable and multi-level privacy-preserving cooperative authentication in distributedm-healthcare cloud computing system. IEEE Trans. Parallel Distrib. Syst. **26**(6), 1693–1703 (2014)

18. Volk, M., Sterle, J., Sedlar, U.: Safety and privacy considerations for mobile application design in digital healthcare. Int. J. Distrib. Sens. Netw. **2015**, 1–12 (2015)
19. Li, X.: Understanding eHealth literacy from a privacy perspective: eHealth literacy and digital privacy skills in American disadvantaged communities. Am. Behav. Sci. **62**(10), 1431–1449 (2018)
20. Edemacu, K., Park, H.K., Jang, B., Kim, J.W.: Privacy provision in collaborative ehealth with attribute-based encryption: survey, challenges and future directions. IEEE Access **7**, 89614–89636 (2019)
21. Omoogun, M., Seeam, P., Ramsurrun, V., Bellekens, X., Seeam, A.: When eHealth meets the internet of things: Pervasive security and privacy challenges. In: 2017 International Conference on Cyber Security and Protection of Digital Services (Cyber Security), pp. 1–7. IEEE (2017)
22. Bhuiyan, M.Z.A., Zaman, M., Wang, G., Wang, T., Wu, J.: Privacy-protected data collection in wireless medical sensor networks. In: 2017 International Conference on Networking, Architecture, and Storage (NAS), pp. 1–2. IEEE (2017)
23. Liu, L.S., Shih, P.C., Hayes, G.R.: Barriers to the adoption and use of personal health record systems. In: Proceedings of the 2011 iConference, pp. 363–370. ACM Seattle, WA, 8–11 February (2011)
24. Romanou, A.: The necessity of the implementation of privacy by design in sectors where data protection concerns arise. Comput. Law Secur. Rev. **34**(1), 99–110 (2018)
25. Sousa, M., et al.: OpenEHR based systems and the general data protection regulation (GDPR). building continents of knowledge in oceans of data: the future of co-created ehealth (2018)
26. Iwaya, L.H., Fischer-Hübner, S., Åhlfeldt, R.M., Martucci, L.A.: mhealth: a privacy threat analysis for public health surveillance systems. In: 2018 IEEE 31st International Symposium on Computer-Based Medical Systems (CBMS), pp. 42–47. IEEE (2018)
27. Sawand, M.A., Khan, N.A.: Privacy and security mechanisms for ehealth monitoring systems. Int. J. Adv. Comput. Sci. Appl. **8**(4), 533–537 (2017)
28. Pirbhulal, S., Samuel, O.W., Wu, W., Sangaiah, A.K., Li, G.: A joint resource-aware and medical data security framework for wearable healthcare systems. Futur. Gener. Comput. Syst. **95**, 382–391 (2019)
29. Al-Janabi, S., Al-Shourbaji, I., Shojafar, M., Shamshirband, S.: Survey of main challenges (security and privacy) in wireless body area networks for healthcare applications. Egypt. Inform. J. **18**(2), 113–122 (2017)
30. Luo, E., Bhuiyan, M.Z.A., Wang, G., Rahman, M.A., Wu, J., Atiquzzaman, M.: Privacyprotector: privacy-protected patient data collection in IoT-based healthcare systems. IEEE Commun. Mag. **56**(2), 163–168 (2018)
31. Cavoukian, A.: Privacy by design [leading edge]. IEEE Technol. Soc. Mag. **31**(4), 18–19 (2012)
32. Lambrinoudakis, C.: The general data protection regulation (GDPR) era: ten steps for compliance of data processors and data controllers. In: Furnell, S., Mouratidis, H., Pernul, G. (eds.) TrustBus 2018. LNCS, vol. 11033, pp. 3–8. Springer, Cham (2018). https://doi.org/10.1007/978-3-319-98385-1_1
33. Tikkinen-Piri, C., Rohunen, A., Markkula, J.: EU general data protection regulation: changes and implications for personal data collecting companies. Comput. Law Secur. Rev. **34**(1), 134–153 (2018)
34. Huth, D.: A pattern catalog for GDPR compliant data protection. In: PoEM Doctoral Consortium, pp. 34–40 (2018)
35. Rubinstein, I.S., Good, N.: Privacy by design: a counterfactual analysis of Google and Facebook privacy incidents. Berkeley Technol. Law J. **28**(2), 1333–1413 (2013)

36. Antignac, T., Scandariato, R., Schneider, G.: Privacy compliance via model transformations. In: 2018 IEEE European Symposium on Security and Privacy Workshops (EuroS&PW), pp. 120–126. IEEE (2018)
37. Rösch, D., Schuster, T., Waidelich, L., Alpers, S.: Privacy control patterns for compliant application of GDPR. In: AMCIS 2019 Proceedings > Information Security and Privacy (SIGSEC) > 27 (2019)
38. Palmirani, M., Martoni, M., Rossi, A., Bartolini, C., Robaldo, L.: Legal ontology for modelling GDPR concepts and norms. In: JURIX, pp. 91–100 (2018)
39. Notario, N., et al.: PRIPARE: integrating privacy best practices into a privacy engineering methodology. In: Proceedings - 2015 IEEE Security and Privacy Workshops, SPW 2015, pp. 151–158 (2015)
40. Kalloniatis, C., Kavakli, E., Gritzalis, S.: Addressing privacy requirements in system design : the PriS method. Requirements Eng. **13**(3), 241–255 (2008)
41. Robol, M., Salnitri, M., Giorgini, P.: Toward GDPR-compliant socio-technical systems: modeling language and reasoning framework. In: Poels, G., Gailly, F., Serral Asensio, E., Snoeck, M. (eds.) PoEM 2017. LNCS, vol. 305, pp. 236–250. Springer, Cham (2017). https://doi.org/10.1007/978-3-319-70241-4_16
42. Kalloniatis, C.: Incorporating privacy in the design of cloud-based systems: a conceptual meta-model. Inf. Comput. Secur. J. **25**(5), 614–633 (2017). https://doi.org/10.1108/ICS-06-2016-0044
43. Kalloniatis, C.: Designing privacy-aware systems in the cloud. In: Fischer-Hübner, S., Lambrinoudakis, C., López, J. (eds.) TRUSTBUS 2015. LNCS, vol. 9264, pp. 113–123. Springer, Cham (2015). https://doi.org/10.1007/978-3-319-22906-5_9
44. Mouratidis, H., Islam, S., Kalloniatis, C., Gritzalis, S.: A framework to support selection of cloud providers based on security and privacy requirements. J. Syst. Softw. **86**(9), 2276–2293 (2013)
45. Carbonaro, G., Leanza, E., McCann, P., Medda, F.: Demographic decline, population aging, and modern financial approaches to urban policy. Int. Reg. Sci. Rev. **41**(2), 210–232 (2018)
46. Sharma, R.: The demographics of stagnation: why people matter for economic growth. Foreign Aff. **95**(2), 18–24 (2016)
47. Petrakaki, D., Hilberg, E., Waring, J.: Between empowerment and self-discipline: governing patients' conduct through technological self-care. Soc Sci Med **213**, 146–153 (2018)

Author Index

© The Editor(s) (if applicable) and The Author(s), under exclusive license
to Springer Nature Switzerland AG 2025
N. Pitropakis and S. Katsikas (Eds.): *Security and Privacy in Smart
Environments*, LNCS 14800, p. 287, 2025.
https://doi.org/10.1007/978-3-031-66708-4